Lecture Notes in Physics

Edited by H. Araki, Kyoto, J. Ehlers, München, K. Hepp, Zürich
R. L. Jaffe, Cambridge, MA, R. Kippenhahn, München, D. Ruelle, Bures-sur-Yvette
H. A. Weidenmüller, Heidelberg, J. Wess, Karlsruhe and J. Zittartz, Köln

Managing Editor: W. Beiglböck

376

D. Berényi G. Hock (Eds.)

High-Energy Ion-Atom Collisions

Proceedings of the 4th Workshop on
High-Energy Ion-Atom Collision Processes
Held in Debrecen, Hungary, 17–19 September 1990

Springer-Verlag
Berlin Heidelberg GmbH

Editors
D. Berényi
G. Hock
Institute of Nuclear Research of the Hungarian Academy of Sciences
P. O. Box 51, H-4001 Debrecen, Hungary

ISBN 978-3-662-13864-9 ISBN 978-3-540-47080-9 (eBook)
DOI 10.1007/978-3-540-47080-9

Originally published by Springer-Verlag Berlin Heidelberg New York in 1991
Softcover reprint of the hardcover 1st edition 1991

2153/3140-543210 – Printed on acid-free paper

Preface

This is the fourth workshop on high-energy ion–atom collision processes organized by the Institute of Nuclear Research of the Hungarian Academy of Sciences *(ATOMKI)* in Debrecen. (The proceedings of the former (third) workshop were also published by Springer-Verlag: Lecture Notes in Physics, Vol. **294**, 1988.)

According to the philosophy of the workshop, invited surveys (two, in general) are given on the most topical subjects, and these are selected with the help of the international advisory committee of the workshop. The surveys are followed in the individual sessions by pertinent contributions (mostly in poster form) on recent results.

Here, the following topics were selected and treated as separate sessions:

- electron capture and loss (including ionization)
- double electron and resonance processes
- electron correlation and post collision interaction effects
- collisions of antiparticles with atoms, etc.

Additionally, a *keynote talk* launched the workshop and a *'hot topics' session* and *concluding remarks* ended it.

In these proceedings, the invited surveys are included together with the relevant invited contributions. In certain cases, other (so-called *poster*) contributions are also included, at the referees' suggestion. The grouping of articles follows the sessions of the workshop. On the first page of each session, the titles and authors of the other contributions not published in the proceedings are additionally listed. All the *abstracts* of the contributions to the workshop (surveys, invited and other contributions) are published as an issue of ATOMKI Report 1990/1. This report has been distributed as usual and is still available. Of course, both the keynote talk and the concluding remarks are included in these proceedings.

Debrecen,
31 October 1990

D. Berényi
G. Hock

SPONSORS

Hungarian Academy of Sciences, *Budapest*
National Committee for Technical Development, *Budapest*
European Physical Society, *Geneva*

SUPPORTED by

TUNGSRAM Co. Ltd., *Budapest*

INTERNATIONAL
ADVISORY BOARD

J.S. Briggs *(Freiburg/Br., FRG)*
C.L. Cocke *(Manhattan, KS, USA)*
S. Datz *(Oak Ridge, TN, USA)*
B. Fastrup *(Århus, Denmark)*
K.-O. Groeneveld *(Frankfurt/M., FRG)*
N.M. Kabachnik *(Moscow, USSR)*
J.H. McGuire *(Manhattan, KS, USA)*
W. Mehlhorn *(Freiburg/Br., FRG)*
P.H. Mokler *(Darmstadt, FRG)*
T. Mukoyama *(Kyoto, Japan)*
I.A. Sellin *(Oak Ridge, TN, USA)*

ORGANIZING COMMITTEE

Chairman: D. Berényi
G. Hock, T. Papp
J. Pálinkás, B. Sulik
Mrs. E. Kovách, Mrs. M. Halász
Mrs. I. Németh

HOST INSTITUTION

Institute of Nuclear Research
of the Hungarian Academy of Sciences
(ATOMKI), Debrecen, Hungary

Contents

Opening

Chairman: J.M. HANSTEEN

Welcome by D. BERÉNYI, Director of the Institute,
Vice-President of the Hung. Acad. Sci.

Keynote: I.A. SELLIN
Newly Appreciated Roles for Electrons in
Ion–Atom Collisions

Welcome

by **D. Berényi**
Vice-President of the Hungarian Academy of Sciences

Ladies and Gentlemen, Dear Friends,

This is the fourth opportunity we have had for getting together to discuss recent issues in the field of high-energy ion–atom collisions.

The first such meeting took place in 1981. The series of meetings began as a large workshop, but now this seems to become a larger, more popular conference, having about one hundred participants and more than seventy contributions altogether.

Welcoming you, may I outline here again the general philosophy of our meetings as they have developed. With the help of the International Advisory Committee several topical subjects are selected, such as electron capture and loss, collision of antiparticles with atoms and multi-electron processes. Then two speakers are invited to survey the topics concerned from their particular viewpoint (e.g. experimental–theoretical, different experimental or theoretical approaches, etc.). After the surveys there are one or two oral contribution(s). The relevant posters are also included in the sessions. The individual sessions are then closed, in general, by the discussion of the topic of the session based on the talks *and* posters.

But now, I think, is the time to start work. First, however, it is my pleasant duty, on behalf of the Hungarian Academy of Sciences and this Institute, to wish you very successful work during the workshop and also a pleasant time during your stay in Debrecen and Hungary.

NEWLY APPRECIATED ROLES FOR ELECTRONS IN ION-ATOM COLLISIONS

Ivan A. Sellin

Department of Physics and Astronomy
University of Tennessee, Knoxville, TN 37996
and
Oak Ridge National Laboratory, Oak Ridge, TN 37831, USA

Abstract

Since the previous Debrecen workshop on High-Energy Ion-Atom Collisions there have been numerous experiments and substantial theoretical developments in the fields of fast ion-atom and ion-solid collisions concerned with explicating the previously largely underappreciated role of electrons as ionizing and exciting agents in such collisions. Examples to be discussed include the double electron ionization problem in He; transfer ionization by protons in He; double excitation in He; backward scattering of electrons in He; the role of electron-electron interaction in determining beta parameters for ELC; projectile K ionization by target electrons; electron spin exchange in transfer excitation; electron impact ionization in crystal channels; resonant coherent excitation in crystal channels; excitation and dielectronic recombination in crystal channels; resonant transfer and excitation; the similarity of recoil ion spectra observed in coincidence with electron capture vs. electron loss; and new research on ion-atom collisions at relativistic energies.

Introduction

The role of electrons as passive, shielding agents in ion-atom collisions, whose effects are largely taken into account by central field approximations and by Pauli antisymmetrization, has a long and successful history in explicating many fascinating collision phenomena among electrons, ions, atoms, and molecules in general, and ion- atom collisions in particular. The active role of electrons as ionizing and exciting agents in their own right has been generally less well appreciated. In thinking over the various possibilities for responding to Prof. Berényi's injunction to keynote speakers – to give "special emphasis to the development in the recent three years, pointing out the most actual problems of the field and also the future tendencies as you observe them"– it seemed to me that recognition of this active participation of electrons as ionizing and exciting agents in the course of ion-atom collisions has been a distinguishing feature of many of the most stimulating papers that have appeared in the literature of the field during these past three years. Hence the choice of title: the term "Newly Appreciated" emphasizes that the processes to be discussed have always been prominent in the physics – what is new is mainly our improved insights into them.

We will divide our discussion, somewhat arbitrarily, into two parts: the first in which electron-electron and electron-ion interactions are internal to one of the two colliding systems, and the second in which these interactions are shared.

Internal Electron-Electron Ionization And Excitation Interactions

We take as a starting point double ionization by photons, since a single photon projectile interacts cleanly with only one target electron. Double ionization is entirely attributable to electron correlation, taken here to mean Coulomb interaction among target electrons, and to include exchange effects. Even for photoionization, which lacks the complexity of the heavy collision partner, other complexity immediately arises: double ionization can arise from direct ejection of two electrons; or from a two-step process where photoionization of an inner shell electron is followed by an Auger transition; or by a two-electron Auger decay following creation of an inner shell vacancy.

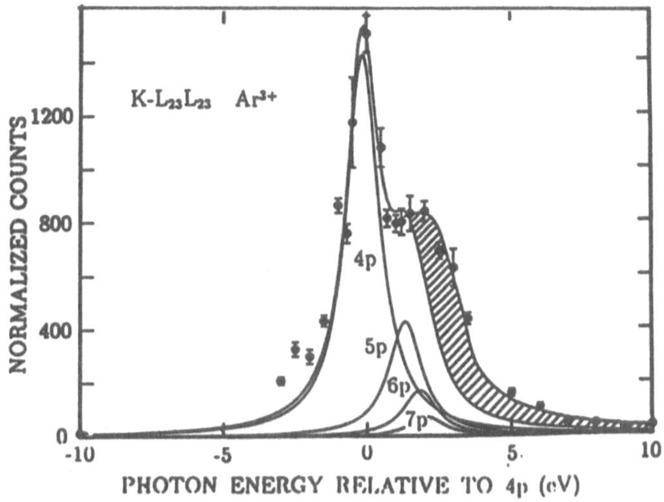

Fig. 1. Decomposition of Ar^{3+} photoion yields observed in coincidence with K-L_{23}-L_{23} autoionization electrons into components resulting from excitation of an Ar K electron into bound *np* levels. The shaded area represents recapture of the photoelectron by post-collision interaction. From J. Levin et al., Ref. 1.

An example of how large internal electron-electron ionization probabilities can be is illustrated by a recent experiment of Levin et al[1], who studied the photoion spectrum resulting from the Auger decay of Ar atoms which had been resonantly photoexcited to the *4p*, *5p*, *6p*, *7p* .... levels of neutral Ar. Analysis of the photoion spectrum reveals that the L-MM decay of each of the two L_{23} vacancies usually formed in the most probable K-LL autoionization decay channel additionally ionizes ≈27% of these Rydberg electrons, with the ionization probability of *7p* electrons reaching $\gtrsim$ 55%! Figure 1 illustrates the excitation function of one of the photoion states observed as a function of photon energy. As the exciting photon energy is raised in energy above the Rydberg resonances to just above ionization threshold, a prominent shoulder in the excitation function is seen, exhibiting the interesting phenomenon of recapture of the photoelectron by post-collision interaction with departing Auger electrons, reducing the charge of the residual ion from the expected, above threshold diagram value from +4 to +3.

Two-Electron Ionization in He

In view of the complex as well as highly probable internal electron-electron interaction phenomena accompanying inner-shell excitation phenomena just noted, even for the simple case of photon - single electron interactions, it is natural that the greatest progress in understanding dynamics of both photon and ion interactions with atoms in which electron correlation is important has concerned the simplest such target, He. Even for He, a number of surprises keep turning up.

For example, it is interesting to compare single and double photoionization cross sections with corresponding charged particle interaction cross sections. In the limit of high photon energies and projectile particle energies (say above 10 times the ionization potential and 10 MeV/u, respectively), the naive expectation is that apart from kinematic factors the particle and photoionization cross sections approach converge to the same values. Because of its specific sensitivity to electron correlation in double ionization, and because common factors drop out, it has become customary to compare R, the ratio of the double to single ionization cross sections in each case rather than the raw cross sections themselves. Yet this ratio is found to lie[2] in the range $R(h\nu) = 4$ to 5%, and R(protons, antiprotons, electrons) in the range 0.2 to 0.4%, more than an order of magnitude different[3−5]. Actually, $R(h\nu)$ is surprisingly ill established experimentally, and is based on scattered data which is remarkably sparse above 200 eV, and could profit from new measurements at higher photon energy[6]. Until the comparatively recent work reported in Refs. 3 - 5, little was firmly established about the asymptotic limit for the charged projectile case as well.

Reading and Ford have attributed the difference in R to the fact that "the ejected electron in a high-energy ion-atom collision moves rather slowly away from the atom. Thus the shakeoff limit is not applicable, and gives a cross section predicted to be an order of magnitude higher than observed..." Here shakeoff refers to the sudden ejection of one electron, leaving the other to collapse from a single-particle orbit in neutral He to an appropriate linear combination of states in He^+, including those shakeoff states lying in the He^{2+} ionization continuum.

The divergent approach to an asymptotic R value common to electrons, positrons, protons and antiprotons at large v_P has a rich experimental and theoretical background much of which almost (but not quite) fits into the three-year span of central interest here. A succinct summary of events has been provided by Heber et al.[5] Following up interesting differences between cross sections for double ionization for equal velocity electrons and protons (cf. Fig. 1, first article, Ref. 6), more recent experiments and theory have successfully sought to compare data for equivelocity protons and antiprotons[3,4] and electrons and positrons[7]. These experiments have shown that the observed differences reflect a large effect associated with projectile charge sign. At projectile velocities corresponding to $\sim$ 1 MeV/u R is found to about be a factor of 2 larger for electrons and antiprotons than it is for positrons and protons. Above 10 MeV/u the R values for all four projectiles appear to level off at about 2 - 3 x 10^{-3}.

A number of interesting possible mechanisms to explain these particle-antiparticle differences were discussed (and some rejected) by Andersen et al.[3] Among these were shakeoff, disregarded for reasons already given; a two-step (second Born approximation) process labelled $TS-1$, in which a first electron "struck" by the incident projectile goes on to knock out a second (scaling as the second power of the projectile charge); another two-step mechanism thought important at lower projectile velocities labelled $TS-2$, consisting of two consecutive projectile-electron encounters in the same collision (scaling as the fourth power of the projectile charge). Since the particle-antiparticle differences reflect an odd power of the charge, two interference effects were considered. Interference between $TS-2$ and shakeoff, or between $TS-1$ and $TS-2$, may account for these differences. In fact, it has recently been shown by Végh and Burgdörfer[3] that $TS-1$ and shakeoff are equivalent (up to a sign).

Subsequently the novel calculations of Reading and Ford, using their so-called forced impulse method, not only produced good qualitative understanding of the magnitude and velocity dependence of R for both protons and antiprotons, but also (with the inclusion of d as well as s and p orbitals in their basis) quantitative agreement with the asymptotic value of R. Figures 2 and 3, drawn respectively from Ref. 4, item 2, and Ref. 5, illustrate these points. The former reference attributes the proton-antiproton difference primarily to an interference between first and second Born amplitudes depending critically on non-dipole transitions, without commenting further on the relative merits of the interference mechanisms identified by Andersen et al. The dotted curve shows the prediction of Ford and Reading multiplied by 1.35. Convergence toward a good understanding of the double ionization problem seemed well on its way until publication of the results of Heber et al., who used beams of N^{7+} ions to study R in the 10 - 30 MeV/u range. In these experiments R was found to remain nearly constant over the velocity range at about 0.01, some 4 - 5 X higher than the high-velocity limit established previously for $q = 1$ projectiles, and also found for 20 MeV/u He projectiles! On this disquieting note we end this discussion of double ionization in He, and pass on to the closely related subjects of transfer ionization by protons in He, and double excitation of He by fast ions.

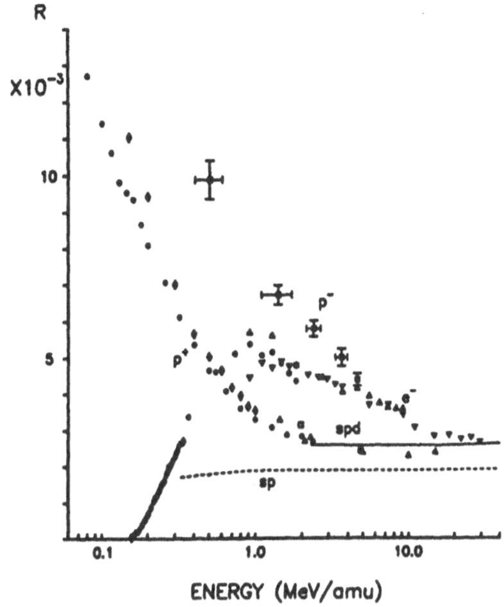

Fig. 2. The ratio R between double and single ionization cross sections for antiprotons, protons, and electrons in He. The broken and full curves labelled *sp* and *spd* are first Born results of Ford and Reading, Ref. 4, item 2. Filled squares, antiproton data from Ref. 3. Other filled symbols, electron data. Open symbols, proton data. References to experimental data are given in Ref. 4, item 2.

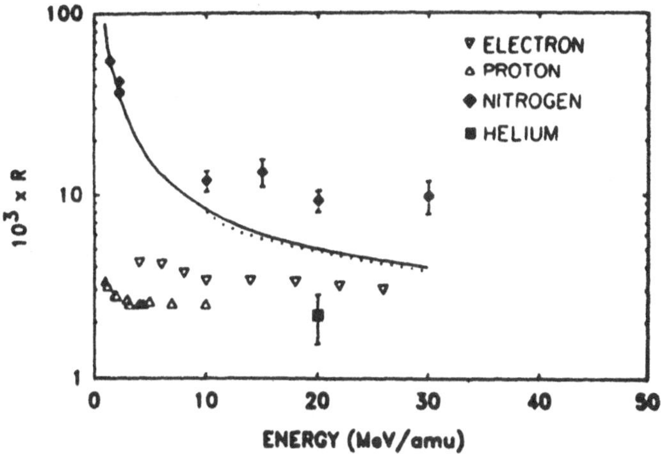

Fig. 3. Comparison of R found by Heber et al. for nitrogen and helium projectiles with other data for electrons, protons, and low- energy nitrogen ions. References to the other data as well as to a semiempirical calculation of H. Knudsen et al. are given in Ref. 5.

Transfer Ionization by Protons in Helium

In a recent experiment Pálinkás[8] et al. investigated the angular distribution of electrons ejected from He near the projectile velocity in coincidence with the capture of the other electron by 1 MeV protons, a special case of double ionization usually referred to as transfer ionization. In an approximate description of the process studied, the captured electron first collides with the proton, and then scatters into a bound state of the projectile through a second collision with the other electron (p-e-e scattering). The signature of the process is a peak near 90° in the angular distribution of the ejected electrons, and is illustrated in Fig. 4. Good agreement for this very small cross section is found with the second Born calculations of Briggs and Taulbjerg.[8]

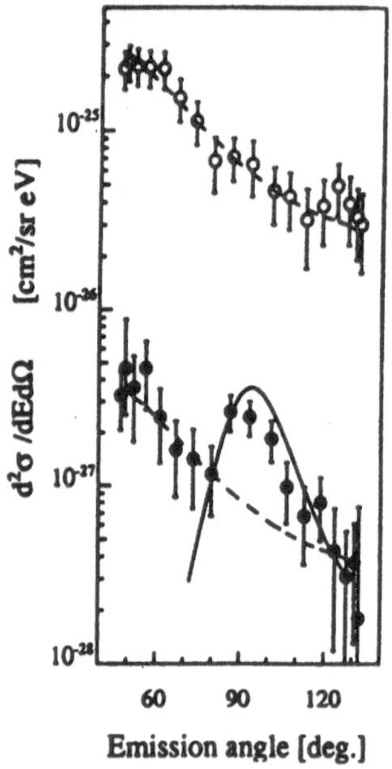

Fig. 4. Doubly differential cross section of electron emission at 600 eV following transfer ionization of Ne (open circles) and He (closed circles) by 1 MeV protons. The solid curve is derived from Briggs and Taulbjerg. From J. Pálinkás et al., and J. Briggs and K. Taulbjerg, Ref. 8.

Double excitation of He by Fast Ions

The kinship of double excitation of He by charged particles to double ionization has been explored recently by J. Giese et al.[9], who note that double excitation leads to better understood states of well defined energies, quantum numbers, and angular distributions, but also the drawback of requiring interpretation of interferences between these resonances and the underlying single ionization continuum. First and second order processes quite analogous to those already discussed for the two-electron ionization problem enter, with corresponding opportunities to study first order processes scaling as the second power of the projectile charge q, second order processes scaling as q^4, and various possible interferences scaling as q^3. Better understanding of these scalings was sought through exploration of the projectile charge q dependence of the electron emission yields from the doubly excited $2s^2(^1S)$, $2s2p(^1P)$, and $2p^2(^1D)$ states of He produced by electrons, protons, C ions (q=4-6), and F ions (q=7-9). The results indicated that excitation to the $2s^2$ and $2p^2(^1D)$ states increase approximately as $\lesssim q^3$, while excitation to the $2s2p$ state varies as approximately q^2. Figure 5 illustrates the total cross section data obtained for the $2p^2(^1D)$ state scaled by q^2, together with overlapping earlier data of Pedersen and Hvelplund[9], and a very recent coupled states calculation of Fritsch and Lin[9]. The Fritsch and Lin calculation is said to specifically treat electron-electron interactions, and is seen to give a much steeper q dependence than the data.

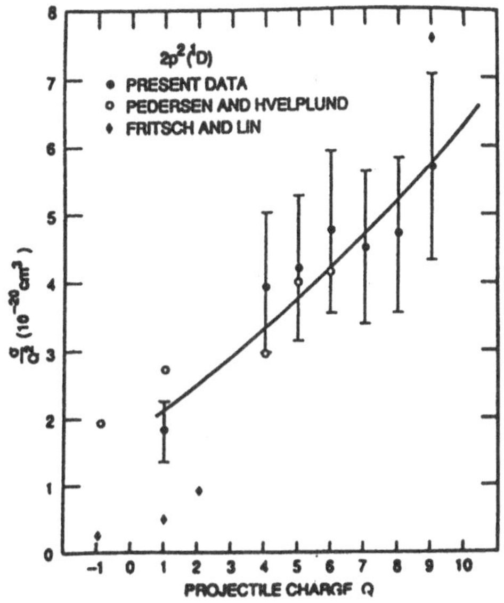

Fig. 5. Total averaged emission cross section for the $2p^2(^1D)$ state scaled by q^2. Open circles represent corrected data of Pedersen and Hvelplund, while the diamonds represent calculations of Fritsch and Lin. From Ref. 9.

Shared Electron-Electron Ionization And Excitation Interactions

We turn now to the some of the many recently observed, highly interesting manifestations of electron-electron and electro~-ion interactions in situations where the excitation or ionization of one partner in an ion-atom collision is attributable to collision interactions with electrons of the other. The first of these to be considered is backward scattering of electrons from projectile ionizing collisions.

Backscattering of Electrons in Projectile Ionizing Collisions

An exploratory study of projectile ionizing collisions (which includes electron loss to continuum, ELC) for ≈ 0.5 MeV/u He^+ on He, Ne, Ar has been undertaken by Köver et al. of ATOMKI in Debrecen, working together with Heil et al.[10] of the University of Frankfurt. A plot of singly differential cross sections $d\sigma/d\Omega$ vs. electron ejection angle is shown in Fig. 6.

The strong deviation of the measured data from two theoretical calculations, one using plane wave Born approximation by Köver, Szabó, Heil et al, and the second by Hartley and Walters[10], is especially evident at the most backward angles. According to calculations in progress by Wang and Burgdörfer[11], the most likely explanation is the increasing prominence of a second order process involving electron - electron inelastic scattering in combination with a hard elastic scattering of the freed electron with the nucleus.

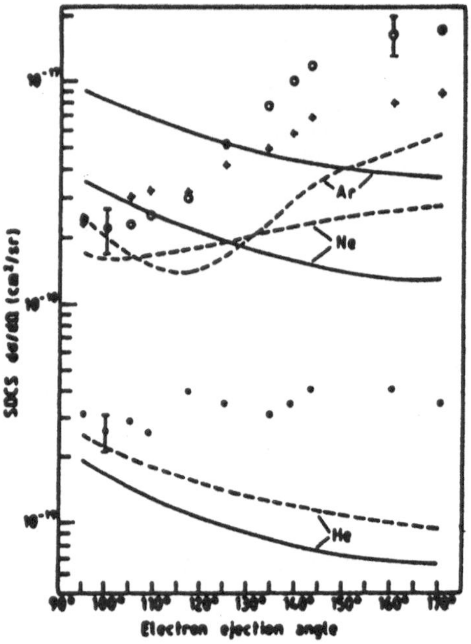

Fig. 6. Singly differential cross section for projectile ionization for 2 MeV He^+ on He, Ne, Ar collisions over the range 90 to 180 deg. From Ref. 10.

Sign Change in Quadrupole Asymmetry Parameter β_2 in He^+ - He ELC Cusps

ELC corresponds to the fraction of projectile ionization electrons emitted within some half-angle of collection Θ_0 of the forward direction. Under some circumstances, e.g. for loss from $He^+(2p_0)$ substates in He^+ - rare gas collisions at sufficiently high velocities for the Born approximation to apply, Burgdörfer et al.[11] found that in the now familiar multipole expansion of the ELC angular distribution, the quadrupole (β_2) component can take on large positive values. For sufficiently large values, a dip or inversion near the tip of the cusp can occur.

For He^+ - He collisions in the range 50 - 150 keV/u, Gulyás et al. find the steep variation of β_2 with beam velocity illustrated in Fig. 7. Since it is a feature of the theory that inelastic electron - electron interactions reduce the value of β_2, its rapid decrease to negative values are thought to reflect a correspondingly speedy onset of the importance of inelastic electron-electron collisions in this ELC process. The decrease observed is considerably more rapid than found in the related theoretical calculations, which are also illustrated in Fig. 7.

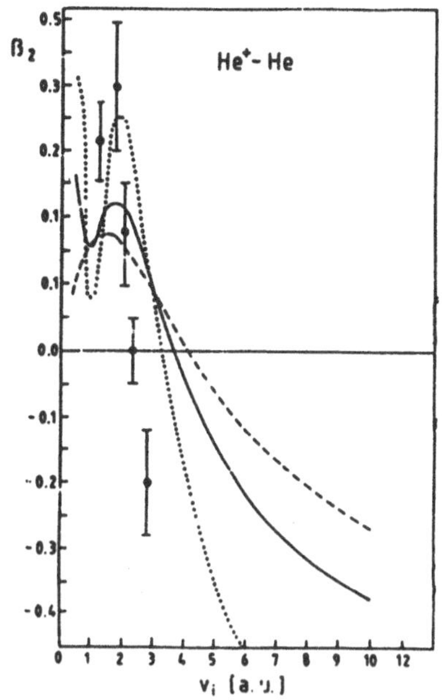

Fig. 7. β_2 parameter for He^+ - He collisions as a function of projectile velocity. Full circles, experiments of Gulyás et al. Theory from Burgdörfer, Szábo et al.: dashed line, initial state of projectile, 1*s*; dotted line, 2*s*; solid line, 80% 1*s* plus 20% 2*s*. From Ref. 11.

Projectile K Ionization by Target Electrons

Very clear illustrations of electron-electron interaction between projectile and target electrons are provided by very recent measurements by Hülskötter et al.[12] of cross sections for projectile K-shell ionization for 0.75 - 3.5 MeV/u C^{5+} and O^{7+} projectiles colliding with H_2 and He targets. Some of the experimental results are displayed in Fig. 8.

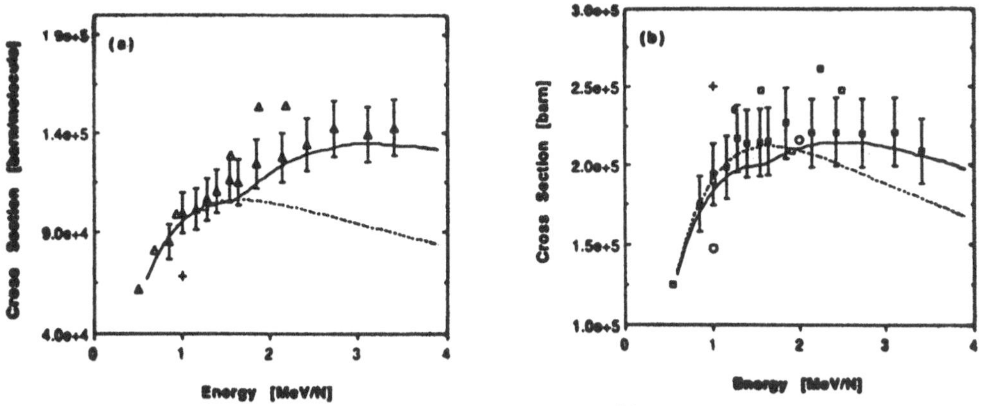

Fig. 8. Projectile ionization cross sections for O^{7+} on (a) H_2, and (b) He. The solid curves give the results of screening-antiscreening calculations described in Ref. 12, and the dashed curves pure PWBA calculations. The solid symbols display the recent experimental results of Hülskötter et al., and the open symbols earlier results from other laboratories (see citations in Ref. 12).

In explaining the observed data, the authors take into account the effect of target electrons by introducing a screened Coulomb interaction between the projectile electron and the target, where however the target electrons not only act coherently as screening agents, but also incoherently as ionizing (antiscreening) agents. The experimental results are found to agree with plane wave Born approximation calculations which take both into account. For energies for which the target electrons have sufficient energy in the projectile frame to ionize the projectile electron, the electron-electron interaction thus leads to a significant observed increase in the total ionization cross section.

Spin Exchange in Transfer Excitation

A direct manifestation of the interaction of projectile and target electrons has been studied by Zouros et al.[13], who measured the production of $1s2s2p\,^4P$ projectile states excited in collisions of $1s^22sO^{5+}$ and F^{6+} with He and H_2. The ^4P state cannot be produced in direct projectile electron - target nucleus excitation, since the necessary electron spin flip is very rare for such low Z ions. However, $1s$ to $2p$ excitation to the quartet P state can proceed by the exchange of the projectile electron with the exciting target electron. Zouros et al. showed that the production cross sections was found to increase sharply with projectile energy above .75 MeV/u. Figure 9 shows data from O^{5+} on He, H_2 collisions. The energy dependence (but not the magnitude) of the thresholdlike behavior of the measured cross sections is thus found to be well described using calculated cross sections[13] for electron impact ionization found in the literature, folded with the momentum distribution or Compton profile of the target electrons. This type of accounting for the target electrons' approximately "free" nature is analogous to the impulse approximation treatment of resonance transfer excitation (RTE), which relates dielectronic recombination, another free electron-ion collision process, to that of RTE occurring in ion-atom collisions.

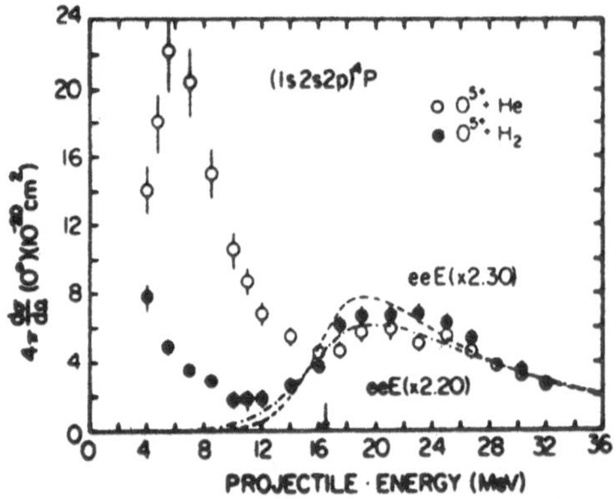

Fig. 9. Cross sections for the production of $1s2s2p\,^4P$ states by $1s \rightarrow 2p$ projectile excitation in collisions of O^{5+} with He and H_2 targets vs projectile energy. Dashed and dashed-dotted lines give corresponding scaled theoretical excitation functions, using theoretical electron impact excitation cross sections folded by the Compton profile of the target. Threshold is marked by an arrow. From Ref. 13.

Electron Impact Ionization and Excitation in Crystal Channels

Collisions between ions and the electrons of a macroscopic "molecule", namely a perfect single crystal, have long been studied in channeling experiments, where most of the interaction occurs with crystal electrons in regions distant from lattice sites, with the projectile ions colliding only gently with those of the lattice. The nearly free electrons found near the centers of channels, spread in energy only by the Fermi distributions of the conduction or valence electrons, has been exploited by Claytor et al.[14] to determine the electron impact ionization cross sections for Be-like to H-like uranium for 222 keV electrons (in the rest frame of the projectile ions). The large density of electrons – $\gtrsim 10^{23}$ per cm^3 – makes possible the measurement of very small cross sections. In these experiments, 405 MeV/u uranium ions were channeled in the < 110 > channel of Si to obtain cross sections of 3.9; 11.0; 16.0, and 31.0 b (+100%, - 50%), respectively, for $1s$, $1s^2$, $2s$, and $2s^2$ electrons. The results for the $1s$ and $1s^2$ cross sections disagree with present theory.(See Table I.)

TABLE I. Electron impact ionization cross sections (b).

Ion	State	Expt.	Pindzola&Buie	Scofield	Younger	Lotz
U^{91+}	$1s$	3.9		1.5	0.8	0.7
U^{90+}	$1s^2$	11.0		3.0	1.7	1.4
U^{89+}-U^{90+}	$2s$	16.0	13.0	29.0	9.4	12.0
U^{88+}-U^{90+}	$2s^2$	31.0	26.0	57.0	19.8	24.0

For original theoretical papers, see Ref. 14.

Resonant Coherent Excitation of Convoy Electrons in Crystal Channels

An axially channeled ion feels the anharmonic periodic potential of the crystal as an oscillatory electric field with a fundamental frequency $\nu = v/d$, where d is the atomic spacing and v the ion velocity. When the frequency (or a higher harmonic) coincides with an excitation energy of the ion, RCE can occur. This effect was first observed through the change in the charge state distribution of emergent ions[15], reflecting the higher probability for electron loss from resonantly excited states of projectile ions. Recently, Iwata et al[15] have observed photons emitted from excited states formed by RCE.

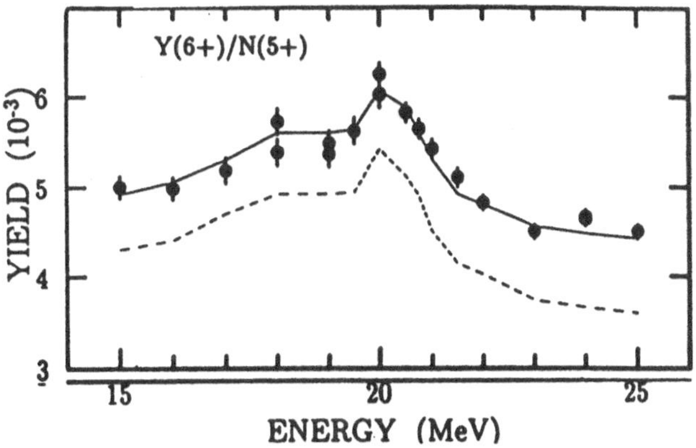

Fig. 10. Beam energy dependence of absolute convoy electron yields measured in coincidence with exit C^{6+} ions, normalized to the number of exit C^{5+} ions. The solid line shows results of a model calculation for ELC, and the dashed line shows the calculated contribution from excited states. From Ref. 16.

Very recently, Kimura et al.[16] have made the first identification of a clear effect of RCE on convoy electron production. Figure 10 shows a large enhancement of the yield of convoy electrons in coincidence with C^{6+} ions traversing the $\langle 100 \rangle$ axis of 160 nm thick Au crystals, normalized to the number of exit C^{5+} ions, as the incident velocity is swept through the second harmonic $1s-2s$ and $2p$ (RCE) resonances near 20 MeV. A detailed rate-equation production model fit to the data, which includes separate treatment of the C^{5+} ground and excited state populations, reveals that while the fraction of all C^{5+} ion in the excited state is $\lesssim 1/3$, these excited ions contribute to $\gtrsim 80\%$ of the convoy yield!

Dielectronic Excitation and Recombination – in Crystal Channels and in Resonant Transfer Excitation

Strictly speaking, dielectronic recombination is a resonant process in which a free electron of carefully selected relative velocity resonantly excites a bound target electron to an excited state, sticks to form a doubly excited state, and radiative stabilization of the recombined atom or ion then occurs. Though it is tempting and timely to discuss the beautifully resolved dielectronic recombination resonances observed in the recent work of Andersen et al.[17], we elect to bypass this interesting subject because it has been so extensively discussed and reviewed in many recent conferences and review articles, because it focusses primarily on free electron-ion collisions, and also because it is to discussed by others (e.g., Tanis and Cocke) in subsequent papers at this Workshop.

In a series of recent experiments, Datz et al.[18] have succeeded in studying dielectronic excitation and recombination in crystal channels. These phenomena were observed for H-like S, Ca, and Ti, and for He-like Ti, traversing the $< 110 >$ and $< 100 >$ channels in Si. Sample data is shown in Fig. 11, which exhibits for Ca^{19+} the yield per injected ion of unresolved $h\nu_1(2\ell n\ell' \rightarrow 1sn\ell')$ plus $h\nu_3(2\ell \rightarrow 1s)$ transitions. Both arise from filling a K hole in an empty K shell. The features correspond to dielectronic excitation of KLL, KLM, KLN, etc., and, at 300 MeV, to $1s \rightarrow 2p$ excitation. The dielectronic recombination resonances are broadened and reduced in amplitude, owing to the small but finite width of the electron Fermi energy distribution (≈ 10 eV).

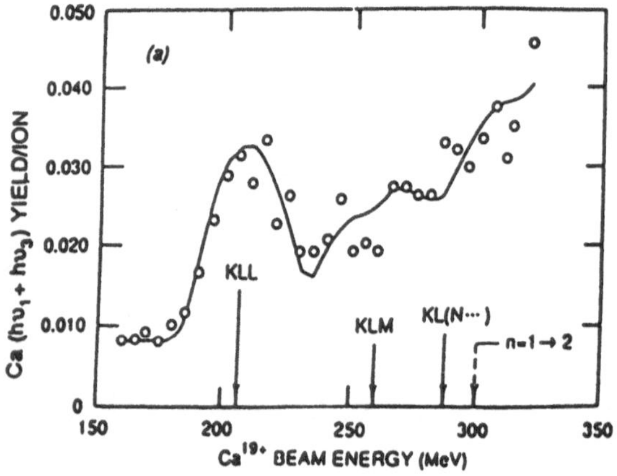

Fig. 11. Yield of $h\nu_1(2\ell n\ell' \rightarrow 1sn\ell')$ plus $h\nu_3(2\ell \rightarrow 1s)$ as a function of Ca^{19+} ion energy incident on a $< 110 >$ channel in a 1.2 μm thick Si crystal. Thresholds for various dielectronic processes are shown. The solid curve corresponds to a computer simulation. From Ref. 18.

That similar phenomena occur in ion-atom collisions, where electron transfer from a target atom is accompanied by simultaneous excitation of a bound projectile electron, has been known for quite some time. Tanis has recently provided two reviews[19]. In this case resonances are broadened more strongly still than for dielectronic recombination of ions in crystals, owing to the folding of the bound electrons' momentum distribution (Compton profile) with the resonances.

Because Tanis is to discuss this subject subsequently in this Workshop, no doubt more expertly than I, it seems best to leave the bulk of discussion about resonant transfer and excitation (RTE) to him. Here we provide only two illustrations of the transfer excitation process. The first is Fig. 12, which presents a schematic overview of the three different known mechanisms which lead to transfer excitation in energetic ion- atom collisions – the resonant mechanism just discussed, in which electron-electron interactions between projectile and target electrons dominate; non-resonant transfer excitation (NTE), which occurs through independent capture by the projectile coincident with projectile electron excitation by the target nucleus; and the so-called uncorrelated transfer excitation (UTE), coming from target electron excitation of a projectile electron, coincident with electron capture by the projectile. In the systems so far investigated, the NTE and UTE processes tend to dominate the TE probability at somewhat lower and higher projectile energies, respectively, than the range for which RTE dominates.

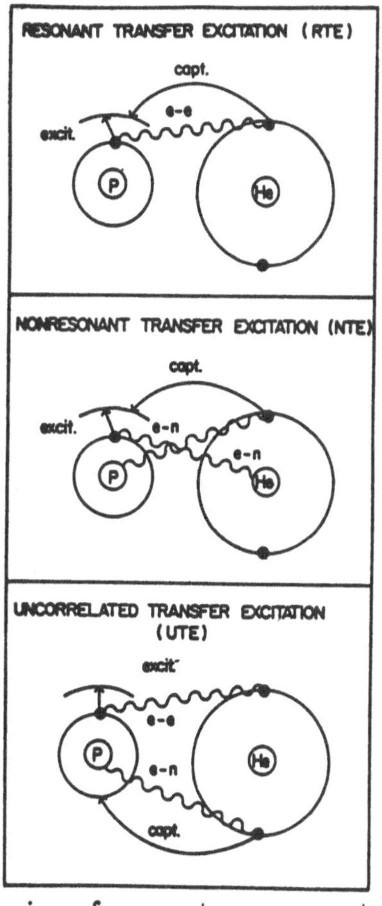

Fig. 12. Schematic overview of resonant, nonresonant, and uncorrelated transfer excitation processes. From J. A. Tanis, presented at the U. S.- Japan Seminar on Exotic and Highly Ionized Ions, Anchorage, June, 1990.

This situation is illustrated in Fig. 13, which shows the various contributions to TE for S^{13+} on He, observed in the x-ray decay channel.

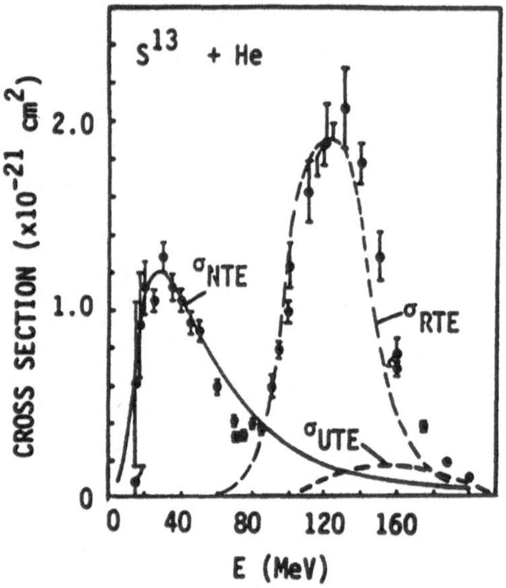

Fig. 13. Transfer excitation cross sections for S^{13+} on He. Data measured by J. Tanis et al. [Phys. Rev. A31, 4040 (1985)] are for S K x-rays coincident with single capture. The upper dashed curve is the calculated RTE cross section X 0.85, and the solid curve is the calculated NTE cross section normalized to the data. The lower dashed curve is the estimated UTE cross section taken from the work of Y. Hahn and H. Ramadan, Nucl. Inst. Meth. B43, 285 (1989). From Ref. 19.

Similarity of Recoil Ion Spectra Observed in Coincidence with Electron Capture vs. Electron Loss

In a paper by Levin et al.[20] concerning the decisive importance of vacancy cascades in determining high recoil ion charge state distributions in nearly symmetric, 0.7 MeV/u Cl on Ar collisions, the recoil ion charge state spectrum was studied in coincidence with single and double, electron capture and loss collisions. The remarkable similarity of the recoil ion spectra for coincident double capture vs. double loss is apparent in Fig. 14.

The enrichment of highly ionized states in the recoil spectrum corresponding to capture of two L-shell electrons proved easy to explain quantitatively by considering the additional autoionization corresponding to two vacancy cascades. The nearly identical appearance of the recoil spectra coincident with double electron loss strongly suggests that loss of two projectile L-shell electrons is highly correlated with ionization of two L-shell target electrons in the same collision. (Similar results were obtained in comparing recoil ion spectra for single capture and loss). Electron-electron collisions among target and projectile electrons giving rise to symmetric ionization of both is a plausible mechanism for accounting for this mutual ionization. Although the fast collision conditions prevailing are well out of the adiabatic collision regime for which molecular orbital collision models are expected to apply, such a mechanism is reminiscent of the promotion of $4f\sigma$MO's into the continuum, well known to produce L-shell vacancies in much lower energy collisions.

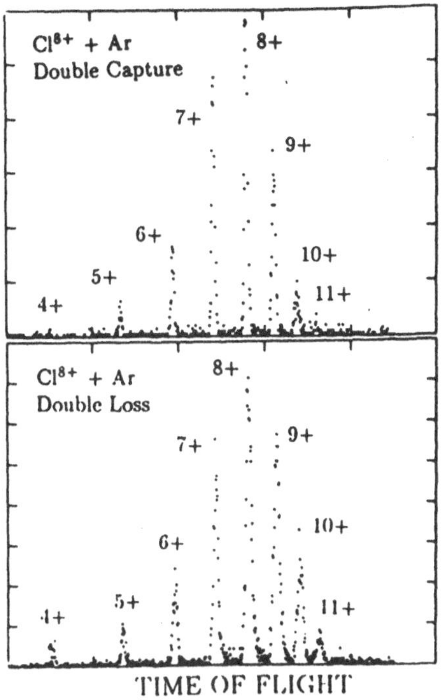

Fig. 14. Time of flight spectra for Ar recoil ions produced by beams of 0.7 MeV/u Cl ions, measured in coincidence with double electron capture, and double electron loss, respectively. Note the remarkable similarity of the recoil ion spectra for double capture vs. double loss. From Ref. 20.

New Research on Ion-Atom Collisions Physics at Relativistic Energies

We conclude with some examples of new experiments in which the role of electrons – and in this case, their positron antiparticles – is again decisive, but in the domain of relativistic atomic collisions. In our choice of examples we draw upon the work of a panel that considered relativistic atomic collisions within a larger workshop on future opportunities in atomic and molecular science held in 1989.[21]

The first of two main areas of research identified as having high potential by the panel lies in the area of fast heavy ion-atom collisions: Because peripheral atomic collisions at TeV/u energies can be viewed as virtual photon-photon collisions owing to Lorentz contraction of the Coulomb fields, the resultant production of single and multiple electron-positron pairs presents new experimental and theoretical challenges. For example, in TeV/u U + U collisions, the projectile field at a target nucleus appears essentially as an intense pulse of photons with energies $\gtrsim$ 100 GeV, not only permitting copious production of electron-positron pairs but also elementary particle pairs (muons, tauons, W's, etc.). While possibilities for studying such collisions do not yet exist, beams of 1 GeV/u U presently exist at GSI and LBL, and of 200 GeV/u S at CERN.

Examples of particular experimental possibilities include:

(1) Study of the differential cross section for electron-positron pair production, including the energy and angular distribution of both electrons and positrons. There seems to be a wide variation in theoretical predictions for these cross sections[21];

(2) Pair production with capture of the electron into a vacant projectile (K) shell. Experimentally, this process can be distinguished from free pair production because of the charge change of the projectile. Though the cross section is thought to be an order

of magnitude or two smaller than that for free pair production, it is also thought to be measurable already at a relativistic time dilation factor $\gamma = 2$;

(3) Radiative electron capture accompanied by electron-positron pair emission. For fast enough collisions, it has long been known that electron capture from a target atom accompanied by photon emission can dominate over mechanical capture. Preliminary calculations[21] indicate that for $\gamma \gtrsim 3$, radiative capture accompanied by electron-positron pair emission may be detectable. If so, a careful discrimination between possibilities (2) and (3) must be made, since in the former charge state change occurs with positron emission alone, while in the latter a pair is emitted.

At the time of writing an initial experiment[22] set up to explore possibilities (1) and (2) is underway at the CERN SPS facility. The experiment is being run with 200 GeV/u S beams on Au, and (for testing scaling and calibration), Al and Pd. Recent calculations by Rhoades-Brown, Bottcher, and Strayer[22] for the cross sections for capture for various symmetric pairs are shown in Fig. 15.

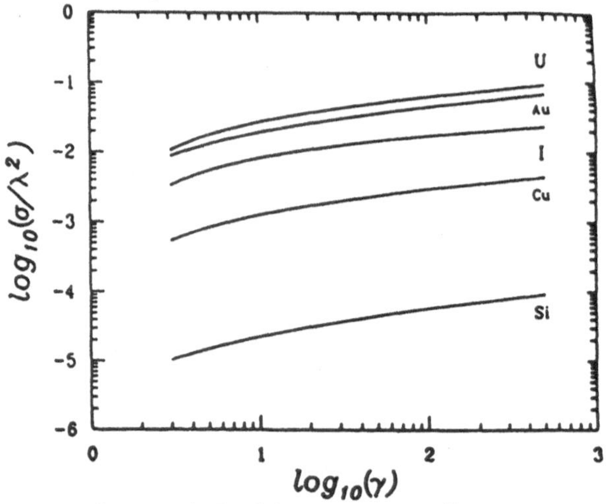

Fig. 15. Capture cross sections scaled with respect to $\lambda^2 = 1.49$ kb, where λ is the rationalized Compton wave length, for the various symmetric collision pairs marked. From Ref. 22.

If the cross section is scaled as Z^2, the cross section for the S on Au collision system at the γ value corresponding to the available center of mass energy is predicted to be $\approx$ 2b.

Work supported in part by the National Science Foundation and by the U.S. Department of Energy, Office of Basic Energy Sciences, Division of Chemical Sciences, under Contract No. DE-AC05-840R21400 with Martin Marietta Energy Systems, Inc.

References

1. J. C. Levin, C. Biedermann, N. Keller, L. Liljeby, C.-S. O, R. T. Short, and I. A. Sellin, Phys. Rev. Lett. 65, 988 (1990).
2. Z. Smit, M. Kregar, and D. Glavic-Cindro, Phys. Rev. A 40, 6303 (1989).
3. L. H. Andersen, P. Hvelplund, H. Knudsen, S. P. Møller, K. Elsener, K.-G. Rensfelt and E. Uggerhøj, Phys. Rev. Lett. 57, 2147 (1986). For a very recent discussion of shakeoff and the so-called TS-1 process, see the paper of L. Végh and J. Burgdörfer, Phys. Rev. A42, 655 (1990).

4. J. F. Reading and A. L. Ford, Phys. Rev. Lett. 58, 543 (1987); J. Phys. B21, L685 (1988).

5. O. Heber, B. Bandong, G. Sampoli, and R. L. Watson, Phys. Rev. Lett. 64, 851 (1990).

6. J. McGuire, private communication; Phys. Rev. Lett. 49, 1153 (1982); Nucl. Inst. Meth. B 10/11, 17 (1985).

7. M. Charlton, L. H. Andersen, L. Brun-Nielsen, B. I. Deutsch, P. Hvelplund, F. M. Jacobsen, H. Knudsen, G. Larrichia, M. R. Poulsen, and J. O. Pedersen, J. Phys. B21, L545 (1988).

8. J. Pálinkás, R. Schuch, H. Cederquist, and O. Gustafsson, Phys. Rev. Lett. 63, 2464 (1989); J. Briggs and K. Taulbjerg, J. Phys. B 12, 2569 (1979).

9. J. Giese, M Schulz, J. Swenson, H. Schöne, M. Benhenni, S. Varghes, C. Vane, P. Dittner, S. Shafroth, and S. Datz, to be published in Phys. Rev. A; corrected data of J. O. P. Pedersen and P. Hvelplund, Phys. Rev. Lett. 62, 2373 (1989); W. Fritsch and C. D. Lin, to be published in Phys. Rev. A.

10. O. Heil, A. Kövér, G. Szabó, L. Gulyás, K. Tökési, J. Kemmler, H. Rothard, D. Berényi, and K. O. Groeneveld, Inst. für Kernforschung der Uni. Frankfurt/M Annual Report No. IKF-47, p. 19 (1987), and J. Phys. B 21, 3231 (1988); H. Hartley and H. Walters, J. Phys. B 20, 3811 (1987); J. Wang and J. Burgdörfer, private communication, and to be published.

11. J. Burgdörfer et al., Phys. Rev. Lett. 51, 374 (1983); J. Burgdörfer, M. Breinig, S. B. Elston, and I. A. Sellin, Phys Rev. A 28, 3277 (1983); L. Gulyás, A. Kövér, G. Szabó, D. Berényi, L. Sarkadi, J. Pálinkás, and T. Vajnai, "Cusp Inversion in the Electron Loss Process", ECAMP - 3, Bordeaux, 1989; J. Burgdörfer and G. Szábo, to be published.

12. H.-P. Hülskötter, W. E. Meyerhof, E. Dillard, and N. Guardala, Phys. Rev. Letters 63, 1938 (1989).

13. T. J. M. Zouros, D. H. Lee, and P. Richard, Phys. Rev. Lett. 62, 2261 (1989); S. J. Goett and D. H. Sampson, At. Data Nucl. Data Tables 29, 535 (1983).

14. N. Claytor, B. Feinberg, H. Gould, C. Bemis, J. Gomez del Campo, C. Ludemann, and C. R. Vane, Phys. Rev. Letters 61, 2081 (1988).

15. S. Datz, C.D. Moak, O.H. Crawford, H.F. Krause, P.F. Dittner, J. Gomez del Campo, J.A. Biggerstaff, P.D. Miller, P. Hvelplund, and H. Knudsen, Phys. Rev. Lett. 40, 843 (1978); Y. Iwata, K. Komaki, Y. Yamazaki, M. Sekiguchi, T. Hattori, T. Hasegawa and F. Fujimoto, Nucl. Inst. Meth. B48, 163 (1990).

16. K. Kimura, J. Gibbons, S. Elston, C. Biedermann, R. DeSerio, N. Keller, J. Levin, M. Breinig, J. Burgdörfer, and I. Sellin, submitted for publication.

17. L. Andersen, P. Hvelplund, H. Knudsen, and P. Kvistgaard, Phys. Rev. Lett. 62, 2656 (1989).

18. S. Datz, C. Vane, P. Dittner, J. Giese, J. Gomez del Campo, N. Jones, H. Krause, P. Miller, M. Schulz, H. Schöne, and T. Rosseel, Nucl. Inst. Meth. B48, 114 (1990).

19. See for example the excellent reviews by J. A. Tanis, "Electron Transfer and Projectile Excitation in Single Collisions", Nucl. Inst. Meth. A262, 52 (1987); and in the Proceedings of the X-90 International Conference on X-rays and Inner Shell Ionization, to be published in Nucl. Inst. Meth.

20. J. Levin, C.-S. O, H. Cederquist, C. Biedermann, and I. Sellin, Rapid Communications, Phys Rev. A38, 2674 (1988).

21. Report of the Panel on Relativistic Atomic Collisions, Future Opportunities in Atomic, Molecular, and Optical Sciences Workshop, November 7, 1989. Panel members: C. Bottcher, H. Bryant, H. Gould, W. Meyerhof, P. Mohr, and W. Smith. See also R. Anholt and H. Gould, in Advances in Atomic and Molecular Physics, D. Bates and B. Bederson, eds., (Academic Press, New York, 1986) Vol. 22, p. 315.

22. Participants include S. Datz, C. Vane, P. Dittner, H. Gould, H. Knudsen, P. Hvelplund, R. Schuch et al. The theoretical calculations displayed in Fig. 15 are from M. J. Rhoades-Brown, C. Bottcher, and M. Strayer, Physical Review A 40, 2831 (1989), and Nucl. Inst. and Meth. B43, 301 (1989).

A Electron Capture and Loss

Chairman: M.W. LUCAS

A.1 Invited Surveys

M.G. MENENDEZ, A. HUETZ and M.M. DUNCAN
Electron Capture, Electron Loss and the Cusp

R.D. DuBOIS
Coincidence Measurements of Electron Capture and Loss in Ion–Atom Collisions

A.2 Invited Contributions

B.L. MOISEIWITSCH
Electron Capture to the Continuum

L. SZÓTÉR
The Possible Role of the Negative Ion Resonances in Cuspology

L. GULYÁS, A. KÖVÉR, GY. SZABÓ, T. VAJNAI and D. BERÉNYI
Testing the Series Expansion Method Used for Studying the 'Cusp' Spectra

T.J.M. ZOUROS, P. RICHARD and D.H. LEE
Z_p– and q–Dependence of 0° Binary Encounter Electrons in 1–2 MeV/u Collisions of Ions with H_2 and He Targets

A.3 Contributions (Posters)

H. FUKUDA, L. VÉGH, I. SHIMAMURA and T. WATANABE
Singly Ionizing Proton–He Collisions at Very Small Scattering Angles

R.D. DuBOIS and S.T. MANSON
Double Differential Cross Sections for He^+ – Atom/Molecule Collisions: Experiment and Theory

SH.D. KUNIKEEV and V.S. SENASHENKO
On the Theory of Electron Capture into Continuum by Neutral Atoms

M. GERETSCHLÄGER, O. BENKA and Ž. ŠMIT
K-Shell Ionization for Collision Systems $0.2 \leq Z_1/Z_2 \leq 0.62$ by Carbon and Oxygen Ions

L. GULYÁS, L. SARKADI, T. VAJNAI, GY. SZABÓ, Á. KÖVÉR, J. PÁLINKÁS and D. BERÉNYI
Cusp Inversion in $He^+ \rightarrow$ He, Ar Collision Systems

P. RYMUZA, Z. SUJKOWSKI, M. CARLEN, J.-Cl. DOUSSE, J. KERN and Ch. RHÊME
L- and M-Shell Ionization Probabilities in Central Collisions Deduced from KX-Ray Satellite Spectra

ELECTRON CAPTURE, ELECTRON LOSS AND THE CUSP

M G Menendez* and A Huetz
Laboratoire de Dynamique Moleculaire et Atomique
Université P & M Curie
Paris, France

M M Duncan
Department of Physics & Astronomy
University of Georgia
Athens, Georgia USA

1. INTRODUCTION

The velocity (energy) spectrum of electrons ejected during fast ion-atom collisions exhibits a sharp cusp in the forward direction. This cusp is centered about the projectile velocity when measured in the extreme forward direction, that is, at and near zero degrees. The cusp has attracted a great deal of interest since its initial observation and its explanation as the capture of target electrons into low-lying continuum states of the projectile by Macek (1970). The primary interest in the cusp, or more accurately, the shape of the cusp, stems from the fact that electrons moving very slowly in the field of the projectile are responsible for it. Thus it is reasonable to expect that analyses of cusps can lead to inferences about the dynamics of the processes that give rise to them. There are three well known processes that give rise to cusps of slightly different shape: (1) (target)-electron capture to the (projectile)-continuum, ECC; (2) (projectile)-electron loss to the (projectile)-continuum, ELC, Drepper and Briggs (1976); (3) transfer ionization, a processes where two (target)-electrons are captured by the projectile, one into a bound state and the other into the continuum, TI, Vane et al (1979).

The issues of atomic dynamics being addressed by detailed studies of the secondary electron cusps produced by various processes have become increasingly sophisticated in the last two decades. Some current questions that may serve to indicate the flavor of some of these issues are:

Is there a length scale for capture to the continuum?

Under what conditions is a two-electron capture process the capture of two correlated electrons?

Is transfer ionization a one-step or two-step process?

Under what conditions are post collisional interaction effects important?

In the next section we present a brief survey of recent work on fast ion-atom collisions where the shape of the cusp and its analysis is central to the elucidation of the physics of the situation. (This survey is not intended to be exhaustive.) In Section 3 we review the usual method used to characterize the cusp and point out a problem when dealing with electrons moving very slowly with respect to the projectile. In the last section we propose a method for dealing with this problem.

2. BRIEF SURVEY

In this section we mention a few recent studies, theoretical and experimental, that make use of the shape of the cusp in order to extract details of the atomic dynamics of the processes. The concentration is on work of the last three years. It shall become obvious that we have taken license in presenting the results of these works; first, in brevity, second, with a bias toward the result(s) that depend primarily on the analysis of the cusp. The results are ordered roughly by process and, in the spirit of some of these findings, no apology is made for the mixing of processes.

ECC

The zero degree cusp produced by bare C and O ions in 1.6 MeV/u collisions with He was carefully analyzed by Knudsen et al (1986). They found that the projectile-frame double differential cross section, expressed as a mutipole expansion, contains an important quadrupole contribution.

Reinhold and Schultz (1990) have calculated the shape of the ECC cusp for bare and structured projectile ions. Their classical trajectory Monte Carlo calculations show that the electron spectra are sensitive to the structure of the projectile.

Comparisons between ECC and TI cusps led Sarkadi et al (1989) to consider the issue of whether short range potentials can support the ECC process. By performing coincidence measurements of the ejected electron with the projectile charge state in the collision systems He /He,Ar the Atomki group uncovered ECC to a neutral projectile.

Jakubassa (1990) has examined the ECC process in short range potentials and finds theoretical evidence for ECC. The theoretical zero degree peak is broader than the experimental one.

Jagutzki et al (1990) have performed impact parameter dependent measurements of the electron spectra for 0.5 MeV/u light ions on Ne and He targets and find that a long-range post-collisional Coulomb focusing effect seems to be important for ECC.

ECC/ELC

Elston et al (1985) performed one of the first detailed comparisons and systematic characterizations of ECC and ELC. They found that ECC displays a dipolar behavior and that ELC shows preferential transverse electron transmission.

In recent coincidence experiments on the collision systems He^+ / He,Ar at 100 KeV/u and below , Köver et al (1989) found the unexpected result that the ECC process not only contributes to the cusp but dominates it below 100KeV/u.

ECC/TI

Using the zero degree cusp produced by highly charged I and U ions at medium velocities on He targets, Datz et al (1990) find that TI can not be viewed as two independent processes; the process is the result of electron-electron correlation.

ELC

Born calculations of the shape and width of the cusp by Burgdörfer (1983) and (1986), respectively, show that the ELC cusp depends strongly on the initial state of the projectile, weakly on the projectile velocity and that whereas the Born approximation is adequate for light targets it is not adequate for heavy targets where the nucleus drives the projectile ionization.

Andersen et al (1986) and Menendez and Duncan (1987) studied the single-ELC process in 1 MeV H^-/atom collisions. Both groups measured the structure on the ELC cusp due to the $^1P^o$ shape resonance in the e-H exit channel. These observations and the good agreement with the calculations of Liu and Starace (1989) provide an example of the sort of detail that is often obtainable from analysis of the cusp.

The situation is not as straightforward for the case of collisional double ionization of H^-. Sorensen et al (1988) and Duncan and Menendez (1990) have measured the cusp due to double-ELC of H^- from fast collision with rare gas atoms: the Aarhus group at zero degrees and the University of Georgia group from zero to three degrees. Although the zero degree cusps of both groups are in excellent agreement, the multipole expansions of the cross section developed from data taken at zero degrees are not consistent with data taken at angles close to zero.

Atan et al (1990) have made multipole expansion analysis of ELC cusps for 1.6-2.8 MeV He^+ on He, Ne and Ar in order to measure details of target dependence on the ELC cusp. In presenting the various multipole expansion parameters, some in agreement with Born calculations and some not, the above authors point out that these experimentally derived values are very sensitive to the detector parameters.

We end this brief survey intentionally on two examples that suggest that there may be difficulties in the analysis of the cusp. Furthermore, we hope that this survey may serve to strenghten the idea that the connection between the dynamics of fast atomic collisions that render electrons moving slowly in the field of the projectile and the experimentally accessible zero degree cusp can be exploited to give useful information. We now turn our attention to the standard technique used to characterize experimental cusps.

3. CHARACTERIZATION OF THE CUSP

The standard and preferred method used for analysis of the cusp utilizes a multipole expansion of the differential cross section in the projectile frame. The method requires that the projectile-frame double differential cross section be transformed to the laboratory frame and then convoluted with the analyzer/detector function. The multipole expansion analysis follows the works of Dettmann et al (1974) for ECC and Briggs and coworkers for ELC; both of these early analyses are based on solid theoretical ground, namely, the threshold behavior of charged particles in a Coulomb field studied by Wigner (1948). Since the cusp electrons of primary interest are those that move very slowly in the field of the projectile we write the double differential cross section in the projectile frame as:

$$(ddcs)_p=(d\sigma/\ dE'd\Omega') = \Sigma_l\ \beta_l(E')\ P_l(\cos\ \theta') \qquad (1)$$

for the azimuthally symmetric case and where θ' is the scattering angle of an electron of energy E'. The expansion parameters $\beta_l(E')$ are often expanded in a Taylor series about the electron speed in the projectile frame, v'. However, we shall keep the energy dependence of β_l explicitly in terms of E' for convenience, generality and because the expansions have nothing to do with the problem we wish to bring forward.

It is not possible to measure $(ddcs)_p$ directly; every electron analyzer/detector system has its own "view" of the interaction region from which the measurable fluxes flow. Thus, it is necessary to develop procedures to find the $(ddcs)_p$ that are consistent with the measured electron distributions. A relationship between the measured yield and the double differential cross section in the laboratory frame, $(ddcs)_L$, is given by

$$\text{Yield} \propto \int\int S(E_L,\Omega_L)\ (ddcs)_L\ dE_L\ d\Omega_L \qquad (2)$$

where all the quantities are in the laboratory system and $S(E_L,\Omega_L)$ is a detector function that contains the details of the interaction region and the analyzer/detector system. The laboratory-frame differential cross sections are related to the projectile-frame differential cross sections by

$$(ddcs)_L = (ddcs)_p\ v_L/v' = (ddcs)_p\ (E_L/E')^{1/2} \qquad (3)$$

where v_L is the speed of the electron in the laboratory frame. Combining (2) and (3) we have

$$\text{Yield} \propto \int\int S(E_L,\Omega_L)\ (E_L/E')^{1/2}\ (ddcs)_p\ dE_L\ d\Omega_L \qquad (4)$$

which tells us to integrate the transformed $(ddcs)_L$ over the "resolution volume". Such a procedure, using a resolution volume in terms of electron velocity instead of energy, was first discussed by Briggs and Day (1980) and utilized by Meckbach et al (1981) in ECC cusp analysis. In addition, it is common practice to take $S(E_L,\Omega_L)=R(E_L)\ G(\theta_L)$ (for cases where there is no Φ dependence) and to make some analytical approximations for $R(E_L)$ and $G(\theta_L)$ in order to perform the integration indicated in (4).

We believe that the use of equation (4) could lead to spurious results for certain important situations, among them, the analysis around the peak of a cusp measured at zero degrees, ie around $v'=0$. We uncovered this problem not by addressing the specific approximations commonly used for $R(E_L)$ and $G(\theta_L)$, for they seem reasonable, but, rather, by questioning the validity of equation (3) for the frame transformation of the (ddcs) for very small projectile-frame electron speeds and detector apertures of finite size.

For apertures of finite size one has for the equality of the outgoing flux in the laboratory and projectile frames

$$(\mathrm{ddcs})_L\,(\delta E_L \delta\Omega_L) = (\mathrm{ddcs})_p\,(\delta E' \delta\Omega') \qquad (5)$$

from which it follows that

$$(\mathrm{ddcs})_p = (\mathrm{ddcs})_L\,(\delta E_L \delta\Omega_L)/(\delta E' \delta\Omega')\;. \qquad (6)$$

We shall call equation (6) the finite form of the (ddcs) transformation. The relationship between the finite form and the differential form , equation (3), is given by

$$\lim_{\delta E_L \to 0,\ \delta\Omega_L \to 0} (\delta E_L \delta\Omega_L)/(\delta E' \delta\Omega') = v'/v_L\,. \qquad (7)$$

For each value of the electron laboratory-frame speed, v_L , contained within the resolution volume of a real analyzer/detector system there is a distribution of projectile-frame *velocities* , $\underline{v}'$. Whereas the solid angle for all the v_L contained in the resolution volume is constant, the solid angles generated by each of the v' are different. These differences need to be taken into account around the peak of the cusp where v' is very small.

We have numerically calculated the quantity $(\delta E_L \delta\Omega_L)/(\delta E' \delta\Omega')$ for comparison with v'/v_L using simple functions: $R=1$ in the range from $E_L - \delta E_L/2$ to $E_L+\delta E_L/2$, zero otherwise; $G=1$ in the range $\theta_L-\delta\theta_L/2$ to $\theta_L+\delta\theta_L/2$, zero otherwise. In all cases we have found that the finite form of the transformation breaks away from the differential form for very small values of v'. For example, for an energy resolution $(\delta E_L/E_L)$ of 0.01 and an angular resolution $(\delta\theta_L)$ of 0.3° the transformation flattens out around $E'=0.1$ eV while the differential form (v'/v_L) plummets to zero. Thus one is alerted to the fact that even if an analyzer/detector system can be specified adequately by energy and angular resolution functions the yield calculated using equation (4) is likely to give specious results for very small values of v' unless the finite form of the transformation is used.

It is noteworthy that the analysis of the zero and near zero degree peaks for the case of the $^1P^o$ shape resonance in single-ELC of H^- did not require the use of points very near the peak of the cusps for the fitting routines. On the other hand, it is believed that the lack of consistency we seem to have regarding the double-ELC data is in large part due to the fact that the fitting procedures used to develop the parameters are especially sensitive to the very low projectile-energy part of the cusp.

Our computer electron trajectory calculations of the angular acceptance of our analyzer/detector system show that the angle distribution functions, the $G(\theta)$, can have unexpected shapes at and very near zero degrees. The use of a calculated numerical function for $G(\theta)$ and the finite form of the transformation factor in equation (4) makes the integration over the resolution volume cumbersome. In the next section we propose a numerical procedure that replaces the integration over the resolution volume by an integration over real space , allows for dealing with the finite size and density distributions of particle beams and does not require a priori assumptions about energy and angular resolutions.

4. A ZERO DEGREE SOLUTION

The analysis of zero degree cusps near the peak requires that a calculated or guessed or otherwise assumed projectile-frame differential cross section be convoluted with a function that correctly describes the analyzer/detector system to obtain a calculated yield for comparison with experimental yields. Equation (4) provides a prescription for convoluting a projectile-frame differential cross section even for very small v' provided that the finite form of the transformation is used. However, integration over the "resolution volume" requires precise knowledge of the analyzer/detector characteristics. Our work on the double-ionization of H^- demanded that we pay very close attention to the analyzer/detector characteristics at zero degrees in order to make the best use of equation (4). We calculated the energy and the angular characteristics of our detection system using a computer program developed over the last three years. For our analyzer/detector system, which can be characterized as having good energy resolution, 1%, and good to excellent angular resolution, from about 2° to 0.3°, we find: (1) the shape of the energy functions to be typical and independent of the detector setting; (2) the shape of the angular functions to be atypical and to depend on the detector setting for the first few degrees. (Details of these numerical results will be presented elsewhere.) The above result on the angular functions further complicates the use of equation (4) for the analysis of a cusp within the first few degrees.

We propose an alternative to the use of equation (4) for the analyses of cusps for the first few degrees. The approach is to make straightforward numerical calculations of the electron trajectories from a statistically significant sample of the interaction region. One is interested in specifying those trajectories that pass through the analyzer/detector system and noting the locations of the trajectories at some plane, for example, the plane of a position sensitive detector.

The yield as a function of the angle of the analyzer/detector, Θ_d, and the voltage of an electrostatic analyzer, V_d, in terms of an integral over the interaction region is given by

$$Y(\Theta_d,V_d)=I^\circ\int dE_L \int (ddcs)_L \ (\delta S_L/R^2)\ N(x,y,z)\ dV \quad (8)$$

where:

I°	is the flux of the incident uniform beam
δS_L	is the surface generated by electrons of energy E_L from a point (x,y,z) that *pass through the analyzer/detector system*
$\delta S_L/R^2$	is the corresponding solid angle
R	is the magnitude of the vector from the point to the centroid of δS_L
N(x,y,z)	is the number density of target scattering centers at (x,y,z)
V	is the volume of the interaction region, it is specified by the incident beam, cross beam, and analyzer/detector.

Our procedure is to numerically integrate (8) over the volume V. The interaction region is sampled by selecting a statistically large number of random points (x,y,z). For a given electron energy E_L and for a selected point (x,y,z) one calculates a large number of trajectories through the analyzer. Each trajectory is labeled by its laboratory-frame parameters E_L, Ω_L as well as its projectile-frame parameters E', Ω'. Some trajectories are detected, some are not; the envelope of those trajectories that pass through the analyzer/detector encloses the surface element δS_L. Each surface element δS_L is readily expressable in terms of smaller surface elements $\delta S_P(E_i', \Omega_i')$. The $\delta S_P(E_i', \Omega_i')$ are taken to be sufficiently small that they can be specified by specific values of (E_i', Ω_i') and that the differential form of the transformation, equation (3), can be used to weight each one by the corresponding value of $(ddcs)_p(E_i')^{-1/2}$. In addition, trajectories coming from a point are weighted by the target number density N(x,y,z). The process is repeated for a statistically significant number of points in the volume V and over the entire range of the energy bandpass of the analyzer. In this way one can evaluate the integral in (8) and account for the characteristics of the analyzer/detection system at the same time. Equation (8) can be rewritten as

$$Y(\Theta_d, V_d) = I^\circ \int (E_L)^{1/2}\, dE_L$$

$$\int \Sigma_i((E_i')^{-1/2} (\Sigma_l \beta_l(E_i') P_l(\theta_i')) (\delta S(E_i', \Omega_i')/R^2))\, N(x,y,z)\, dV \quad . (9)$$

Although the numerical integration of equation (9) is tedious it is straightforward and does not require any assumptions about the analyzer/detector system. We have programmed the differential equation for the trajectories through our cylindrical mirror analyzer region in order to deal with trajectories with an angular momentum about the coaxis. We have also included subroutines to calculate the N(x,y,z) due to the cross beam.

In summary, the cusp at zero degrees contains important information about fast atomic collisions that give rise to final states where one or more electrons are left moving slowly in the field of the projectile. Our studies indicate that the use of the differential form of the transformation of the (ddcs) and/or simple analytical functions to describe the angular properties of detector systems at and near zero degrees is likely to lead to inconsistencies. Our recommendation is clear: cusps at zero and near zero degrees need to be treated carefully!

* Permanent address: Department of Physics and Astronomy, The University of Georgia, Athens, Georgia

REFERENCES

Andersen L H, Frost M, Hvelplund P, Knudsen H and Datz S 1984 Phys. Rev. Lett. **52** 518-21

Atan H, Steckelmacher W and Lucas M W 1990 J. Phys. B: At . Mol. Opt. Phys. **23** 2579-2593

Briggs J and Day M H 1980 J. Phys. B: At. Mol. Opt. Phys **13** 4797-810

Datz S, Hippler R, Andersen L H, Dittner P F, Knudsen H, Krause H F, Miller P D, Pepmiller P L, Rosseel T, Schuch R, Stolterfoht N, Yamazaki Y and Vane C R 1990 Phys. Rev. **41** 3559-3571

Drepper F and Briggs J S 1976 J Phys. B: At. Mol. Opt. Phys. **7** 267-87

Duncan M M and Menendez M G 1990 Phys. Rev. A **41** 2858

Elston S B, Berry S D, Burgdorfer J, Sellin I A, Brenig M, DeSerio R, Gonzalez-Lepera C E, Liljeby L, Groeneveld K-O, Hofman D, Koschar P and Nemirovsky I B E 1985 Phys. Rev. Lett. **55** 2281-2284

Jagutski O, Koch R, Skutlartz, Kelbch C and Schmidt-Böcking H 1990 submitted to J. Phys. B: At. Mol. Opt. Phys.

Jakubassa-Amundsen D H 1989 J. Phys. B: At. Mol. Opt. Phys. **22** 3989-99

Knudsen H, Andersen L H and Jensen K E 1986 J. Phys. B: At. Mol. Opt. Phys. **19** 3341

Köver A, Sarkadi L, Palinkas J, Berenyi D, Szabo Gy, Vajnai T, Heil O, Groeneveld K-O, Gibons J and Sellin I A 1989 J. Phys. B: At. Mol. Opt. Phys. **22** 1595-602

Liu C R and Starace A F 1989 Phys. Rev. A **40** 4926

Macek J 1970 Phys. Rev. A **1** 235-41

Meckbach W, Nemirovsky I B and Garabotti C R 1981 Phys. Rev. A **24** 1793-802

Menendez M G and Duncan M M 1987 Phys. Rev. A **36** 1653

Reinhold C O and Schultz D R 1989 J. Phys. B: At. Mol. Phys. **22** L565

Sarkadi L, Palinkas J, Köver A, Berenyi D and Vajnai T 1989 Phys. Rev. Lett. **62** 527-30

Sorensen J, Andersen L A, and Nielsen L B 1988 J. Phys. B: At. Mol. Opt. Phys. **21** 847-58

Vane C R, Sellin I A, Elston S B, Suter M, Thoe R S, Alton G D, Berry S D and Glass G A 1979 Phys. Rev. Lett. **43** 1388-91

Wigner E P 1948 Phys. Rev. **73** 1002

Coincidence Measurements of Electron Capture and Loss in Ion-Atom Collisions

R.D. DuBois
Pacific Northwest Laboratory
P.O. Box 999
Richland, WA 99352 USA

Abstract:

Collisions between fast, fully stripped projectiles and atomic targets predominantly result in target electrons being ejected to the continuum. For fast partially stripped projectiles which bring weakly bound electrons into the collision, projectile ionization can also contribute to the observed electron spectra. At lower impact velocities, electron capture by the projectile ion becomes important and higher order processes, often referred to as transfer ionization, can be a significant source of free electrons. In recent years, coincidence techniques have been used to evaluate the relative importance of electron capture and loss in free electron production, to separate the capture and loss contributions from those resulting from target ionization alone, and to provide more detailed information about electron capture and loss mechanisms than is available from total cross section measurements. A brief survey of these experiments will be presented.

Introduction:

Fast charged particles traversing gaseous media gradually lose energy and slow down as they interact with individual target atoms. The energy transferred to the target as a result of these interactions often results in ionization of a bound target electron; excitation to discrete excited states can also occur. If the projectile also contains loosely bound electrons, projectile ionization can be an additional source of free electrons. Thus, studying the ejected electron spectra can provide information about the interaction and energy loss processes.

For the purpose of this paper, ionizing collisions will arbitrarily be classified as follows. 1) Fast, fully stripped ion impact where target ionization dominates, 2) lower velocity collisions involving fully, or partially, stripped ion impact where electron capture becomes important, and 3) fast collisions, involving projectile ions having loosely bound electrons of their own, where electron loss from the projectile is important.

For category 1, numerous experimental and theoretical studies have demonstrated that the first Born approximation can adequately describe single target ionization resulting from fully stripped ion impact. Typically the theoretical predictions are poorer for heavier targets because of the increased probability for multiple electron emission and because more sophisticated target wave functions are required. In general, however, our understanding of these collisions is rather good.

Category 2, electron capture by the projectile, becomes important at lower impact velocities, i.e. where the impact velocity is approximately equal to, or smaller than, the velocity of the bound target electron. Simple electron capture results in target ionization, but higher order processes, such as transfer ionization, can liberate additional target electron(s) to the continuum. Our present theoretical understanding of this process is much poorer than for fast, fully stripped ion impact because perturbative techniques such as the Born approximation are inappropriate at these low velocities. More sophisticated theoretical treatments are required. However these processes are attractive from a theoretical viewpoint since they involve a two electron transition with one of the final states being a discrete bound state while the other is a continuum state.

For interactions where the projectile brings loosely bound electrons of its own into the collision, category 3, electron loss tends to be important. Although the collision velocities are sufficiently large that perturbative techniques can be applied, the theoretical description must consider ionization of the target **and** of the projectile **with, and without**, simultaneous excitation of the other collision partner. Each of these processes needs to be modeled and calculated independently and then summed to yield information about the observed electron emission spectra. In addition, for both target and projectile ionization, the interaction involves a screened nuclear charge. Hence modeling these collisions and comparing with experimental data is often complicated.

Therefore, understanding collisions where electron capture or loss occur is one of the next major steps in extending our knowledge of ion-atom collisions. This requires experimental data capable of a) steering the theoretical models in the proper direction by indicating the relative importance of the various ionization processes and b) testing the models in order to identify their successes and failures. Sometimes it is possible to use total cross section or differential electron emission measurements obtained using non-coincidence techniques for these purposes. However, in recent years coincidence techniques have been applied in order to investigate electron capture and loss mechanisms in greater detail. For a summary of the work done in this field prior to 1980 see reference 1.

In this paper examples of various coincidence methods will be presented in order to demonstrate what information can be derived from each. I wish to emphasize that the examples cited were chosen for illustrative purposes and I do not mean to imply that they represent the only research efforts in this field. The experiments discussed are intended to demonstrate what additional knowledge can be derived using these techniques and also to serve as a starting point for obtaining references and information in this field.

Because of space limitations, examples will be restricted to those pertaining to the emission of free electrons. One technique that has been applied toward this end is the study of coincidences between charge state analyzed projectile ions and either recoil ions or emitted electrons. Another technique is to study coincidences between scattered projectile ions and again either recoil ions or emitted electrons. From these measurements a better understanding of these ionizing collisions is gradually emerging.

Electron Capture:

As previously stated, for projectile velocities comparable to or less than the velocity of the bound target electrons, electron transfer from the target to the projectile becomes important. For example, total cross section measurements for H^+ and He^+ impact on various atomic and molecular targets [2-4] have shown that the probability of electron capture can exceed the total ionization cross section (meaning the electron production cross section) at low impact velocities. When the capture cross section is comparable to or larger than the electron production cross section, higher order effects, such as transfer ionization, can then become relatively important in the production of free electrons.

a) Projectile ion-recoil ion coincidences:

At the Pacific Northwest Laboratory we have used the charge state analyzed projectile ion-recoil ion coincidence technique to study light ion impact on various atomic targets.[4-6] These studies showed that the relative number of free electrons produced via the electron capture process, in other words the transfer ionization process, was larger for targets possessing more loosely bound electrons, e.g. Ar, Kr, than for those with fewer or more tightly bound electrons, e.g. He, Ne. However, even for H^+ and helium ion impact, transfer ionization was found to be a significant source of electrons in slow collisions. This is demonstrated in Fig. 1.

Even larger probabilities were found for heavier or more highly charged projectile ions. This can also be seen in Fig. 1 but a better example involves slow highly charged ion impact on helium [7] where transfer ionization can dominate in the production of free electrons. The Leningrad and Belfast groups have also used projectile ion-recoil ion coincidence techniques to study these processes [8-10].

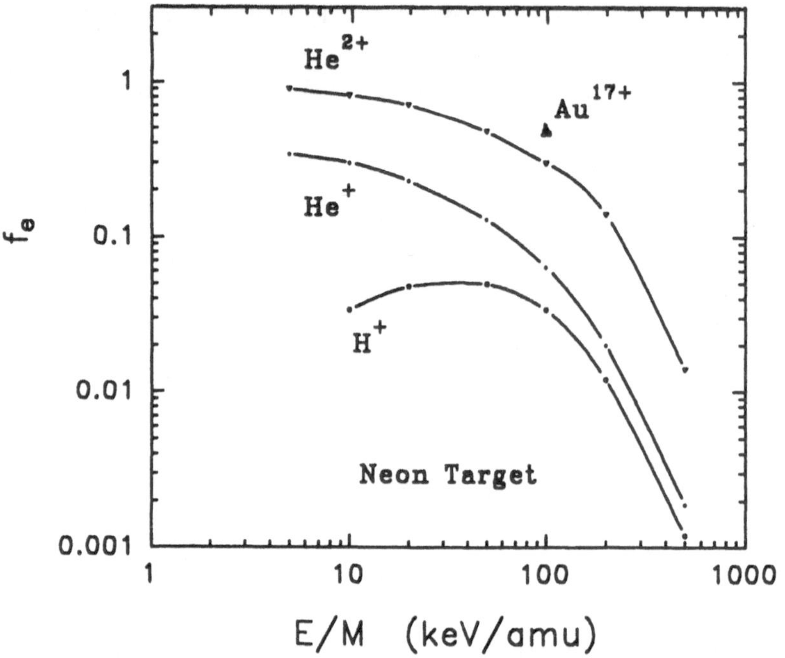

Fig. 1 The fraction of the total electron production resulting from higher order electron capture processes for various projectiles colliding with neon. H^+ impact, ref. 6; He^+ impact, ref. 4; He^{++} impact, ref. 5; Au^{17+} impact, ref. 7.

b) Projectile ion-emitted electron coincidences:

Transfer ionization can be modeled in various ways. The simplest model uses an independent particle approximation where the transfer of one target electron to the projectile is treated independently of the ionization of a second electron to the continuum, the total probability then being the product of the capture probability, $P_c(b)$, and the ionization probability, $P_i(b)$, where b is the impact parameter. A second model assumes that electron-electron interactions are important, the Thomas double scattering model being a prime example. [11] In order to distinguish between these two models, differential electron spectra, measured in coincidence with charge state analyzed projectile ions, are useful.

In Fig. 2 doubly differential cross sections (DDCS) for the total electron production and for electrons originating from transfer ionization (TI) are shown. [12] Although more extensive data are required to determine the mechanism involved, these data indicate that the relative fraction of the total free electron production obtained from data as discussed in Fig. 1 is also reflected in the differential electron spectra. For example for 90° electron emission in neon, the TI DDCS are roughly a factor of ten smaller than the total DDCS whereas when integrated over all emission angles and electron energies the TI contribution is roughly 5% of the total electron emission.

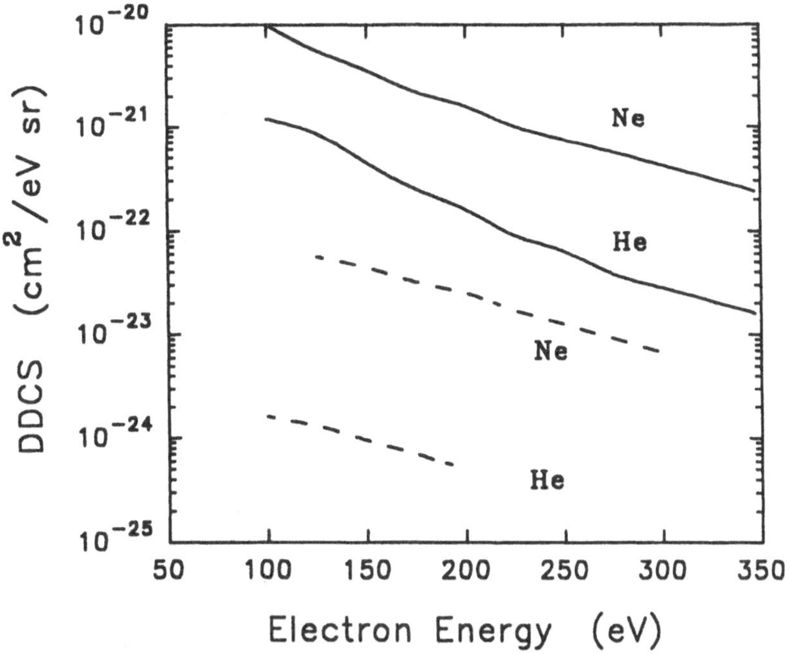

Fig. 2 Doubly differential cross sections for electron emission occurring in 300 keV H^+ - He, Ne collisions. Solid lines, total electron emission; dashed lines, electron emission resulting from electron capturing collisions. Data are from ref. 12.

In another experiment where coincidences between emitted electrons and projectiles that captured an electron from the target were investigated, Pálinkás et al. [13] recently demonstrated that the Thomas double scattering mechanism is responsible for the TI events occurring in fast proton-Ar collisions.

There also exists another electron capture mechanism which can result in the production of free electrons-namely electron capture to the continuum (ECC). In this process a target electron is "captured" into a continuum state of the projectile. This results in an enhancement of the number of electrons having a direction and velocity matching that of the projectile ion. The TI process can also produce electrons in the forward direction having an enhanced probability for velocities matching that of the projectile. By using charge analyzed projectile ion-zero degree emitted electron coincidence techniques, it is possible to separate these two processes from each other and from a third process that occurs when a loosely bound electron is lost from a projectile ion. This latter process, which will be discussed in a following section, is electron loss to the continuum (ELC).

As an example of the ECC, TI and ELC processes, consider He^+ impact. An ECC process will result in a zero degree electron - He^+ coincidence. TI will yield an electron in coincidence with a neutral helium atom and ELC will yield an electron-ionized projectile (He^{++}) coincidence. Fig. 3 shows the relative importance of these three processes for 75 keV/amu He^+ - He collisions. These data were obtained in a collaborative effort between the Debrecen and Univ. Frankfurt groups. [14] In this example, transfer ionization is considerably less important than is capture to the continuum. However other studies using highly charged ions [15,16] showed TI to be the dominating contributor to the cusp peak.

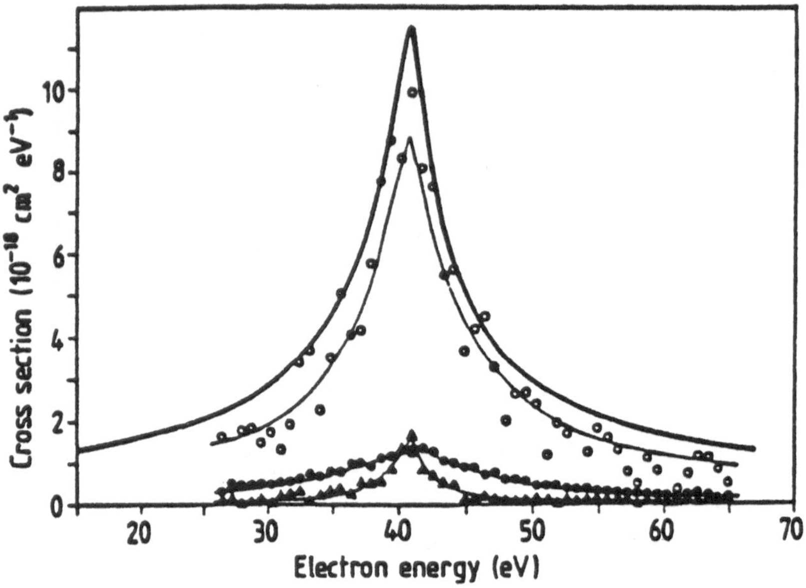

Fig. 3 Zero degree cusp electron production in 75 keV/amu He^+ - He collisions indicating the relative importance of the ECC, open circles, TI, filled triangles, and ELC, filled circles, mechanisms to the total electron production, heavy solid line. Data from ref. 14.

c) **Scattered projectile ion-recoil ion or emitted electron coincidences:**

Another method of studying the mechanisms leading to electron emission associated with electron capture is to investigate the cross sections as a function of impact parameter. This can be done by measuring the recoil ion production as a function of the projectile ion scattering angle. According to the classical Thomas scattering mechanism, a signature of these events would be an enhancement in the differential scattering probability at 0.55 mrad in proton-helium collisions and, in addition, the production of a doubly charged helium ion. In Aarhus, using this method a strong enhancement in the amount of doubly ionized target ions at the critical angle was found. [17] These data, shown in Fig. 4, were interpreted as evidence of the double scattering Thomas mechanism in the TI process although other interpretations are possible.

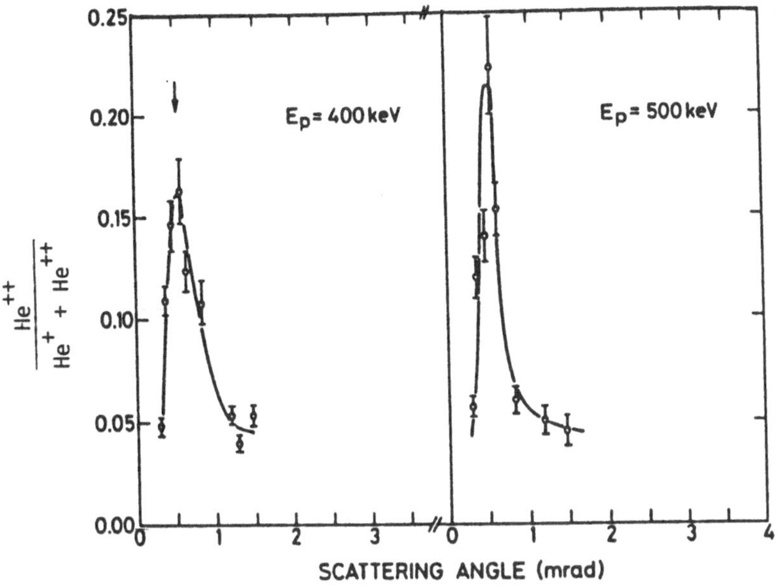

Fig. 4 Double to single ionization probabilities as a function of projectile scattering angle in $H^+ + He \rightarrow H + He^{q+} + (q-1)e^-$ collisions. Data from ref. 17.

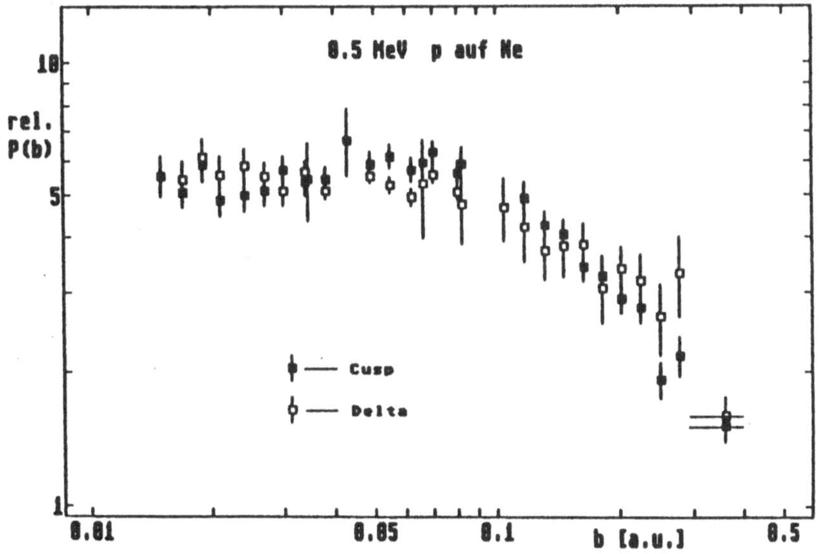

Fig. 5 Impact parameter dependencies of the cusp electron production and continuum target electrons with similar energies measured for 500 keV H^+ - Ne collisions. Data from ref. 19.

By measuring the scattering angles of charge state analyzed projectile ions in coincidence with zero degree electron emission, the groups at Kansas State

University [18] and the University of Frankfurt [19] have studied the ECC and TI processes. Heavy (fluorine) and light (protons, helium ions) projectiles were used at the two laboratories respectively. For light ion impact, one of the results of this work was the demonstration that cusp electrons and continuum (delta) electrons of similar energies have identical impact parameter dependencies (see Fig. 5). This confirms that ECC cusp electrons are delta electrons that are swept in the forward direction due to the coulomb attraction of the projectile ion. From these examples, it should be clear that the use of coincidence methods has greatly contributed to our present understanding of ionization associated with electron capture processes. Certainly other interesting studies have been performed but the examples and references cited here should provide the interested reader with an overview of the field.

Electron loss:

Let us now turn our attention to collisions involving projectiles containing loosely bound electrons of their own. For these collisions an additional electron production mechanism, namely projectile ionization or electron loss to the continuum (ELC), exists. In these collisions either the target, or the projectile, can be ionized and, in either case, the collision partner can remain in the ground state or be excited to a discrete or continuum state. Also in both cases the coulomb interaction involves a screened nuclear charge. Thus interpreting the observed electron spectrum is vastly more complicated than for fully stripped ion impact.

In recent years several experiments have been performed where coincidence methods have been used to separate and identify the various ionization processes occurring in these collisions. Examples of a few of these experiments are given below.

a) Projectile ion-recoil ion coincidences:

Using projectile ion-recoil ion coincidence techniques in our laboratory, we have investigated the relative importance of the electron loss channels toward to total electron production resulting from light ion impact. [4] In Fig. 6 the total electron loss cross sections along with the results of the coincidence measurements are shown for He^+ impact on helium and argon. The nomenclature used for the cross sections is that the number indicates the final target charge state, e.g. 1,2,.. imply that the target was singly, doubly, et cetera, ionized in a collision where the He^+ projectile was also ionized. Due to the coincidence technique employed, pure electron loss by the projectile cannot be investigated since it does not form a target recoil ion. However by summing the cross sections (∇) for simultaneous projectile-target ionization (where the number is non zero) and comparing with total loss cross sections obtained from a separate experiment, the pure electron loss cross section (dashed curve labeled by 0) can be deduced.

As can be seen in Fig. 6, for a light target such as helium roughly half of the time ionization of the projectile results in target ionization as well. For a heavier target, the sum of the coincidence cross sections equals the total loss cross section implying that pure projectile ionization is an unlikely event; collisions where projectile ionization occur <u>always</u> cause target ionization.

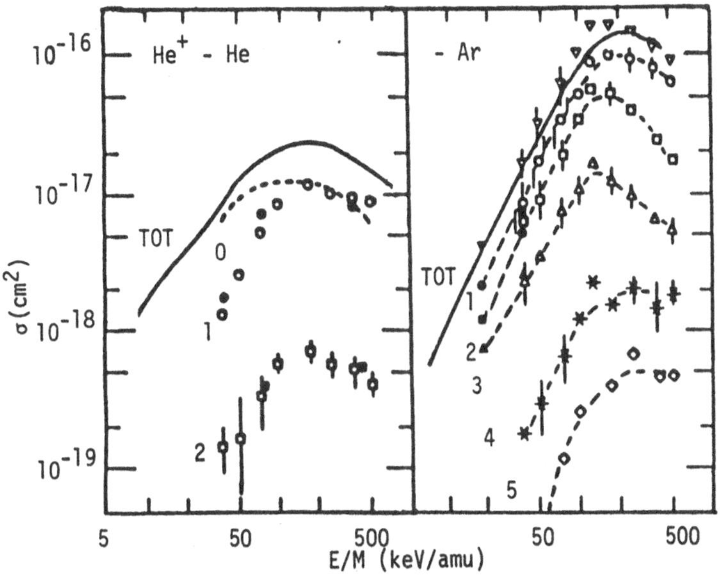

Fig. 6 Cross sections for ionization occurring in He^+ - He, Ar collisions. Solid curves, total cross sections for electron loss by the projectile taken from refs. 3 and 20. Dashed curves though symbols, cross sections for ionization of both the projectile and the target in the collision. For argon, ∇ denotes the sum of the cross sections for simultaneous projectile-target ionization. Data taken from ref. 4. The number denotes the degree of target ionization. For the helium target the curve denoted by zero is for pure electron loss by the projectile. See text for details.

Projectile ion-recoil ion coincidence data also can be used to determine the relative importance of the electron loss channels toward the total electron production. At higher impact energies, electron capture is relatively unimportant. Thus the total electron production comes from 1) ejecting an electron from the target, or 2) ejecting an electron from the projectile, or 3) ejecting electron from both the target and the projectile. The latter case will be called simultaneous ionization although there is no experimental or theoretical evidence implying that this is a correlated or uncorrelated process.

Examples of the relative importance of these three processes for He^+ impact on helium and argon are shown in Fig. 7. The contribution of projectile electrons, process 2, (designated by P) are seen to be roughly independent of the target for fast collisions. At lower impact energies there is a greater contribution for the lighter target because the number of loosely bound target electrons is larger for the heavier target and hence the target ionization contribution increases.

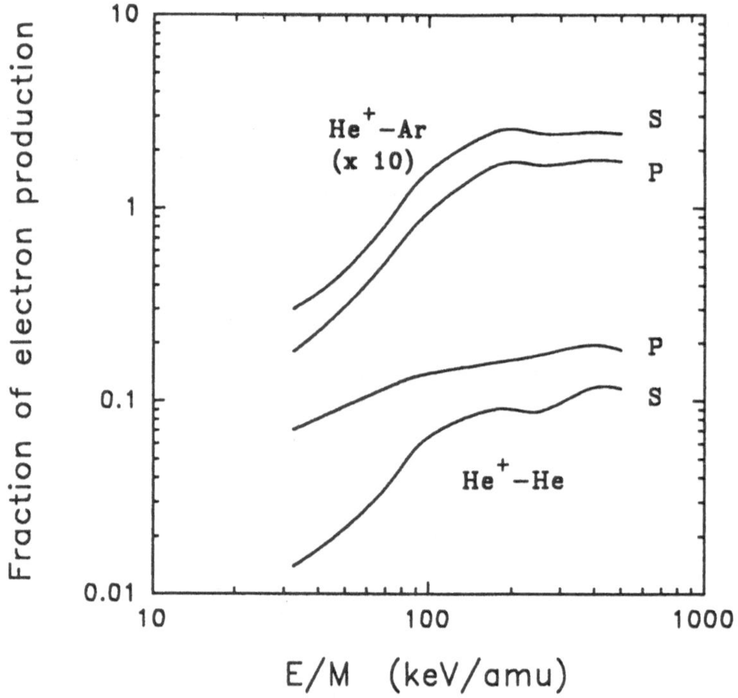

Fig. 7 The relative importance of electrons originating from the projectile (P) and those originating from the target in collisions where the projectile is also ionized (S) for He^+ impact on helium and argon targets. The argon data have been shifted by a factor of 10 for display purposes only. Data from ref. 4.

The remainder of the electrons originate from the target. However, a portion of these electrons are emitted in collisions where the projectile is also ionized, process 3 as described above. This process is of interest since a two electron transition has taken place. The contribution of these electrons (designated by S in Fig. 7) is relatively important for both light and heavy targets. The combined processes, 2 and 3, indicate that, at higher impact energies a third to nearly half of the emitted electrons result from collisions where electron loss by the projectile occurs.

b) Projectile ion-emitted electron coincidences:

Additional details of the projectile and simultaneous ionization processes can be obtained by studying the differential electron emission spectra in coincidence with projectile ions that have lost an electron during the collision. These data can be used to test our theoretical understanding of these processes. Several years ago we reported data for He^+ - He collisions where electron emission angles of 20° and 30° were studied. [21] Comparisons of the experimental data with first Born calculations indicated that the theoretical treatment of target and projectile ionization were reasonably good but the theoretical model used severely underestimated simultaneous ionization events. In a recent study, [22] we extended this study to other targets. For an argon target, a wider range of emission angles was studied. In this case very poor agreement between experiment and theory was found.

In order to clarify the situation, additional experimental work has recently been completed at the University of Frankfurt. [23] Fast neutral hydrogen and helium beams were collided with a helium target in order to provide data for benchmark testing of theory. An example of these data are shown in Fig. 8. The large peak centered near 500 eV is due to projectile ionization since the majority of the projectile electrons are emitted with small kinetic energies in the moving projectile reference frame. Hence, in the laboratory frame, they appear as a peak centered on an electron velocity equal to the projectile velocity. The coincidence measurement (solid diamonds) confirms that the electrons in this peak are associated with ionization of the projectile. In addition the coincidence measurement indicates that a significant number of the electrons emitted with energies smaller and larger than this peak are also associated with projectile ionization. These electrons were originally identified by DuBois and Manson [21] as being target electrons that were emitted in simultaneous ionization events.

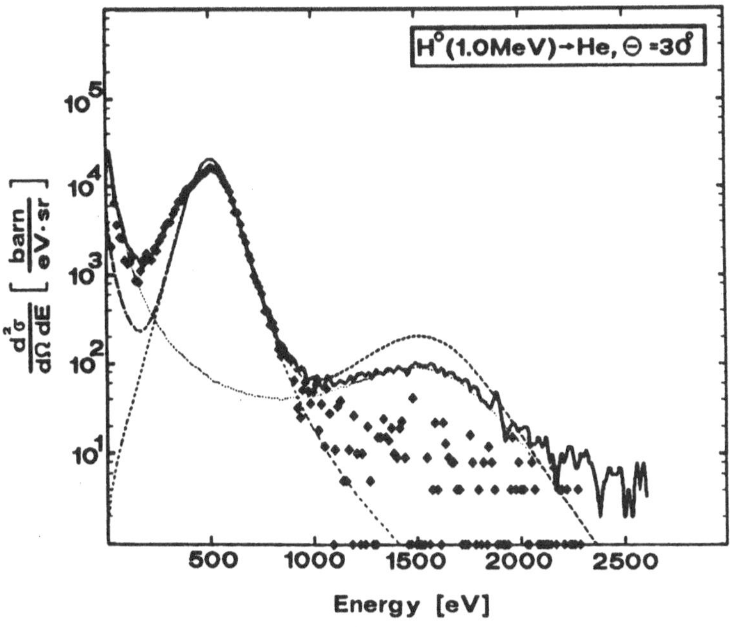

Fig. 8 DDCS for electron emission in 1 MeV H - He collisions. Experiment: solid line, total electron emission at 30°; ♦, electron-ionized projectile coincidence data. PWBA theory: -- -- --, total electron emission; - - - -, projectile ionization, ..., target ionization for proton impact. Data from ref. 23.

The dotted and dashed curves in Fig. 8 are theoretically predicted cross sections obtained using a simple PWBA calculation. Comparisons between experiment and theory for other impact energies and angles indicated that projectile and target ionization are handled reasonably well by the PWBA theory. Thus, it was concluded that the discrepancies previously noted for He^+ - Ar collisions are due to inadequate wave functions used in the calculation. Additional details and data for neutral hydrogen and helium impact can be obtained from another paper in this set of proceedings.

Coincidence measurements between ionized projectile ions and zero degree electron emission (cusp electrons) have also been studied by the Debrecen [14] and ORNL groups. [1] Data from reference 14 are shown in Fig. 3 of this paper.

c) Scattered projectile ion-emitted electron coincidences:

At the University of Frankfurt, impact parameter measurements of the ELC process have recently been performed. [19] Electron emission at 180° was investigated for He^+ - Ne. A constant ionization probability as a function of impact parameter was found (see Fig. 9). This is understandable since electron loss at 180° results from direct collisions between the projectile electrons with the target nucleus without influencing the projectile trajectory.

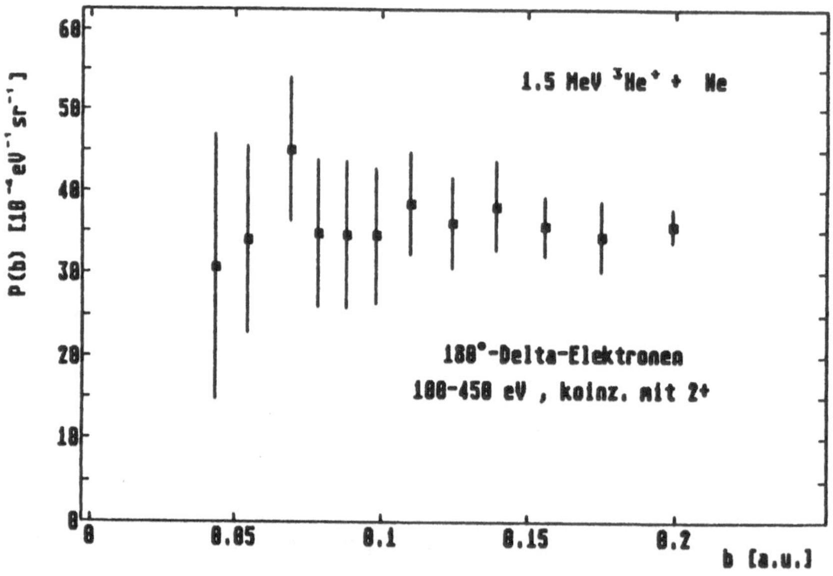

Fig. 9 The impact parameter dependence electron loss from the projectile observed at 180° in 500 keV/amu He^+ - Ne collisions. Data from ref. 19.

Conclusion:

Although this brief survey by no means covers all investigations of the electron capture and loss processes, it should provide a flavor of the range of experimental methods being used to unravel the mechanisms involved. Clearly the next generation of experiments will be even more sophisticated. But the studies already undertaken or in progress have provided many new insights into these complicated processes. All of the examples cited are useful in providing details about these processes; however more information is obtainable from experiments where the differential electron emission or the impact parameter dependencies are measured. Hopefully additional studies of these types will be performed in the near future.

Acknowledgments: This work supported by the Office of Health and Environmental Research (OHER), U.S. Department of Energy, under contract No. DE-AC06-76RLO 1830.

References:

1. M. Breinig, S.B. Elston, S. Huldt, L. Liljeby, C.R. Vane, S.D. Berry, G.A. Glass, M. Schauer, I.A. Sellin, G.D. Alton, S. Datz, S. Overbury, R. Laubert and M. Suter, Phys Rev A 25, 3015 (1982).
2. M.E. Rudd, R.D. DuBois, L.H. Toburen, C.A. Ratcliffe and T.V. Goffe, Phys Rev A 28, 3244 (1983).
3. M.E. Rudd, T.V. Goffe, A. Itoh and R.D. DuBois, Phys Rev A 32, 829 (1985).
4. R.D. DuBois, Phys Rev A 39, 4440 (1989).
5. R.D. DuBois, Phys Rev A 36, 2585 (1987).
6. R.D. DuBois, Phys Rev Lett 52, 2348 (1984).
7. H. Damsgaard, H.K. Haugen, P. Hvelplund and H. Knudsen, Phys Rev A 27, 112 (1983).
8. V.V. Afrosimov, Y.A. Mamaev, M.N. Panov and V. Uroshevich, Zh Tekh Fiz 37, 717 (1967), [Sov Phys Tech Phys 12, 512 (1967)].
9. V.V. Afrosimov, Y.A. Mamaev, M.N. Panov and N.V. Fedorenko, Zh Tekh Fiz 39, 159 (1969), [Sov Phys Tech Phys 14, 109 (1969)].
10. M.B. Shah, P. McCallion and H.B. Gilbody, J Phys B 22, 3037 (1989).
11. L.H. Thomas, Proc Roy Soc London 114, 561 (1927).
12. R. Hippler, G. Schiwietz and J. Bossler, Phys Rev A 35, 485 (1987).
13. J. Pálinkás, R. Schuch, H. Cederquist and O. Gustafsson, Phys Rev Lett 63, 2464 (1989).
14. A. Kövér, L. Sarkadi, J. Pálinkás, D. Berényi, Gy Szabó, T. Vajnai, O. Heil, K.-O. Groeneveld, J. Gibbons and I.A. Sellin, J Phys B 22, 1595 (1989).
15. S. Datz, C. Bottcher, L.H. Andersen, P. Hvelplund and H. Knudsen, Nucl. Inst and Meth B10/11, 116 (1985).
16. T.A. Underwood, M. Breinig, C.C. Gaither III and J. Freyou, Phys Rev A 38, 6138 (1988).
17. E. Horsdal, B. Jensen and K.O. Nielsen, Phys Rev Lett 57, 1414 (1986).
18. A. Skutlartz, S. Hagmann and H. Schmidt-Böcking, J Phys B 21, 3609 (1988).
19. O. Jagutzki, Diplomarbeit (University of Frankfurt) (1989).
20. I.S. Dmitriev, V.S. Nikolaev, L.N. Fateeva and Ya.A. Teplova, Sov Phys JETP 15, 11 (1962).
21. R.D. DuBois, Phys Rev Lett 57, 1130 (1986).
22. R.D. DuBois and S.T. Manson, Phys Rev A 42, 1222 (1990).
23. O. Heil, R. Maier, M. Kuzel, K.-O. Groeneveld and R.D. DuBois, submitted to Phys Rev Lett (1990).

ELECTRON CAPTURE TO THE CONTINUUM

B L Moiseiwitsch
Department of Applied Mathematics and Theoretical Physics
The Queen's University of Belfast, Northern Ireland

Abstract. The second–order Oppenheimer–Brinkman–Kramers approximation is used to obtain a simple analytical formula, evaluated to the lowest order in the fine structure constant α in the numerator, for the differential cross section for electron capture to the continuum (ECC) by incident bare ions having velocity v from target hydrogenic atomic systems. Both non–relativistic and relativistic forms are derived. Comparison of the theory with the experimental data of Dahl (1985) and Andersen et al (1986) for H^+, He^{2+} + He collisions is reasonably satisfactory for energies > 50 keV/amu. However, theory shows that although the velocity dependence obtained by Andersen et al is $v^{-11.3\pm0.2}$ in the range of impact energies 1–2.6 MeV/amu, this does not mean that the asymptotic v^{-11} velocity dependence given by the non–relativistic second–order OBK cross section is nearly attained. In fact it is shown that this cannot happen until an energy > 500 MeV/amu is reached where the effect of relativity produces a significant change in the energy fall off. Also a modification of the second–order OBK approximation obtained by Shakeshaft and Spruch (1978) has been expanded to first order in the atomic number Z_P of the projectile ion to get simple formulas for the yield of continuum electrons and the cusp asymmetry factor β. The accordance with the data of Andersen et al (1986) is fair at energies > 0.5 MeV/amu/Z_P but the situation is uncertain at lower energies since there is disagreement between different experimental groups and CDW theory results in smaller values of β than the OBK2 theory.

Meckbach et al (1981),Dahl (1985),Andersen et al (1986) and Gulyas et al (1986) have performed experimental investigations of electron capture to the continuum (ECC) from target helium atoms by incident H^+ and He^{2+} ions.

Andersen et al (1986) remark that no proper higher–order Born calculation has been carried out which can be related to their work.

In this paper we report an analysis of ECC using the second–order Oppenheimer–Brinkman–Kramers (OBK) approximation evaluated to the leading order in the fine structure constant α. Since the experimental work referenced above is for atomic particles with very low atomic numbers this approximation should be valid as long as the energy is sufficiently high.

Let us consider an incident bare ion having atomic number Z_P and possessing velocity v referred to the nucleus of the target hydrogenic ion having atomic number Z_T.

If the electrons captured to the continuum have velocity v_e and the acceptance angle is θ_0 referred to the target frame, a non–relativistic analysis using the second–order OBK approximation yields for the differential cross section

$$d\sigma/dv_e = (2^{17}/5)\pi a_0^2 Z_T^5 Z_P^3 v_0^{-2}(v_0/v)^{12} f(v)[1+Z_T^2(v_0/v)^2]^{-10}$$
$$\times (v_e/v)[\{(v_e-v)^2+v_e v\theta_0^2\}^{1/2}-|v_e-v|] \quad (1)$$

where

$$f(v) = 1-275/384+5\ln 3/512+(5/2^{11})(\pi/Z_T)(v/v_0) \quad (2)$$

and $v_0 = e^2/\hbar$.

In the limit $v_e \to v$ we obtain

$$d\sigma/dv_e \to 2\pi mvB_0^0\theta_0 \quad (3)$$

where

$$B_0^0 = (2^{16}/5)a_0^2 Z_T^5 Z_P^3 (mv_0^{\;2})^{-1}(v_0/v)^{12} f(v)[1+Z_T^2(v_0/v)^2]^{-10} \quad (4)$$

We see that B_0^0/Z_P^3 is independent of Z_P.

A relativistic generalization of this differential cross section formula has also been derived.

In figure 1 we make a comparison between the values of B_0^0/Z_P^3 determined experimentally by Dahl (1985) for H^+ + He and by Andersen et al (1986) for H^+, He^{2+} + He collisions, and the curve calculated using (4) taking the variationally determined value Z_T=27/16 and multiplying by 2 to allow for the two electrons in the He atom.

The agreement between the experimental data and the OBK2 calculations is reasonably satisfactory for energies > 50 keV/amu.

We also show the OBK1 approximation curve obtained by setting f(v) = 1 in (4). It is about a factor of 3 larger than the OBK2 curve.

Andersen et al (1986) derive a $v^{-11.3\pm0.2}$ velocity dependence for B_0^0 in the range of energies 1–2.6 MeV/amu from their experimental data. This fall off with velocity corresponds to an E^{-n} with n=5.65±0.1 energy dependence which can be compared with the values of the index n given by OBK theory for He collisions in table 1.

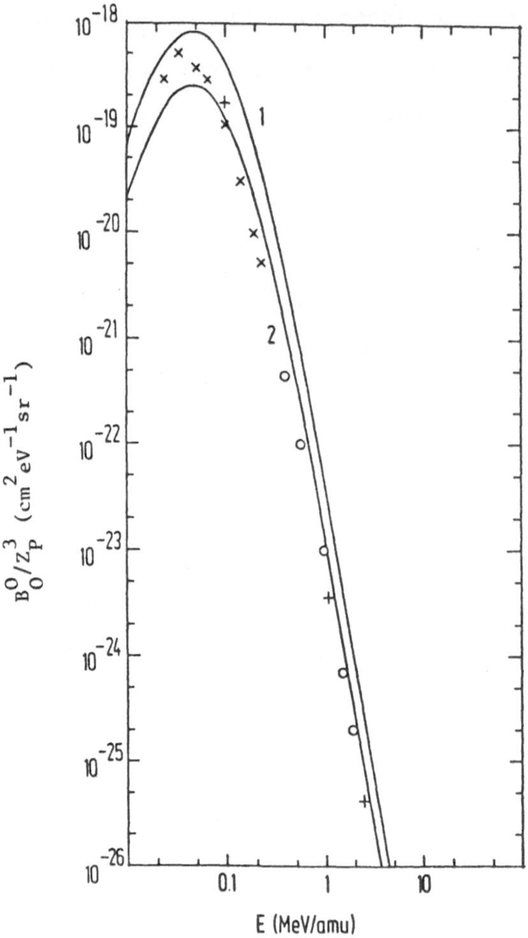

Figure 1. B_0^0/Z_P^3 (in units of $cm^2eV^{-1}sr^{-1}$) as a function of energy E (Mev/amu) for H^+ and He^{2+} + He collisions. Curve 1 : calculated using OBK1 approximation, curve 2 : calculated using OBK2 approximation. Experimental data : × , H^+ + He collisions, o , He^{2+} + He collisions Andersen et al (1986); + , H^+ + He collisions Dahl (1985).

We see that although there is approximate accordance between the experimentally derived index n in the range 1–2.6 MeV/amu and the OBK2 value n=5.60 at 2 Mev/amu, we note that the OBK2 value of n increases to 5.81 at 10–20 MeV/amu and then starts to decrease reaching 5.66 at 500 MeV/amu It is only for energies >500 MeV/amu that the true OBK2 asymptotic limit 5.5 is approached. However, for energies >100 MeV/amu relativistic effects begin to become important and the index n is changed significantly so that at 500 MeV/amu the ROBK2 approximation gives n=4.69, as can be seen from table 1.

Table 1. The rate of decay E^{-n} of B_0^0 with energy E for ECC from He atoms.

Projectile energy E(Mev/amu)	values of index n non–relativistic OBK1	OBK2	relativistic ROBK1	ROBK2
0.1	1.84	1.83		
0.2	3.38	3.35		
0.5	4.75	4.72		
1.0	5.33	5.29		
2	5.66	5.60		
5	5.86	5.77	5.85	5.77
10	5.93	5.81	5.91	5.80
20	5.96	5.81	5.92	5.78
50	5.99	5.78	5.88	5.69
100	5.99	5.75	5.79	5.55
200	6.00	5.71	5.60	5.30
500	6.00	5.66	5.13	4.69
1000	6.00	5.62	4.52	4.04
2000	6.00	5.59	3.72	3.47

OBK1: first–order non–relativistic OBK approximation
OBK2: second–order non–relativistic OBK approximation
ROBK1: first–order relativistic OBK approximation
ROBK2: second–order relativistic OBK approximation

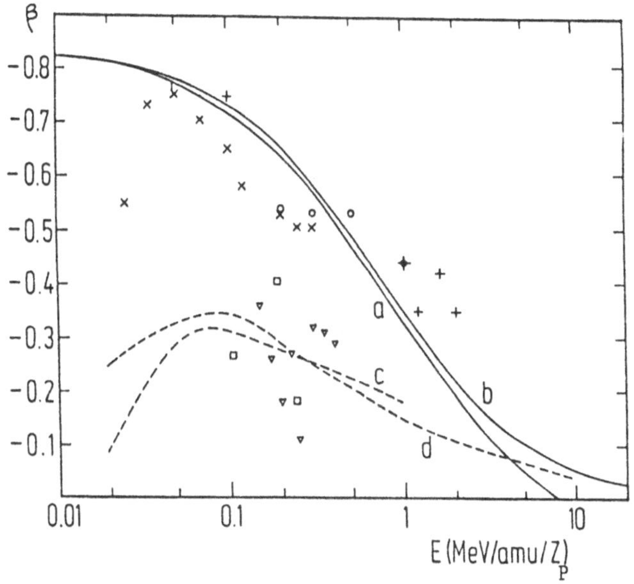

Figure 2. Cusp asymmetry parameter $\beta = B_1^0/B_0^0$ calculated using (a) OBK2 formula, (b) OBK2 formula ignoring Thomas peak, (c) CDW approximation (Burgdorfer 1986), (d) CDW approximation (Crothers and McCann 1987). Experimental data: Meckbach et al (1981) H^+ □ ; Dahl (1985) H^+ × ; Andersen et al (1986) H^+ + , He^{2+} o ; Gulyas et al (1986) He^{2+} ∇ .

The above theory produces a cusp shape for the electron yield which has only a slight asymmetry and which is in the wrong sense. In order to obtain the observed sense of asymmetry it is necessary to carry the OBK2 analysis to higher order. This can be achieved to first order in $Z_P(v_0/v)^2$ by using a similar approach to that introduced by Shakeshaft and Spruch (1978).

As $v_e \to v$ it can be shown that we may write $(2\pi mv)^{-1} d\sigma(\theta_0)/dv_e \to B_0^0\theta_0 - B_1^0\theta_0{}^2/4$ where $\beta = B_1^0/B_0^0$ is the cusp asymmetry parameter for which we have derived a first order simple closed analytical formula.

In figure 2 we compare the OBK2 curves for β obtained from our first order analysis with the experimental points obtained by Meckbach et al (1981), Dahl (1985), Andersen et al (1986) and Gulyas et al (1986). There is reasonable agreement between our theory and the experimental data of Dahl (1985) and Andersen et al (1986) for energies above about 50 keV/amu/Z_P but not with the data of Meckbach et al (1981) and Gulyas et al (1986) for H^+ and He^{2+} + He collisions and also not with the CDW calculations of Burgdorfer (1986) and Crothers and McCann (1987) which give much lower values of β below 0.5 MeV/amu/Z_P. Naturally it should be kept in mind that at such low energies the OBK2 approximation may not be reliable.

At 1.2 MeV/amu/Z_P the value of β found by Andersen et al (1986) for H^+ and He^{2+} + He ECC collisions is −0.35. The value of β calculated in the present work is −0.28 which is about twice that calculated by Crothers and McCann (1987) using CDW theory.

The presence of the Thomas peak in the OBK differential cross section causes the curve for β to change sign at 8 MeV/amu/Z_P. If the Thomas peak is ignored we obtain the curve (b) for β which is negative at all energies and agrees with the CDW calculations of Crothers and McCann (1987) at energies $\gtrsim$ 5 MeV/amu/Z_P.

[Added at this Workshop: L Gulyas, A Kover, Gy Szabo, T Vajnai, D Berenyi (1990, 4th Workshop) obtain for H^+ + He collisions β values of −0.65, −0.59, −0.54 with an error of ±0.05 at 200, 250, 300 keV energies respectively, in satisfactory accordance with the data of Dahl (1985) and Andersen et al (1986), and with the theoretical β curves presented here in Figure 2.]

References.

Andersen L H, Jensen K E and Knudsen H 1986 J Phys B: At. Mol. Phys. 19 L161
Burgdorfer J 1986 Phys Rev A 33 1578
Crothers D S F and McCann J F 1987 J Phys B: At. Mol. Phys. 20 L19
Dahl P 1985 J Phys B: At. Mol. Phys. 18 1181
Gulyas L, Szabo Gy., Berenyi D, Kover A, Groeneveld K O,
Hoffmann D and Burkhard M 1986 Phys Rev A 34 2751
Meckbach W, Nemirowsky I B and Garibotti C R 1981 Phys Rev A 24 1793
Shakeshaft R and Spruch L 1978 Phys. Rev. Lett. 41 1037

The Possible Role of the Negative Ion Resonances in Cuspology

László Szótér

Department of Physics, University of Miskolc H-3515 Miskolc, Hungary

The observation of the forward peak for neutral projectiles is the most outstanding experimental finding of the cuspology in the last few years. In the cases of 75 keV/amu $H^o \rightarrow Ar^o$ and $He^o \rightarrow He^o, Ar^o$ collisions surprisingly large, sharp and symmetric peaks were registered [1,2] in the energy spectrum of the forward detected electrons appearing at electron velocities equal to the velocity of the projectiles. Since the yield of these cusp electrons has been measured in coincidence with the outgoing neutral projectile atoms, the authors concluded that the well known "electron capture to the continuum" (ECC) [3] can arise not only in the case of ionic projectiles [4,5], but in the case of atomic projectiles as well.

To our knowledge, only the next three articles has been published up till now attempting to explain the above mentioned unexpected finding. In the calculations with the peaked impulse approximation it is found [6] that the forward peak for neutral projectiles is considerably broader than for charged projectiles, which is in qualitative contradiction with the above experimental results [1,2]. On the basis of a new theoretical model it is shown [7] that a metastable state of the He^o projectile may support a very low-energy virtual state, causing a sharp forward ECC peak. Finally, an independent "structured projectile" model was proposed by us [8] to explain the origin of this cusp. This model is enlightened here in more details than in the earlier work [8]. We consider the reaction scheme (the asterisks denote the excited electrons of the projectiles):

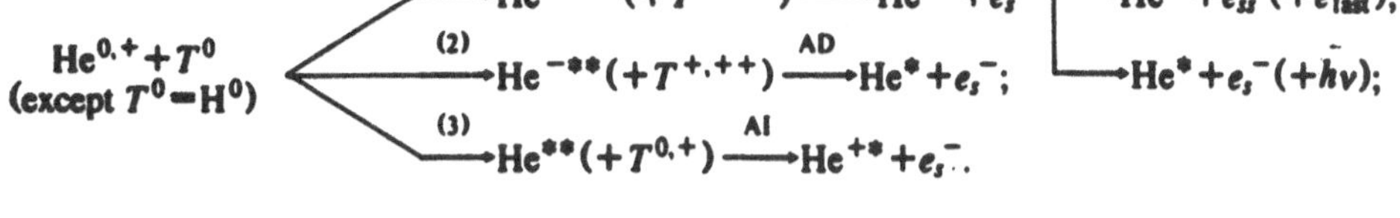

This model is based on the assumption that the electrons emitted by the intermediate short-lived (typically 10^{-13} - 10^{-15} s) autodetaching (AD) doubly and triply excited negative ion states (NIS) [9] may contribute to the cusp production with a non-negligible probability. The NIS of the projectiles are thought to be produced as a result of capturing one or two electrons from the T^o target atoms.

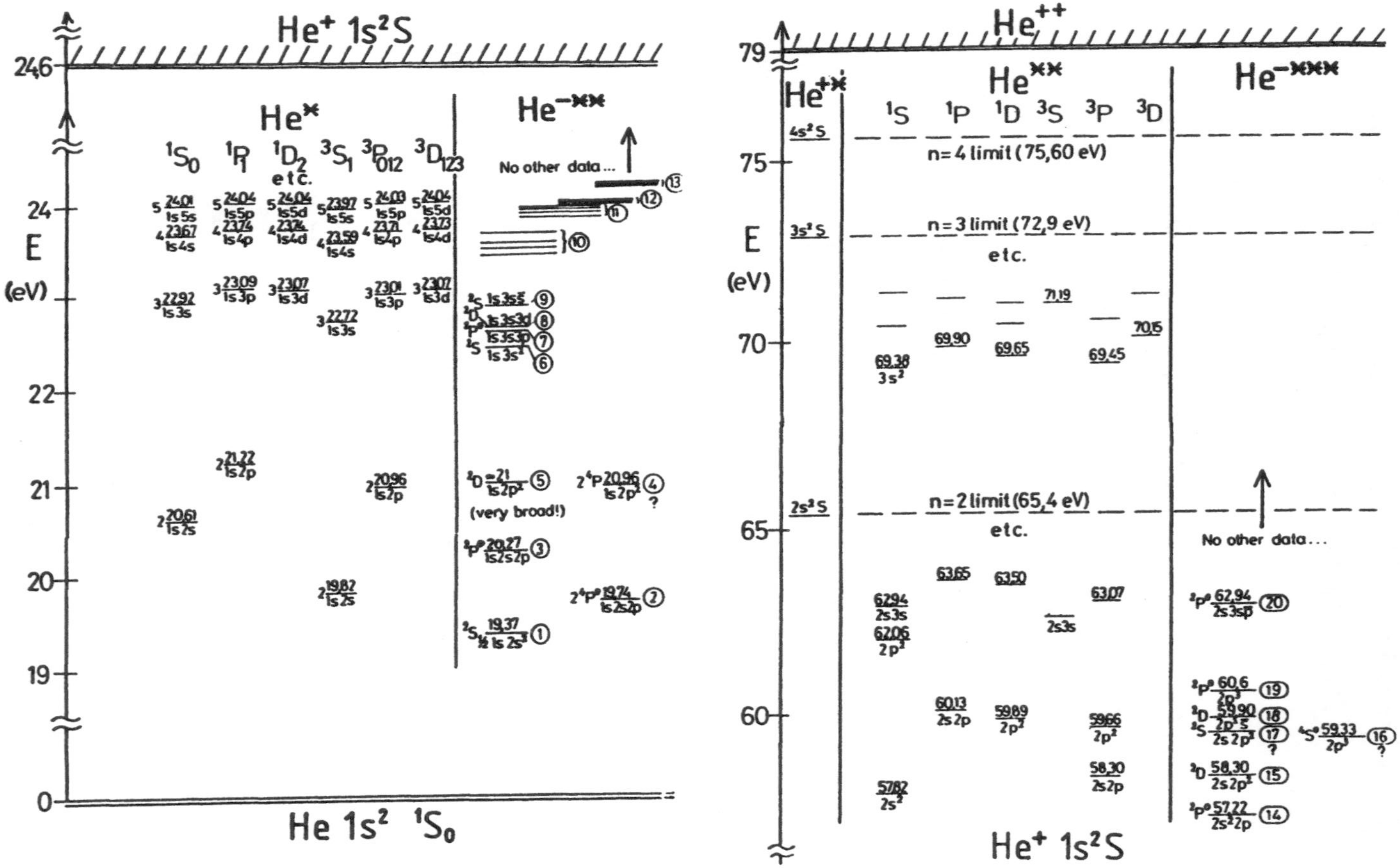

Fig. 1. The energy level diagram for the all kind of helium-states.

For completeness, the autoionizing (AI) atomic states are also involved [10] as intermediate states because their lifetimes – as in the case of the NIS – are also too short to be detected directly (i.e. they decay far before the projectile reaches the detector). Note that this scheme includes only those indirect processes which can principally lead to the release of sufficiently slow electrons from the projectiles. For this reason one must generally consider only those high-lying states which can produce free electrons at energies which fall into the cusp's energy region.

Due to the lack of the relevant cross-sections for the large number of involved intermediate states, no convincing proof of this model can be given at present (see also below). Nevertheless, this model can provide a plausible qualitative explanation. We compiled the level characteristics, shown in Fig. 1, in order to make the picture more clear. Here the most important data for all known NIS of helium [9, 11-14] are included and their energy positions are also illustrated, compared with the other kinds of helium states. Even from the analysis of only the well established NIS of helium, one can find more than twenty lines in the electron spectrum, with energies less than 0.25 eV in the rest frame of the outgoing He^o projectile. Table 1. shows only few examples of these low-energy lines. These electrons could contribute to the measured cusps [1,2,4] between 35–41 and 41–48 eV in the case of their backward and forward ejection, respectively. The maxima were observed at 41 eV. It is remarkable that channel 1 contains a process which can act as a "pumping" process for electrons shifting them towards the energy of the cusp's center. As a result of the post collision interaction (PCI) (e.g. [14]) between the autodetaching and the subsequently autoionizing electrons, the slow AD electron (e_s^-) suffers a further slowing down (e_{ss}^-). The channel 3 may give rise to lines with energies less than 0.25 eV due to transitions of $nln'l'$ AI states [15] into (n-1) states of He^{+*} for $n \geq 5$.

Note, that according to this model each processes of the ECC, the electron loss to the continuum (ELC) [2] and the transfer ionization (TI) [2] can simultaneously occur in the different channels. In the case of the incoming He^o projectile both channels 1 and 2 can contribute to the ($0 \rightarrow 0$) ECC cusp, and, the ($0 \rightarrow +$) ELC cusp can be influenced simultaneously by both channels 1 and 3. (The "incoming $\rightarrow$ outgoing" charges of the projectiles are indicated in parenthesis.) For the incoming He^+ projectiles the similar picture arises: The ($+ \rightarrow +$) ECC cusp can be effected by both channels 1 and 3 and the ($+ \rightarrow 0$) TI peak can contain yield from two different processes again, indicated in the channels 1 and 2. It must be also emphasized here that the possible yields of the different channels to any of the above processes (ELC, ECC or TI) cannot be separated even in coincidence measurements [1,2,4].

Thus, here we are faced with a quite complicated situation when the cross-sections for processes involved in the reaction scheme are large enough indeed to have a non-negligible contribution to the cusp production. But on the ground of the available relevant experimental and theoretical data, it is impossible at the moment to estimate even the order of magnitude of the cross-sections of processes taking place via the high-lying and short-lived NIS of helium considered here.

We could not find in the literature any cross-sections characterizing the production of these helium NIS in the energy range in question. Only in the case of the very low energy (~1 keV/amu) atom (ion)–atom collisions there are very scarce results which indicate the production of certain NIS [16].

Table 1. Some of the low-energy ($\leq$ 0.25 eV) electron spectrum lines coming from the autodetaching of the negative-ion states (NIS) of helium.

The designation of the NIS (see Fig.1)	Refer- ences	The autodetaching He^{-**} or He^{-***} negative ion states			The resulting He^{*} or He^{**} atomic states (Energy, eV)		The energy of the free electrons (eV)
5	[9, 11]	$1s2p^2\,^2D$	($\simeq$ 21)	$\rightarrow$	$1s2p\,^3P$	(20.96)	0–0.2
9	[11, 12]	$1s3s\bar{s}\,^2S$	(22.88)	$\rightarrow$	$1s3s\,^3S$	(22.72)	0.16
10	[12]	$1s4s\bar{s}\,^2S$	(23.67)	$\rightarrow$	$1s4s\,^3S$	(23.59)	0.08
11	[12]	$1s5s^2\,^2S$	(23.86)	$\rightarrow$	$1s4s\,^1S$	(23.67)	0.19
11	[12]	$1s5s5p\,^2P$	(23.91)	$\rightarrow$	$1s4s\,^1S$	(23.67)	0.24
12	[12]	$1s6s\bar{s}\,^2S$	(24.18)	$\rightarrow$	$1s6s\,^3S$	(24.17)	0.01
15	[13, 14]	$2s2p^2\,^2D$	(58.30)	$\rightarrow$	$2s2p\,^3P$	(58.30)	<0.01
18	[13, 14]	$2p^2\bar{s}\,^2D$	(59.90)	$\rightarrow$	$2p^2\,^1D$	(59.89)	$\leq$0.01
20	[14]	$2s3s\bar{p}\,^2P$	(62.94)	$\rightarrow$	$2s3s\,^1S$	(62.94)	<0.01

The quantitative estimation of the yield from the indirect processes here proposed might be a complicated theoretical problem, because it requires the summarized yield of a large number of individual intermediate states. The mechanism of the above phenomena proposed by this model needs further investigations.

Below we propose also *experiments* which do not require any change in the equipments used earlier [1,2,4]. (i) A cardinal question here is that whether the decay of NIS can give a considerable electron yield under the conditions of [1,2,4] or not, it can be tried to detect some of the characteristic electron-spectrum lines. Since in the vicinity of the maxima of cusps the strong overlap of these lines is expected, it would be expedient to search firstly those well resolvable lines, which are located far from the cusp's region (e.g. the decay of the $1s2s^2\,^2S$ NIS of helium provides 19.73 eV electrons in the projectile's frame; one could attempt to detect them at 117 eV in the case of 75 keV/amu projectiles). (ii) It can be also tried to detect directly in these collisions the long- lived He^- ions, because the determination of their yield promises indirect information about the rate of the metastable atoms in the incoming beam. In the collisions of ground-state atoms the high-spin $1s2s2p\,^4P$ metastable NIS is not easily accessible, but not forbidden when one of the atoms in the 2^3S metastable state. The rate of the metastable atoms in the primary beam is crucial for the model [7], but not for the model given here. (iii) Furthermore, it is desirable to carry out the measurements of the ECC cusp for He^{2+} projectiles, but – at long last – in a *coincidence* regime. It is the only case of projectiles with nuclear charge Z=2 when our model plays no role. The result from this experiment would be useful to compare the significance of this model with the 'pure' ECC mechanism in it's original sence [3]. (iv) It will also wortwhile to extend the main ideas of the proposals (i) and (iii) to hydrogen projectiles as the simplest system of the states.

The model is probably not restricted to cuspology and seems to play a role in a large field of atom(ion)–atom collisions. Additionally, this model suggests a reconsidered method for the investigations of the negative–ion resonances with much higher energy resolution than in the low energy electron–atom collisions where these resonances have been almost exclusively studied earlier.

Acknowledgements

I am indebted to E.P. Sabad and G. Hock for valuable discussions and to I. Mariscsák and J. Kovács for technical help.

References

1. Á. Kövér, L. Sarkadi, J. Pálinkás, L. Gulyás, Gy. Szabó, T. Vajnai, D. Berényi, O. Heil, K.-O. Groeneveld, J. Gibbons and I.A. Sellin: Nucl. Inst. Meth. **B42** 463 (1989)
2. L. Sarkadi, J. Pálinkás, Á. Kövér, D. Berényi and T. Vajnai: Phys. Rev. Lett. **62** 527 (1989)
3. J. Macek: Phys. Rev. A **1** 235 (1970)
4. Á. Kövér, L. Sarkadi, J. Pálinkás, D. Berényi, Gy. Szabó, T. Vajnai, O. Heil, K.-O. Groeneveld, J. Gibbons and I.A. Sellin: J.Phys. B **22** 1595 (1989)
5. C. O. Reinhold and D.R. Schultz: J. Phys. B **22** L565 (1989)
6. D.H. Jakubaßa-Amundsen: J. Phys. B **22** 3989 (1989)
7. R. O. Barrachina: J. Phys. B **23** 2321 (1990)
8. L. Szótér: Phys. Rev. Lett. **64** 2835 (1990)
9. For a review, see e.g. H.S.W. Massey: *Negative Ions*, Cambridge Univ. Press, 1976
10. N. Stolterfoht: Phys. Rep. **146** 315 (1987)
11. F.H. Read: Phys. Scripta **27** 103 (1983)
12. S.J. Buckman, P. Hammond, F.H. Read and G.C. King: J. Phys. B **16** 4039 (1983)
13. P.J.M. van der Burgt and H.G.M. Heideman: J. Phys. B **18** L 755 (1985)
14. P.J.M. van der Burgt, J. van Eck and H.G.M. Heideman: J.Phys. B **19** 2015 (1986)
15. Y.K. Ho: Z. Phys. D **11** 277 (1989)
16. G. Gerber, R. Morgenstern and A. Niehaus: J. Phys. B **6** 493 (1973)

Testing the Series Expansion Method Used for Studying the 'Cusp' Spectra

L. Gulyás, [1] Á. Kövér, [1] Gy. Szabó, [1] T. Vajnai [2] and

D. Berényi [1]

[1]Institute of Nuclear Research of the Hungarian Academy of Sciences, H-4001 Debrecen, P.O.Box 51, Hungary
[2]University of Miskolc, Physics Department, H-3515 Miskolc, Hungary

Abstract: We have investigated the electron capture to the continuum of projectile resulting from 200–300 keV $H^+ - He$ collisions. These data are fitted to a generic expression of Meckbach *et al.* [Phys. Rev. A **24** 1793 (1981)]. It was found to be of crucial importance the adequate determination of the spectrometer transmission function which strongly influences the result of the fitting. The asymmetry ratio depends linearly while the fitting (shape) parameters do not depend on the angular acceptance for the measured cusp peaks. These observations are in contradiction to the results of Oswald *et al.* [Phys. Rev. Lett. **62** 1114 (1989)].

1 Introduction

Since its discovery, the ETC cusp peak (Electron Transfer to the Continuum states of the projectile) in the secondary electron spectrum has attracted great interest both experimentally and theoretically [1]. The ETC peak consists of the electrons derived from ELC (Electron Loss to the Continuum) and ECC (Electron Capture to the Continuum) mechanisms in general. The peak appears in the laboratory frame at the forward direction and at electron velocity v where $v = v_i$ (v_i is the velocity of the projectile ion). It is well known that the shape of the observed ETC peak depends strongly on the experimental conditions (velocity and angular resolution of the spectrometer, etc.) [2]. For the analysis of the measured electron yield, $Q(v,\vartheta)$, (ϑ is the ejection angle of the electron) Meckbach *et al.* [3] proposed a method. The method is based on the series expansion of the cross- section:

$$Q(v,\vartheta) = \sum_{n,j} B_{nj} \int (v')^{(n-1)} P_j(cos\vartheta') S(v,\Omega) dv \qquad (1)$$

where P_j are the Legendre polynomials, the prime labels the quantities in the projectile frame. B_{nj} are the fitting parameters and $S(v, \Omega)$ is the transmission function of the spectrometer. It can be seen from Eq. 1 that the B_{nj} parameters are independent from the experimental conditions. In spite of this, in most cases the deviations among the B_{nj} values determined from different experiments are much higher than the experimental errors [4].

A few years ago we have decided with the group in Bariloche, Argentina to study the cusp produced in the same process with different spectrometers at the same experimental conditions and evaluate the spectra mutually in order to try to find an explanation for the above discrepancies. Namely to measure the ECC peak in the 200–300 keV $H^+ \rightarrow He$ collision system at 0° ejection angle for a given ϑ_o angular acceptance angle. Meanwhile Oswald *et al.* [5] found that the asymmetry ratio, the partial width of the cusp measured left and right direction from top of the peak, (Γ_L/Γ_R) and the parameters B_{nj} depend markedly on the ϑ_o (see Fig 1). These are in contradiction with the earlier result of Meckbach *et al.* [3] found at a different collision system. That is why we have decided to study the cusp-shape at the above collision system at different angular acceptance angles: $\vartheta_o = 0.3^o, 0.6^o, 1.1^o, 1.4^o, 2.0^o$. In the following we report on the result of this work which have been done by the group of ATOMKI, Debrecen.

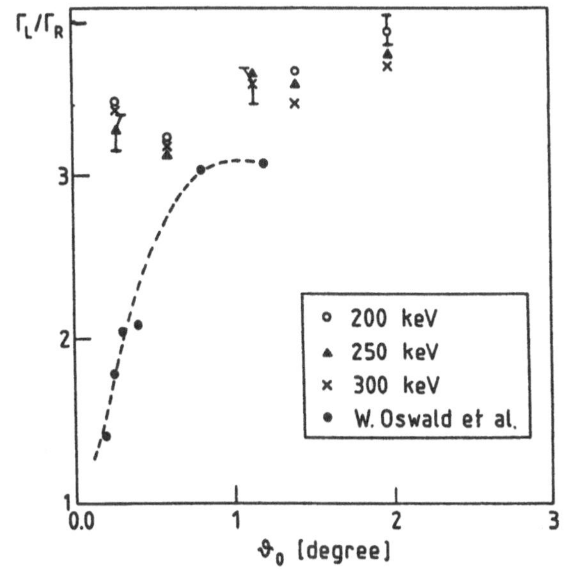

Fig. 1. ETC peak-shape asymmetry ratios Γ_L/Γ_R as a function of angular acceptance angles of the spectrometer ϑ_o for 40-MeV O^{8+}-Ne (•) and H^+- He (o, ▲, x) collision system.

2 Instrumentation and procedure

The 1.5 MV Van de Graaff accelerator facility of ATOMKI was used to render 200–300 keV H^+ impact ions. The electron spectra were taken by a distorted-field double-pass cylindrical mirror electron spectrometer (ESA-13). Here we describe only the pairs of aperture system of the spectrometer which determines the velocity ($R = \Delta v/v$) and angular acceptance of the spectrometer (other information about the arrangement is given in Ref. [6]). Two interchangeable orifices were placed before the detector see Fig. 2(a) (A,B=0.4, 0.5, 0.7, 1.0, 2.0, 2.5, 3.4 mm diameter). The electrons enter the aperture B and passing through the A reach the channel electron multiplier placed behind it. The diameter of orifices B were always larger than that of orifices A which were not larger than 1 mm. These pairs of apertures, which have a distance 50 mm between them was placed in the measuring system on the way that the focal point of the spectrometer coincide with the orifice A. A given angular resolution of the spectrometer was defined by the suitable choosing of the orifices A and B and its actual values were calculated from the geometry described in Fig. 2(a,b). The velocity resolution of the spectrometers was R=0.0025.

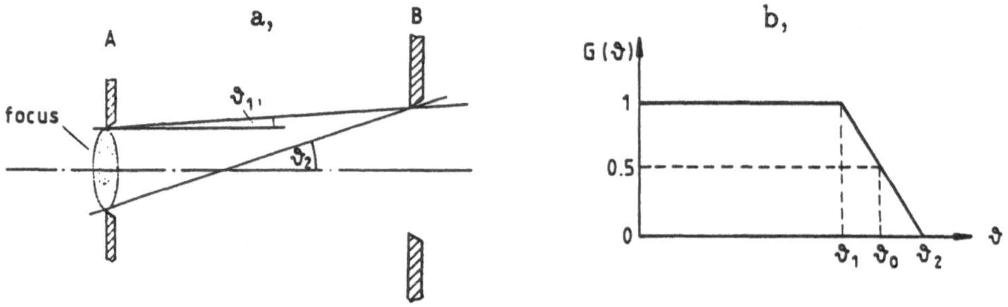

Fig. 2. The pairs of apertures determining the angular acceptance of the spectrometer (a), and the determined G(ϑ) angular acceptance function (b), see the text.

In calculating the integrals in Eq. 1 we assumed that the transmission function can be taken as a product of speed and angular functions, i.e. $S(v, \Omega) = F(v)G(\vartheta)$ where $G(\vartheta)$ is a trapezoidal function, which can be characterized by ϑ_1 and ϑ_2 see Fig. 2(b) (S is independent from ϕ because of the azimuthal symmetry of the system). With this separation of the variables, integrals in Eq. 1 could be calculated analytically as is described in Refs. [3,7]. Analyzing the spectra, the Q function was calculated up to the n=0,1 and j=0,1,2 terms.

3 Results

In analyzing the shape and asymmetry of the measured ETC spectra we determined first the (Γ_L/Γ_R) ratios, see Fig. 1. It can be seen from the figure that there is an essential difference between the results of Oswald *et al.* [3] and ours, especially at lower ϑ_o values. After this, we analyzed our data by means of the procedure described above and we found that the B_{nj} parameters depend strongly on the actual values of the ϑ_o, see Fig. 3, where the asymmetry parameters $\beta_1 = B_{01}/B_{00}$ are shown at 300 keV impact energy (the spectra measured at different ϑ_o angles were fitted separately).

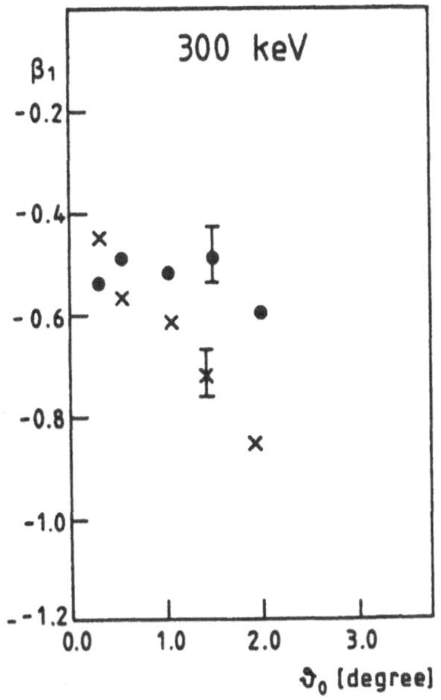

Fig. 3. Asymmetry parameter $\beta_1 = B_{01}/B_{00}$ as a function of the acceptance angle ϑ_o of the spectrometer. $\vartheta_{1,2}$ were fixed (×) and free (•) in the fitting.

This is surprising because we found for the (Γ_L/Γ_R) ratio that it depends linearly on ϑ_o if we calculate cusp distributions at different ϑ_o with the same B_{nj} parameters according to Eq. 1. It is a result which is similar to our result shown in Fig. 1.

In order to find the reason for this contradiction it was evident to examine the role of the spectrometer transmission function in the result of the fitting. It is clear from Fig. 2 that the geometrical description, which is an approximation of the real situation can be destroyed by lot of reasons. We do not enter into the details of these reasons here, we give only one example. It is supposed that the orifice A is placed at the focal point of the spectrometer and the distribution of electrons are homogeneous in this plane which can not be realized at any time because of the fluctuation of the accelerator beam. To try to take into account these effects we estimated a $\Delta\vartheta_i$ error of the ϑ_i values in describing the $G(\vartheta)$ function. After this we make a new fit on the way that the ϑ_i values were free parameters within its errors. A similar study were done by Atan *et al.* [8] at a given ϑ_0 angle. The determined B_{nj} values were nearly the same for the different values of ϑ, Fig. 3 represent the β_1 values. In this way a mean B_{nj} values could be determined. The errors are calculated from the relative deviations. The most important β_1 asymmetry parameters are as follows: -0.65; -0.59; -0.54 with an error of $\pm$ 0.05 at 200; 250; 300; keV energies.

In conclusion, the crucial role of the spectrometer transmission function should be emphasized since it essentially affects the result of the fitting. The different authors use different type of spectrometers and all of them use approximations in determining the $S(v, \Omega)$ function. The errors introduced by these approximations could be the reason of the deviations among the different experimental results. However we think that it would be very important if others carried out similar investigations on this collision system in order to confirm the above statement.

References

1. Forward Electron Ejection in Ion-Atom Collisions, Lecture Note in Physics, Vol. **213**, eds. K.-O. Groeneveld, W. Meckbach and I.A. Sellin (Springer, Berlin 1984)
2. K.R. Chiu, W. Meckbach, G. Sanchez Sarmiento and Wm. McGowan: J. Phys. B **12** L147 (1979)
3. W. Meckbach, I.B. Nemirovsky and C.R. Garibotti: Phys. Rev. A **24** 1793 (1981)
4. D. Berényi, L. Gulyás, Á. Kövér: J. de Phys. **48** C9-231 (1987)
5. W. Oswald, R. Schramm and H.-D. Betz: Phys. Rev. Lett. **62** 1114 (1989)
6. Á. Kövér , Gy. Szabó, D. Berényi, L. Gulyás, I. Csernyi, K.-O. Groeneveld, D. Hofmann, P. Koschar and M. Burkhard: J. Phys. B **19** 1187 (1986)
7. Y.C. Yu and G. Lapicki: Phys. Rev. A **36** 4710 (1987)
8 H. Atan, W. Steckelmacher and M. W. Lucas: J. Phys. B **23** 2579 (1990)

Z_p- and q-dependence of 0° binary encounter electrons in 1–2 MeV/u collisions of ions with H_2 and He targets

T.J.M. Zouros[§], P. Richard and D.H. Lee

[§] Physics Dept., University of Crete and Research Center of Crete, Iraklion, Greece

J.R. Macdonald Laboratory, Kansas State University, Manhattan, KS 66506, USA.

Binary encounter electrons (BEe) ([1] – [3]) are target electrons ionized through direct, hard collisions with energetic projectiles, giving rise to a broad energy distribution. The study of BEe, particularly at 0°, can give us a detailed understanding of small impact parameter collision dynamics and screening considerations, since most of the BEe are produced in collisions with the ions at distances well within the K-shell of the ion. Here, we report on recent 0° results for BEe production in energetic 1–2 MeV/u collisions of bare [4] and structured [5] ions with H_2 targets. Results for He targets were found to be qualitatively similar to those of H_2 [4, 5].

In the impulse approximation [4, 6] (IA) approach, valid for $V >> v$, v and V being the target electron and projectile velocities, respectively, the BEe production process is described within the projectile frame, as the elastic scattering of target electrons from the charged ion. The BEe double differential cross-section (DDCS) can be evaluated within the IA from the following general formula [4],

$$\left(\frac{d\sigma}{d\Omega}\right)_{BEe,q} = \sum_i \int \frac{d\sigma}{d\Omega_i}(\epsilon, \theta, q) \cdot |\psi_i(\mathbf{p}_i)|^2 d^3p_i, \tag{1}$$

The index i refers to the i-th active target electron. The elastic differential cross-section (EDCS), $\frac{d\sigma}{d\Omega}(\epsilon, \theta, q)$, is that for a free electron of energy ϵ scattering from an ion of charge state q into the angle θ. The integral is over the momentum distribution of the target electron $|\psi_i(\mathbf{p})|^2$, where $\mathbf{p}$ refers to the electron's momentum around the target.

Differentiating Eq. 1 with respect to ϵ and introducing the Compton profile $J(p_z)$ to represent the momentum distribution of the target electrons along the z-axis (defined along the beam direction), we utilize the "quadratic model" of the IA (terms quadratic [4] in p_z are maintained together with the ionization energy E_I) to obtain for the BEe DDCS[4, 7]:

$$\left(\frac{d^2\sigma}{d\epsilon d\Omega}\right)_{BEe,q} = \frac{d\sigma}{d\Omega}(\epsilon, \theta, q) \cdot \frac{J(p_z)}{V + \frac{p_z}{m}} \tag{2}$$

with ϵ, the energy of the impinging target electron, given by:

$$\epsilon = \frac{1}{2}mV^2 + V \cdot p_z + \frac{p_z^2}{2m} - E_I \tag{3}$$

where m is the mass of the electron. The only unknown quantity in Eq. 2 is the EDCS, $\frac{d\sigma}{d\Omega}(\epsilon, \theta, q)$, which can be calculated independently. We note that zero degree observation ($\theta_{Lab} = 0°$) corresponds to scattering through $\theta = 180°$ in the projectile rest frame [4].

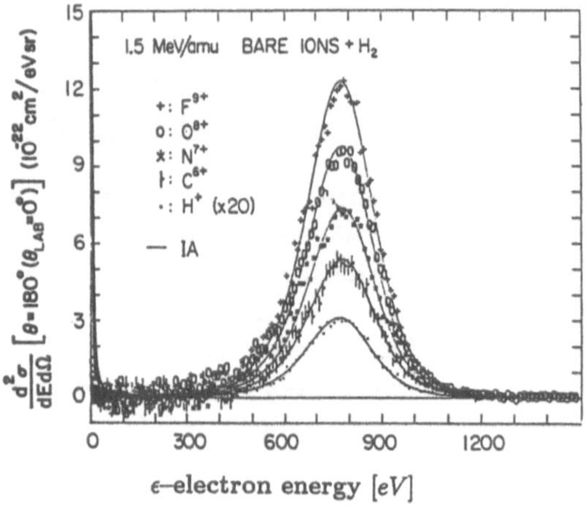

Fig. 1

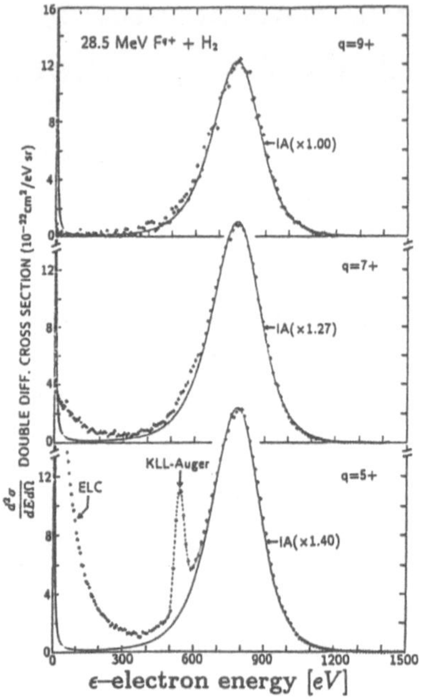

Fig. 2

Figure 1: BEe spectra (projectile frame) for five different bare projectiles. Typical statistical error bars are shown for the case of C^{6+}. Solid lines: IA for each projectile. For protons, both data and IA have been multiplied by 20.

Figure 2: Data: Measured BEe spectra [5] (projectile frame) for collisions of 28.5 MeV $F^{(9,7,5)+}$ with H_2. Solid lines: IA calculation for *bare* F^{9+} BEe production scaled to data by multiplying with numbers in parentheses.

Case I: Bare Projectiles

For bare projectiles, $\frac{d\sigma}{d\Omega}(\epsilon, \theta, q = Z_p)$ is just equal to the Rutherford cross–section (in a.u.):

$$\frac{d\sigma}{d\Omega}(\epsilon, \theta, q = Z_p) = \frac{Z_p^2}{16\epsilon^2 sin^4\frac{\theta}{2}} \tag{4}$$

where Z_p is the projectile nuclear charge. The IA BEe DDCS calculated using Eqs. 2–4 are compared to experimental BEe DDCS in Figs. 1 and 2. As can be seen from Fig. 1, for bare ions, excellent agreement between theory and experiment is observed, thus establishing the validity of the IA treatment for the production of BEe DDCS at zero degrees by bare ions at these fast collision energies.

Case II: Structured Projectiles

For structured projectiles, the additional projectile electrons are expected to screen the projectile nuclear charge. This effect must be included in the calculation of the EDCS. Measured BEe DDCS as a function of q are shown in Fig. 2. The additional electrons in screening the nuclear charge actually increase the BEe yield at 0°. This result was originally rather surprising as it seemed to go against conventional screening expectations [5], and contrary to older q–dependence results [3, 8] for BEe production at non–zero

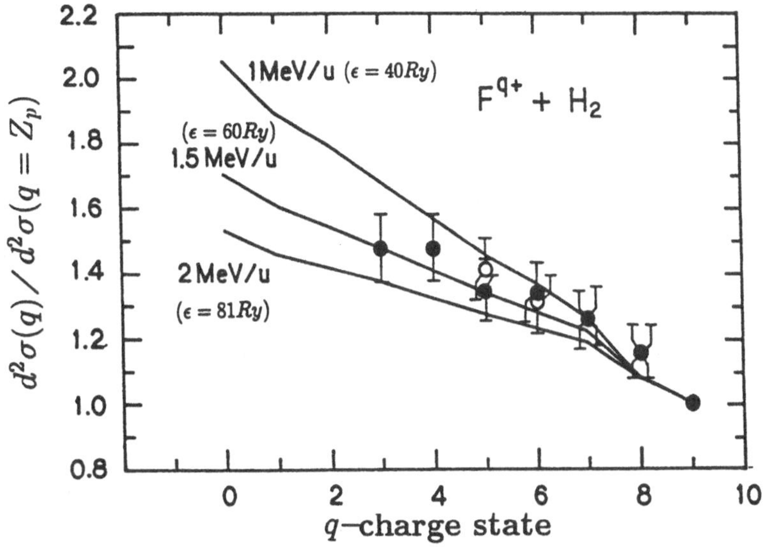

Figure 3: Projectile charge state dependence of BEe production relative to bare ion for collisions of 19 MeV $F^{(3-9)+}$ (dark circles) and 28.5 MeV $F^{(5-9)+}$ (open circles) with H_2 targets. Solid lines: Recent calculations from Ref. [9]. Data from Ref. [5]. Within parentheses are included the energies ϵ of electrons of the same velocity as the projectile.

angles for 30 MeV $O^{q+} + O_2$. The ratios of the BEe DDCS for F^{q+} ($q = 3 - 9$) to that of bare F^{9+}, $d^2\sigma(q)/d^2\sigma(q = Z_p)$, are plotted in Fig. 3 for the various collision systems.

The enhancement of the BEe yields with decreasing q is clearly observed. Similar enhancements were also observed between 1.5 MeV/u O^{8+} and O^{4+}; N^{7+} and N^{3+}; and C^{6+} and C^{3+} in collisions with a H_2 target (not shown in figures).

The effective charge model [3, 8] used the Rutherford EDCS (Eq. 4), but with an effective charge $Z_{eff} = Z_p - S_q(r)$ (where r is given as a function of ϵ) rather than Z_p (for more details see Refs. [3] and [8]). This model is seen to be inappropriate for our data since the screening function $S_q(r)$ is always positive and therefore the BEe DDCS will always decrease with decreasing q, contrary to the observed behavior for 0° measurements.

Recently, a number of authors ([9]–[12]) have explained this "anomalous" q– dependence [5] by computing $\frac{d\sigma}{d\Omega}(\epsilon, \theta, q)$ directly for the scattering of an electron by a structured ion within the non-Coulombic potential resulting from the screening of the projectile nuclear charge by the additional electrons. In fact, the effective scattering potential used is of the form $V_{eff}(r) = -\frac{Z_{eff}(r)}{r}$, with $Z_{eff}(r)$ as already given. Reinhold *et al.* [9] and Shingal *et al.* [10] computed $S_q(r)$ using Hartree–Fock wave functions to obtain the EDCS quantum–mechanically. Some of their results are shown in the following figures.

In Fig. 4 the impact parameter and EDCS are plotted [9] as a function of the scattering angle θ. In Fig. 5 the ratio of EDCS, $R = d\sigma(q)/d\sigma(q = Z_p)$, is plotted [10] as a function of energy ϵ and angle θ.

In Fig. 5(a), it is seen that for ϵ between about $40-80Ry$ (600 - 1200 eV) and $\theta = 180°$, R doesn't vary much with ϵ for different values of q. This explains the relatively good fits of the bare ion IA BEe DDCS to the BEe DDCS for the different charge states of F^{q+} as shown in Fig. 2.

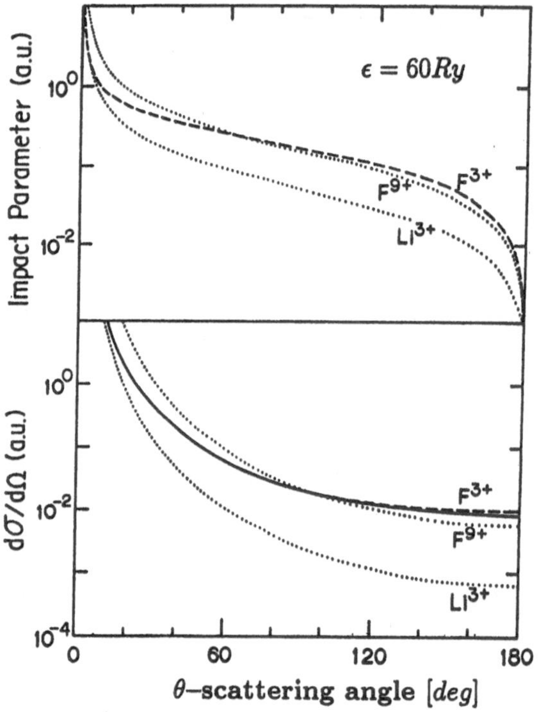

Figure 4: The classical deflection function (top) and the EDCS, $\frac{d\sigma}{d\Omega}(\epsilon, \theta, q)$, (bottom) for scattering of an electron with energy $\epsilon = 815\ eV$ (60 Ry) from fully and partially stripped ions. The solid and dashed curves correspond to the quantum mechanical and classical results for $e^- + F^{3+}$, respectively (adapted from Ref. [9]).

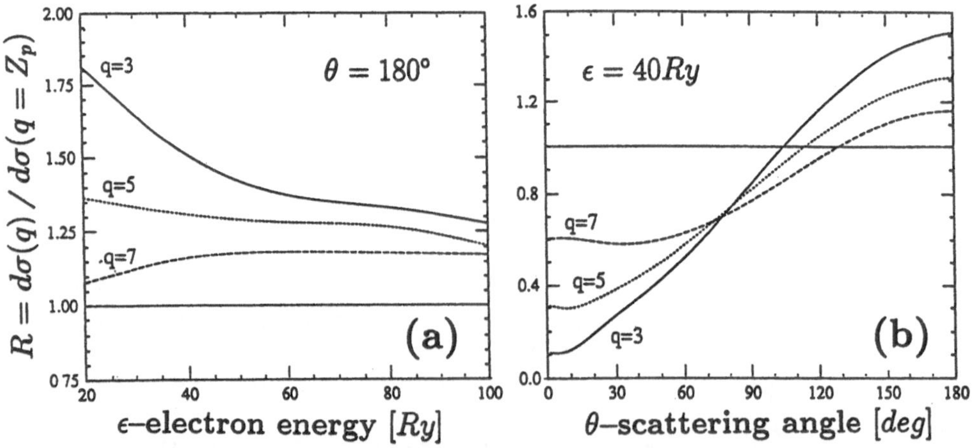

Figure 5: Variation of the calculated ratio R of differential cross-sections for $e^- + F^{q+}(q = 3, 5, 7)$ collisions to that for $e^- + F^{9+}$ collisions as a function of (a) the electron scattering energy ϵ for $\theta = 180°$ (from Ref. [9]) and (b) the scattering angle θ for $\epsilon = 40Ry$. (adapted from Ref. [10].)

In Fig. 5(b), we again note the enhancement of the EDCS with decreasing charge state for $\theta = 180°$. This enhancement continues down to an angle $\theta = 75°$ ($\theta_{Lab} = 45°$) below which the q–dependence is reversed, R decreasing or staying moderately constant with increasing values of q. This result tends to support the BEe q–dependence observed [1, 3, 8] for 30 MeV $O^{q+} + O_2$ at various angles $\theta_{Lab} \geq \theta_{Lab} = 30°$. However, the O_2 target is rather complicated and a much simpler comparison could be made if H_2 or He targets are used to study the angular and q–dependence of BEe production. Such a systematic study of BEe production has still to be undertaken.

Future angular dependence studies in combination with q-dependence studies of BEe production would be useful in further testing the new calculations, eventually leading to a full quantitative description of screening phenomena for electron-ion scattering.

We greatfully acknowledge discussions with J. Reading, N. Stolterfoht, B. Sulik and K. Taulbjerg. In particular, we would like to thank Rajiv Shingal and Carlos Reinhold for allowing us to use their results prior to publication. This work was partially supported by the Division of Chemical Sciences, Office of Basic Energy Sciences, Office of Energy Research, U.S. Department of Energy.

References

[1] N.Stolterfoht, D.Schneider, R.Burch, H.Wieman, and J.S. Risley, Phys. Rev. Lett. 33, 59 (1974).

[2] M.E. Rudd and J.H. Macek, *Case Studies in Atomic Physics* 3, 47 (1972) and references therein; T.F.M. Bonsen and L. Vriens, Physica 47, 307 (1970); M.E. Rudd, L.H. Toburen and N. Stolterfoht, At. Data and Nucl. Data Tables 23, 405 (1979); S.T. Manson, L.H. Toburen, D.H. Madison and N. Stolterfoht, Phys. Rev. A12, 60 (1975); P.D. Fainstein, V.H. Ponce and R.D. Rivarola, J. Phys. B 22, 1207 (1989).

[3] N.Stolterfoht, in *Structure and Collisions of Ions and Atoms*, edited by I.A. Sellin, (Springer-Verlag Press, 1978), pp. 155-199.

[4] D.H. Lee, P. Richard, T.J.M. Zouros, J.M. Sanders, J.L. Shinpaugh and H. Hidmi, Phys. Rev. A41 4816 (1990).

[5] P. Richard, D.H. Lee, T.J.M. Zouros, J.M. Sanders and J.L. Shinpaugh, J. Phys. B23, L213 (1990).

[6] D. Brandt, Phys. Rev. A27, 1314 (1983).

[7] P. Richard, Invited Talk, "X-90, 15th International Conf. on X-Ray and Inner–Shell Processes," Knoxville, Tenn. July 9-13, 1990.

[8] L.H. Toburen, N. Stolterfoht, P. Ziem and D. Schneider, Phys. Rev. A24, 1741 (1981).

[9] C.O. Reinhold, D.R. Schultz and R.E. Olson, submitted to J. Phys. B 1990.

[10] R. Shingal, Z. Chen, K.R. Karim, C.D. Lin and C.P. Bhalla, submitted to J. Phys. B 1990.

[11] J. Reading and T. B. Quinteros, to be published in Nucl. Instrum. and Meth. 1990

[12] B. Sulik, private communication and to be published.

B Two-Center Problems and Saddle-Point Phenomena

Chairman: C. GARIBOTTI

B.1 Invited Survey

R.E. OLSON, C.O. REINHOLD and D.R. SCHULTZ
Two-Center Effects Displayed in Ionized Electron Spectra

B.2 Invited Contributions

R. MAIER, O. HEIL, R.D. DuBOIS, M. KUZEL and K.O. GROENEVELD
New Aspects of Simultaneous Projectile and Target Ionization in
H^0(0.5 MeV/1.0 MeV)→He - Collisions

B.3 Contributions (Posters)

S. SUAREZ, C. GARIBOTTI, W. MECKBACH, G. BERNARDI,
P. FOCKE, R. PREGLIASCO and G. SIGAUD
Two-Center Effects in the Ionization of Ne by H^+ and $^3He^{2+}$ Impact at Intermediate Energies

K. TŐKÉSI and G. HOCK
Classical Trajectories during Atomic Collisions

TWO-CENTER EFFECTS IN ION-ATOM COLLISIONS

R. E. Olson, C. O. Reinhold and D. R. Schultz
Physics Department, University of Missouri-Rolla
Rolla, MO 65401

ABSTRACT

In this paper, certain aspects of ion-atom collisions at intermediate energies ($E \sim 50$ - 1000 keV/u) which involve so called two-center effects will be discussed. In particular, work on the fundamental system H^+ + He will be emphasized, because of the availability of experimental data and the tractability of theoretical calculations using a variety of methods. Furthermore, this review will concentrate on features of the doubly differential cross section for ionization in such ion-atom collisions because they dramatically reflect the importance of the interactions of the ejected electrons with both the projectile and target ionic centers. Comparisons are made with the results of single-center treatments such as the first-order Born approximation in order to underscore the significance of two-center effects. Works considering antiproton-impact and the interchange of electrons between ionic centers in multiply-charged ion - many-electron atom collisions will also be discussed. Finally, the failure of the Born picture in predicting the role of projectile nuclear charge screening in collisions involving partially stripped ions is described in terms of its relationship to observed anomalies in the binary peak magnitude.

INTRODUCTION

The great importance of two-center effects in ion-atom collisions has only been brought to light in the last few years. Originally, the basis for describing collisional phenomena at intermediate energies was the view implicit in the first-order Born approximation. This theory uses a single-center, i.e. target-centered, wave function in order to obtain the collision cross section. In fact, the first-order Born approximation does a very reasonable job when compared to experiment in predicting total cross sections for ionization. Further, in the late 1960's and early 1970's, electron-capture-to-the-continuum (ECC) spectra, in which ionized electrons escape with velocities very nearly equal to the projectile velocity, were first observed [1,2] and explained [3,4] in terms of basis sets centered on both the target and the projectile. To that point, it was thought that the majority of ionized electrons were either directly associated with the target or the projectile.

It was not until 1983, in a theoretical study [5] of the doubly differential ionization cross section which sought to describe the origin of ECC electrons, that it was noted that a much larger flux of electrons were found with velocities approximately half that of the incident projectile (v/2) than were found in the ECC region. In 1986, a subsequent work showed that the v/2, or saddle point, component vastly exceeded the ECC contribution to the total ionization cross section. In fact, with the poor angular and energy resolution of the first calculation [6] using the classical-trajectory Monte Carlo (CTMC) method, the ECC cross section was not resolved.

Later in 1986, a paper by Meckbach and colleagues [7] on "ridge electrons" illustrated the importance of two-center effects experimentally. This work was shown to have suffered from an experimental error [8], but should be credited with motivating much of the investigation which ensued. This 1986 paper was followed in quick succession by a similar study on H^+ + He collisions [9], which pointed out the existence of experimental data from Rudd and colleagues [10,11] from the 1960's, that confirmed the v/2 effect. Moreover, even at high energies with highly-charged ions, Stolterfoht and coworkers [12] found deviations from the predictions of the first-order Born approximation for 25 MeV/u Mo^{40+} ions colliding with He, demonstrating the significance of two-center effects. One year later, Irby and coworkers [13] displayed data to indicate that the saddle point mechanism, a mechanism in which the force cancellation due to the projectile and target charges allows an electron to be ionized, can account for the shape of the ejected electron spectrum for H^+ and He^{2+} colliding with He. In the following section, we review some of the work contributing to the recent advances in our understanding of ion-atom collisions.

RESULTS

The signature of saddle point electrons is not well established for impact ionization in heavy particle ionization. Irby et al [13] in 1988 pointed out from the classical potentials for H^+ + He and He^{2+} + He ionizing collisions, there should be a shift in the maximum of the cross section since the saddle point velocity, v_{sp}, scales as

$$v_{sp} = v_p / [\, 1 + (Z_p / Z_t)^{0.5}\,], \tag{1}$$

where v_p is the projectile velocity and Z_p and Z_t are the nuclear charges of the projectile and target, respectively. Indeed, the experimental data showed the expected shift in

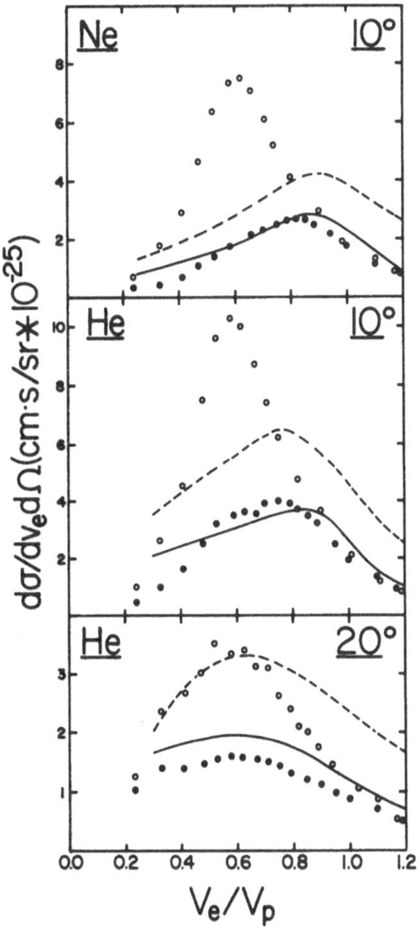

Figure 1. Doubly differential cross sections for projectile energies of 100 keV/u. Solid lines represent the H^+ data of Bernardi et al [14,16]; dashed lines are their He^{2+} results. The data by Gay et al [15] for H^+ and He^{2+} are given by the closed and open circles, respectively. In the bottom two graphs, the H^+ data have been multiplied by 2. In the upper graph, the Gay et al He^{2+} results have been multiplied by 0.5. The relative experimental results of Bernardi et al for Ne have been normalized to their CDW-EIS calculations.

the cross section maximum for He^{2+} projectiles when compared to H^+. However, recent experimental work by Bernardi et al [14] displays no such shift. To remove the discrepancy, Gay et al [15] have remeasured H^+ and He^{2+} on He, Ne and Ar using a different apparatus, that of Rudd, and still find a significant shift in the cross section maxima. A comparison of the experimental results is shown in Fig. 1. To add to the controversy, neither the CTMC [13] nor the continuum-distorted-wave - eikonial-initial-state (CDW-EIS) [14] theoretical methods have been able to obtain a shift. Both of these theories inherently include two-center effects. Thus, a secure experimental benchmark of saddle point electrons has yet to be found.

Two-center effects are readily discernable in calculations describing the electron emission from He from collisions with protons and antiprotons. In Fig. 2 are displayed the singly differential cross sections in angle for the 100 keV collision energy [17]. The CTMC calculations clearly show dramatic differences in the ionized

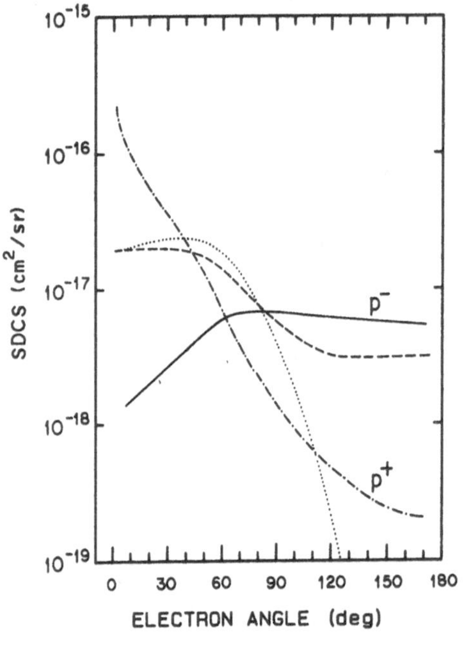

Figure 2. Single differential cross sections for ejection of electrons in 100 keV collisions of protons and antiprotons with He. Dot-dashed and solid lines are CTMC calculations for protons and antiprotons, respectively. Dashed and dotted lines are first-Born and binary-encounter results, respectively.

spectra for the two systems, even though the total cross sections are nearly identical. In contrast, the binary encounter theory and the Born approximation, which are single center theories with a q^2 dependence on the cross sections, differ significantly in shape and magnitude. Calculations using the CDW-EIS method by Fainstein and coworkers [18] supported the CTMC findings.

Examining the doubly differential cross sections in angle and energy for the same systems, Fig. 3, shows that the Born approximation does reasonably well for high ejection energies and proton collisions, but fails to describe the ECC cusp for protons, and the anticusp for antiprotons [17,19]. An incorrect angular dependence is also predicted for low ejection energies. Again, two-center effects manifest themselves in

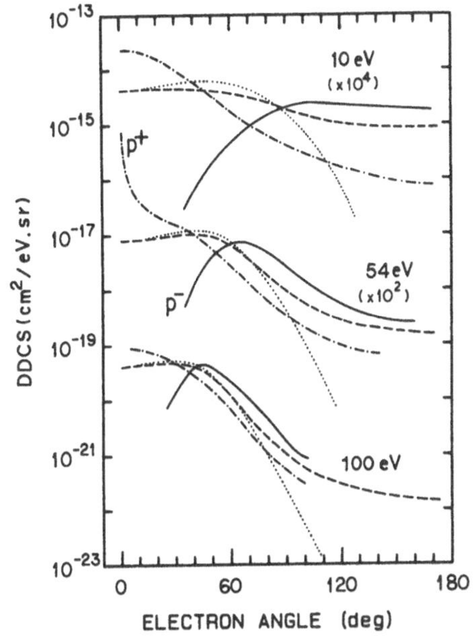

Figure 3. Doubly differential cross sections for 100 keV proton and antiproton collisions on He. The notation is the same as in Fig. 2.

a clear fashion in the electronic spectra. CDW-EIS calculations are in general accord with the CTMC results [20].

One would expect that two-center effects would be less prominent at higher energies, and a single-center theory such as the Born approximation would gain in validity. Such an argument is supported by the fact that the Born approximation predicts the total cross section with good accuracy for a perturbing collision such as for a 500 keV H^+ on He. This conjecture is not valid if the doubly differential cross sections for ionization are examined in detail [19]. At small ejection angles such as 10° or 30° and low ejection energies, the single center Born approximation underestimates the cross sections by about an order-of-magnitude, Fig. 4. In fact, the low energy electrons produced in "soft" collisions are greatly underestimated. Likewise, the Born approximation with q^2 scaling does an extremely poor job in predicting antiproton results and anticusp structure.

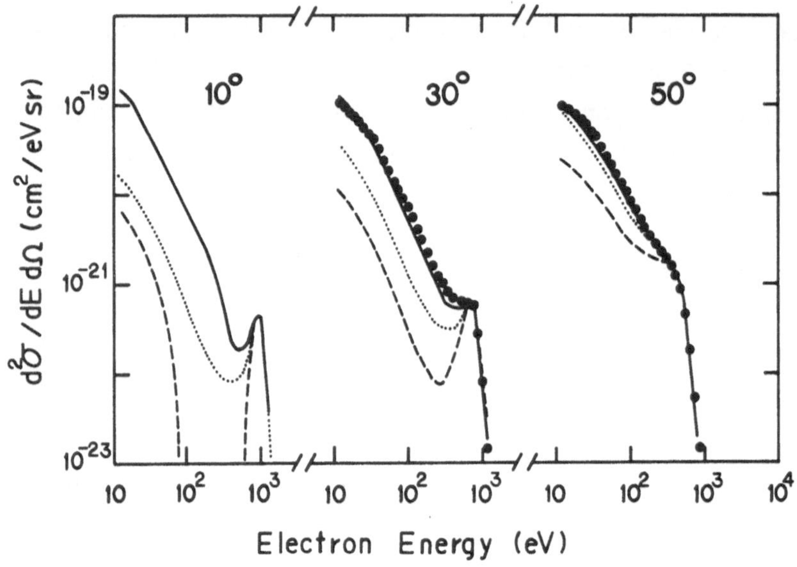

Figure 4. Doubly differential cross sections for electron ejection in 500 keV proton (solid lines) and antiproton (dashed lines) collisions with He calculated using the CTMC method. The dotted lines are the cross sections given by the binary-encounter theory. The solid circles are experimental results of Rudd et al [21] normalized to the total cross sections of Shah and Gilbody [22].

However, one aspect of the electronic structure is accurately portrayed by the single-center Born approximation. This is at the binary peak where the magnitude and shape of the cross section is accurately predicted, Fig. 4. In fact, even the antiproton results are correctly given by the Born approximation at the binary peak. Recent measurements by Petersen et al [23] for He^{2+}, C^{6+} and O^{8+} impact on He show that at the binary peak, the Born approximation and its scaling is valid. These latter measurements confirm early observations made by Toburen and coworkers [24,25] for H^+ and He^{2+} projectiles. Thus, it is safe to say that the electrons giving rise to the binary peak structure can be viewed as arising from a binary collision between the projectile and an essentially "free" electron.

For an accurate description of the ECC peak, two center effects must be incorporated in theoretical methods. Experimentally, Bernardi et al [16] provided an experimental study of ECC cusp structure for both H^+ and He^{2+} colliding with He. Within the Born approximation, the ratio of the cross sections for He^{2+} as compared to H^+ should scale as the charges squared, or as a factor of 4. In fact, at the ECC peak near 0°, there were significant departures from the expected scaling, Fig. 5. Moreover, systematic deviations occur over the whole angular region [17,26].

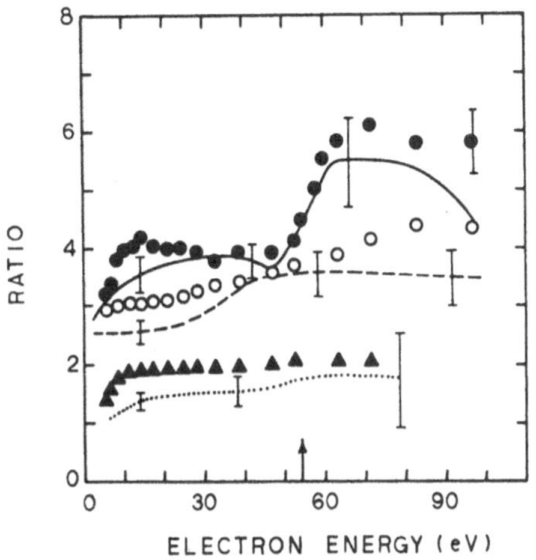

Figure 5. Ratio between the doubly differential cross sections for 100 keV/u collisions of H^+ and He^{2+} with He for specific electron ejection angles. Solid circles, open circles and solid triangles are the data of Bernardi et al [16] for 0°, 30° and 70°, respectively. Solid, dashed and dotted lines are CTMC calculations for 5°, 30° and 70°, respectively.

An excellent review of theoretical approaches with higher-order corrections as applied to ECC structure has been given by Jakubaßa-Amundsen at the New York ICPEAC [27]. Research on deviations from the expected cusp shapes for partially-stripped ion impact has been recently investigated by Reinhold and coworkers for He^+ and C^+ ion impact [28,29].

Direct observation of simultaneous electron removal from both nuclear centers have been made using coincidence techniques by Freyou et al [30] and Kelbch et al [31]. Experimentally, both the projectile charge and the recoiling target charge are determined along with the cross section for the process. Calculations have been made by Schultz and colleagues [32] for the F^{6+} + Ne collision system using a n-body CTMC method with L-shell electrons on both nuclear centers. In Fig. 6 the comparison between the theoretical and the experimental work is shown. In general, there is very good agreement between theory and experiment, especially considering this is a true n-body problem. In the Schultz et al paper, the authors point to an important

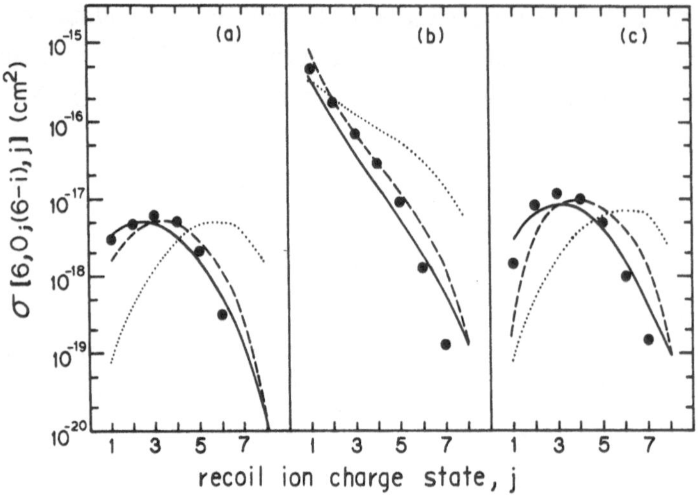

Fig. 6. Coincident charge state total cross sections for 15 MeV F^{6+} + Ne collisions. The solid circles are the experimental data, solid lines are the CTMC method without post-collision autoionization considered, dashed lines are with autoionization, and the dotted lines are independent particle model calculations. (a) Single ionization of F^{6+} in coincidence with multiple ionization of Ne. (b) Multiple ionization of Ne. (c) Single electron capture to F^{6+} in coincidence with multiple ionization of Ne.

collisional process that effects collisions of partially-stripped multiply-charged ions and many-electron atomic targets. This is electron-interchange-to-the-continuum (EIC) where the electrons exchange their nuclear centers during the collision. A cusp for this process is found in the same region of electron energy and angle as that in the ECC and electron-loss-to-the-continuum mechanisms.

Studies of binary peak structure and its scaling with projectile charge state has been of considerable recent interest. Although the cross section magnitude at the binary peak is due to only a binary interaction between the projectile and the target electron, the magnitudes of these cross sections also have not been found to scale with expected q^2 or Z^2 dependencies for partially stripped ion impact. The motivation for binary peak studies was prompted by measurements of Richard et al on the $F^{q+} + H_2$ system [33]. When the incident charge of the flourine was increased from 3+ to fully stripped 9+, the magnitude of the binary peak decreased, rather than increase as expected from q^2 scaling or be constant if one considered the binary peak arises from a very close collision between the electron and the projectile nucleus.

Recent calculations [34] have reproduced the experimental results, Fig. 7. The

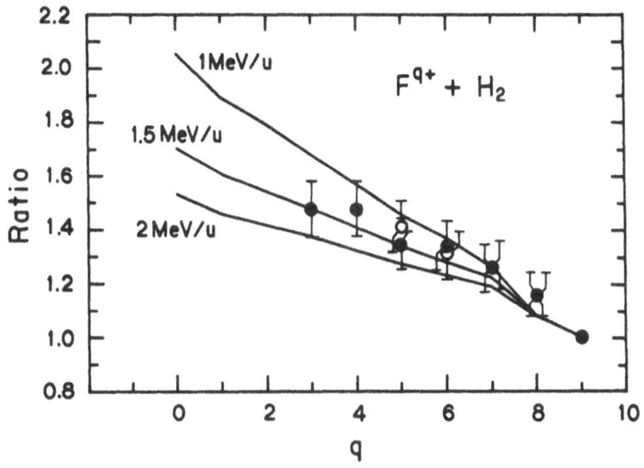

Figure 7. Ratio of the binary peak magnitudes for F^{q+} ions compared to that for fully-stripped F^{9+} for electrons ejected to 0°. The theoretical calculations are compared to the absolute data of Richard et al [33] for 1 MeV/u (solid circles) and 1.5 MeV/u (open circles).

impulse approximation was employed with a screened potential for the F^{q+} ion determined from Hartree-Fock calculations. These calculations clearly described another type of departure from q^2 scaling, that being due to the electronic structure of the projectile. Earlier theoretical work from the same group [35] on 1.4 MeV/u U^{q+} + Ar collisions, q = 11+ to 92+, showed that the binary peak magnitude can be expected to vary by several orders of magnitude from expected q^2 scaling. This effect is accentuated in heavy, partially stripped ion collisions.

CONCLUDING REMARKS

In this short overview of recent progress in ion-atom collisions, we have tried to emphasize the importance of two-center effects in describing the scattering dynamics. These effects clearly manifest themselves in ionized electron spectra and their departures from predictions made using the single-center first Born approximation. The departures are quite global, and are observed in some aspects of almost every collision system, from low electron ejection energies and angles to ECC peak structure. Only for binary peak structure with fully stripped ions, can a single center approach be considered to be valid. Even for this latter case, there are surprises when partially stripped projectile ions are employed. Thus, the study of heavy-ion collision dynamics is a true two-center, n-body problem that challenges the development of new theoretical and experimental techniques.

ACKNOWLEDGEMENT

This work was supported by the Office of Fusion Energy of the U.S. Department of Energy.

REFERENCES

1. Harrison, K. G., and Lucas, M. W., Phys. Lett.**33A**, 142 (1970).
2. Crooks, G. B., and Rudd, M. E., Phys. Rev. Lett. **25**, 1599 (1970).
3. Salin, A., J. Phys. B **2**, 1225 (1969).
4. Macek, J. H., Phys. Rev. A **1**, 235 (1970).
5. Olson, R. E., Phys. Rev. A **27**, 1871 (1983).
6. Olson, R. E., Phys. Rev. A **33**, 4397 (1986).
7. Meckbach, W., Focke, P. J., Goni, R., Suarez, S., Macek, J., and Menendez, M. G., Phys. Rev. Lett. **57**, 1587 (1986).

8. Bernardi, G., Focke, P., Suarez, S., and Meckbach, W., Lecture Notes in Physics, ed. by D. Berenyi and G. Hock (Springer-Verlag, 1988) Vol. 294, pp. 295-9.
9. Olson, R. E., Gay, T. J., Berry, H. G., Hale, E. B., and Irby, V. D., Phys. Rev. Lett. **59**, 36 (1987).
10. Rudd, M. E. and Jorgensen Jr., T., Phys. Rev., **131**, 666 (1963).
11. Rudd, M. E., Sautter, C. A. and Bailey, C. L., Phys. Rev. **151**, 20 (1966).
12. Stolterfoht, N., Schneider, D., Tanis, J., Altevogt, H., Salin, A., Fainstein, P. D., Rivarola, R., Grandin, J. P., Scheurer, J. N., Andriamonje, S., Bertault, D., and Chemin, J. F., Europhys. Lett. **4**, 899 (1987).
13. Irby, V. D., Gay, T. J., Edwards, J. Wm., Hale, E. B., McKenzie, M. L. and Olson, R. E., Phys. Rev. A **37**, 3612 (1988).
14. Bernardi, G., Fainstein, P., Garibotti, C. R. and Suarez, S., J. Phys. B **23**, L139 (1990).
15. Gay, T. J., Gealy, M. W. and Rudd, M. E., J. Phys. B (submitted).
16. Bernardi, G. C., Suarez, S., Fainstein, P. D., Garibotti, C. R., Meckbach, W., and Focke, P., Phys. Rev. A **40**, 6863 (1989).
17. Reinhold, C. O. and Olson, R. E., Phys. Rev. A **39**, 3861 (1989).
18. Fainstein, P. D., Ponce, V. H., and Rivarola, R. D., J. Phys. B **22**, L559 (1989).
19. Reinhold, C. O. and Schultz, D. R., Phys. Rev. A **40**, 7373 (1989).
20. Fainstein, P. D., Ponce, V. H., and Rivarola, R. D., J. Phys. B **21**, 2989 (1988).
21. Rudd, M. E., Toburen, L. H., and Stolterfoht, N., At. Data Nucl. Data Tables **18**, 413 (1976).
22. Shah, M. B., and Gilbody, H. B., J. Phys. B **18**, 899 (1985).
23. Petersen, J. O. P., Hvelplund, P., Petersen, A. G., and Fainstein, P. D., J. Phys. B (in press).
24. Toburen, L. H., and Wilson, W. E., Phys. Rev. A **19**, 2214 (1979).
25. Toburen, L. H., Stolterfoht, N., Ziem, P., and Schneider, D., Phys. Rev. A **24**, 1741 (1981).
26. Reinhold, C. O. and Olson, R. E., J. Phys. B **22**, L39 (1989).
27. Jakubaßa-Amundsen, D. H., The Physics of Electronic and Atomic Collisions, AIP Conf. Proc. 205 (American Institute of Physics, 1989) pp. 358-65.
28. Reinhold, C. O., and Schultz, D. R., J. Phys. B **22**, L565 (1989).
29. Reinhold, C. O., Schultz, D. R., Olson, R. E., Toburen, L. H., and DuBois, R. D., J. Phys. B **23**, L297 (1990).
30. Freyou, J., Breinig, M., Gaither III, C. C., and Underwood, T. A., Phys. Rev. A **41**, 1315 (1990).

31. Kelbch, S., Cocke, C. L., Hagmann, S., Horbatsch, M., Kelbch, C., Koch, R., Schmidt-Böcking, H., and Ullrich, J., J. Phys. B **23**, 1277 (1990).

32. Schultz, D. R., Olson, R. E., Reinhold, C. O., Kelbch, S., Kelbch, C., and Schmidt-Böcking, H., J. Phys. B (accepted).

33. Richard, P., Lee, D. H., Zouros, T. J. M., Sanders, J. M., and Shinpaugh, J. L., J. Phys. B 23, L213 (1990).

34. Reinhold, C. O., Schultz, D. R., and Olson, R. E., J. Phys. B (accepted).

35. Olson, R. E., Reinhold, C. O., and Schultz, D. R., J. Phys. B (accepted).

NEW ASPECTS OF SIMULTANEOUS PROJECTILE AND TARGET IONIZATION IN H^0(0.5MeV/1.0MeV)-->He - COLLISIONS[1]

R. Maier, O. Heil, R.D. DuBois#, M. Kuzel, K.O. Groeneveld
Institut für Kernphysik, J.W.Goethe-Universität Frankfurt,
August-Euler-Straße 6, D-6000 Frankfurt 90, Germany

Abstract

Absolute doubly differential electron emission cross sections for H^0(0.5MeV/1.0MeV)->He collisions have been measured non-coincidently and in coincidence with the outgoing H^+. A comparison is given between the coincident and non-coincident cross sections for H^0 and H^+ impact and PWBA-calculations of the DDCS for electron emission. It is shown that the contribution of simultaneous projectile and target ionizations to the coincident electron emission cross sections is very important. There is some evidence that a great part of these simultaneous ionizations can be explained by uncorrelated simultaneous projectile and target ionizations.

Introduction

Since the contributions of single target and projectile ionization to the **d**oubly **d**ifferential **c**ross **s**ections (DDCS) for electron emission are well known from many theoretical and experimental studies in ion-atom-collisions [1, 2, 3, 4] it is of basic interest to examine the contribution of simultaneous projectile and target ionization processes in the case of projectiles with one or more electrons [5]. In ionization by fast neutral particles the **E**lectron **L**oss to **C**ontinuum (ELC) dominate the electron emission cross sections in the forward direction, since electron capture processes are neglegible. This offers a good chance for experimental and theoretical investigations of the pure electron-loss

1 Supportet by BMFT/Bonn under contract number 06 OF 110 /Ti 476 Gr

permanent address: Pacific Northwest Laboratory, Richland, Washington 99352

process in atom-atom-collisions. Therefore experiments with H^0 projectiles have been performed. Earlier DDCS studies of H^0->Ar collisions [6] have demonstrated that the contribution of simultaneous projectile and target ionization dominates the coincident electron emission cross sections in the low energy region ($E<1/2E_{Cusp}$), while model calculations show a strongly decreasing cross section down to 0eV.

Experiment

The simplest atom-atom-collision system that can be treated experimentally with reasonable effort is H^0->He.

The H^0 beam of this experiment was produced by dissociating H_2^+ molecular ions in a gas filled charge exchange cell and cleaning of charged components by electrostatic field. The remainig H^0 beam (contamination < 2%) then traversed an effusive helium gas target.

Emitted electrons were energy analyzed by an electrostatical cylinder mirror spectrometer described in [7] and detected by a channeltron. For the coincident measurements the projectiles were electrostatically charge analyzed and then detected by a secondary electron emission type detector with counting rates up to $7.0\cdot 10^5$Hz with a detection efficiency of about 50% [8]. Because of the high incoming beam current needed to obtain sufficient counting rates, the coincident measurements could be carried out only with the outgoing H^+ but not with H^0. No information was available about the final target charge state. In the non-coincident measurements the projectile beam current was measured via a Faraday-cup.

Our experimental data of non-coincident electron emission have been placed on an absolute scale via a normalization procedure using the absolute cross sections given by Rudd, Toburen, and Stolterfoht [9] for H^+->He collisions. The coincidence spectra were normalized to the singles spectra at the maximum of the ELC-peak.

Theory

Because of the good agreement between calculations according to the Plane-Wave-Born-Approximation (PWBA) and experiments for H^0->He collisions only these PWBA-calculations shell be considered here in some detail.

The present PWBA-calculations according to the Plane-Wave-Born-Approximation (PWBA) were carried out as described in [1]. The transition matrix element in the PWBA can be written as:

$$T_{fi}=\langle\phi_f\psi_f|V|\phi_i\psi_i\rangle \quad \text{with} \quad V(R,r) = \frac{Z_1Z_2}{R} - Z_1\sum_{j=1}^{n} \frac{1}{|R-r_j|}$$

and $\langle\phi_f|V|\phi_i\rangle = 4\pi N_f N_i q^{-2} e^2 \sum_j \exp(iq\cdot r_j)$

For the system $H^+ \rightarrow H^0$ the cross section can be evaluated analytically according to [10]:

$$d\sigma = n \frac{2^8 Z_e^6 m_p^2}{\pi q^2} \frac{\exp\{-(2Z_e/k_e)\arctan[2Z_e/(Z_e^2+q^2-k_e^2)]\}}{[1-\exp(-2\pi Z_e/k_e)](Z_e^2+q^2+k_e^2-2qk_e\cos\delta)^4}$$

$$* [(q-k_e\cos\delta)^2+Z_e^2\cos^2\delta]/[(Z_e^2+q^2-k_e^2)^2+4Z_e^2k_e^2)\,d\Omega_p d\Omega_e dE,$$

Z_e = $E_b^{1/2}$, E_b=binding energy in Rydbergs
n = number of electrons with binding energy E_b
q = momentum transfer to the electron
k_e = momentum of outgoing electron
$\cos\delta$ = qk_e/qk_e
m_p = proton mass

In our calculations this cross section has been scaled with the square of the effective projectile charge Z_{eff} and the binding energies of the target electrons and then integrated over all projectile emission angles. The evaluation of electron loss cross sections has been carried out by calculating the cross section in the projectile centred system and then transforming these cross section to the laboratory system.

In this model electron-electron-scattering (Doubly-Inelastic(DI)-term) has been taken into account only by the evaluation of an effective projectile charge Z_{eff}, which depends on the momentum transfer to the ejected electron assuming electronic screening and the possibility of projectile excitation as described in [11].

Results

Fig. 1 and 2 show the DDCS for ionization by H^+(1.0MeV) on He and H^0(1.0MeV) impact on He in the ELC-region, where projectile ionization processes dominate at all angles. The difference between the coincidence and the singles data is in the low energy region ($E_e<1/2E_{Cusp}$) 50 to 100%, while the discrepancy in the Binary-Encounter-region (BE) can be estimated to be roughly a factor 10.

A similar behaviour is shown by the data for H^0,H^+(0.5MeV) impact as demonstrated in fig. 3 and 4. Here the influence of the screening of the H^0 projectile by its electron is prominent in the low energy part ($E_e<1/2E_{Cusp}$).

The PWBA-calculations agree with the experiments quite well in the upper energy region ($E_e>E_{Cusp}$), while they underestimate the electron emission cross sections in the lower energy region ($E_e<1/2E_{Cusp}$) by a factor of roughly 5 to 10. In the BE-region the DDCS given by the PWBA-calculations is too large. This is caused by a poor estimation of the influence of doubly-inelastic processes such as electron-electron-scattering in this energy region.

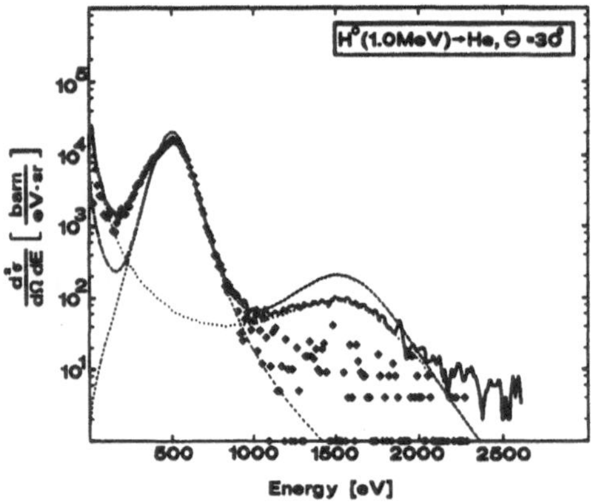

Fig. 1: Experiment and PWBA-calculation for H^0/H^+(1.0MeV) impact on He at 30°:
—— H^0->He, e^- single
◆ e^- coincident
.... H^+->He, e^- single
- - PWBA-calculation of total e^- emission for H^0(0.5MeV)->He
-·- PWBA-calculation of electron loss

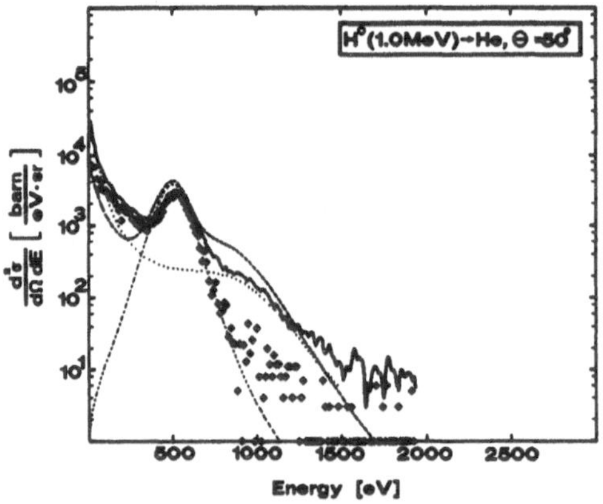

Fig. 2: Experiment and PWBA-calculation for H^0/H^+(1.0MeV) impact on He at 50°: symbols as described in fig. 1.

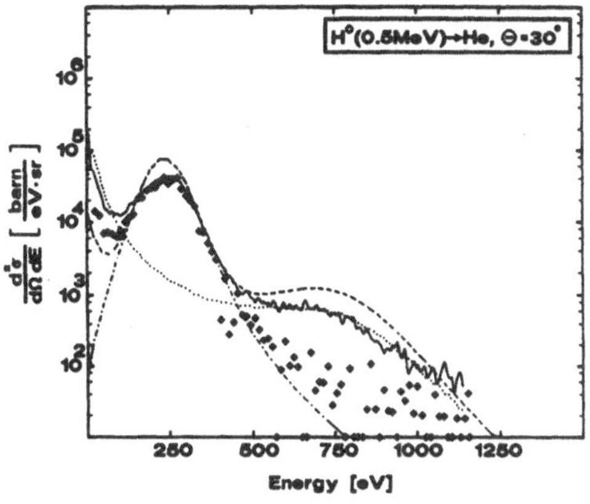

Fig. 3: Experiment and PWBA-calculation for H^0/H^+(0.5 MeV) impact on He at 30°: symbols as described in fig. 1.

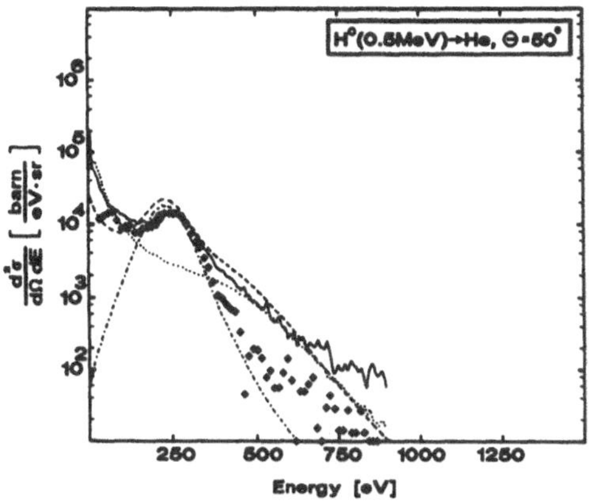

Fig. 4: Experiment and PWBA-calculation for H^0/H^+(0.5MeV) impact on He at 50°: symbols as described in fig. 1.

Discussion

There are three basic processes leading to electron emission in atom-atom-collisions:

1) Scattering of a projectil electron by the screened Coulomb potential of the target nucleus (pure electron loss)

2) Scattering of a target electron by the screened Coulomb potential of the projectile nucleus (pure target ionization)

3) Electron-electron-scattering (projectile and target ionization or excitation by doubly inelastic processes)

The PWBA-calculations presented have taken into account only the pure electron loss (1) and the pure target ionization (2) processes. Doubly inelestic processes (3) are represented only in the evaluation of the projectile form factor. A better evaluation of the DI-term is given by Hartley and Walters [3]. In their calculation simultaneous projectile and target ionization is refered to electron-electron-scattering processes, where projectile and target ionization are correlated. To obtain cross sections for doubly inelastic processes from experiments it is necessary to estimate the probability of uncorrelated simultaneous projectile and target ionization due to the processes (1) and (2).

Data of DuBois [12] show that the total cross sections for target ionization are:

$\sigma_{01}=3.5\cdot10^{-17}cm^2$ for H^+(0.5MeV)->He
$\sigma_{01}=2.0\cdot10^{-17}cm^2$ for H^+(1.0MeV)->He

The total cross sections for electron loss in H^0(0.5MeV/1.0MeV)->He collisions are presented in [13, 14]:

$\sigma^{01}=2.82\cdot10^{-17}cm^2$ for the total loss in H^0(0.5MeV)->He collisions
$\sigma_+^{01}=1.22\cdot10^{-17}cm^2$ for simultaneous projectile and target ionization in H^0(0.5MeV)->He collisions
$\sigma^{01}=1.8\cdot10^{-17}cm^2$ for projectile ionization in H^0(1.0MeV)->He collisions (see Ref. [14]).

In order to estimate roughly the projectile and target ionization probabilities we can compare these cross sections with the geometric dimensions of the helium atom. The geometric cross section for Helium can be estimated from the charge density distribution as $\sigma_G(He)\approx7\cdot10^{-17}cm^2$. From this we get as a rough estimation for the projectile ionization probability $P_{IP}=\sigma^{01}/\sigma_G\approx0.4$ for H^0(0.5MeV) impact and $P_{IP}=\sigma^{01}/\sigma_G\approx0.25$ for H^0(1.0MeV) impact. The target ionization probabilities can be estimated similarly as $P_{IT}=\sigma_{01}/\sigma_G\approx0.5$ for H^+(0.5MeV) impact and $P_{IT}=\sigma_{01}/\sigma_G\approx0.3$ for H^+(1.0MeV) impact. The contribution of simultaneous ionization processes to the DDCS of electron emission can be therefore in the order of 25 to 40%.

The different behaviour of the experimental data in the high energy region indicates that the probability of projectile ionization simultanoeous with the emission of a BE-electron is much lower than the probability of projectile ionization simultaneous with emission of a slow secondary electron. This fact suggests that there is an impact parameter dependence in the projectile ionization.

Conclusion

It has been shown that uncorrelated simultaneous projectile and target ionization processes can contribute significantly to the DDCS of electron emission in atom-atom-collisions because of the very high projectile ionization probabilities in the collision systems presented here. Existing model calculations of the DDCS in H^0->He collisions take into account only electron-electron-scattering (DI-term) as contribution to the simultaneous projectile and target ionization. Thus a comparison between experiments and model calculations of the DDCS for electron emission in atom-atom-collisions is very difficult and generally limited to the upper energy regime.
The different behaviour of the contributions of simultaneous projectile and target ionization processes in the low and high energy region of the spectra suggests an impact parameter dependence of the projectile ionization cross section.

References:

[1] M.E.Rudd, J.H.Macek, Case studies in Atomic Physics 3, 47 (1972); Eds. E.W.McDaniel, M.C.McDowell, North-Holland Publ. Corp., Amsterdam (1972)

[2] M.E. Rudd, Phys. Rev. A20, 787 (1979)

[3] H.M. Hartley, H.R.J. Walters, J.Phys. B20 (1987), 3811-3831

[4] D.H. Jakubaßa, J. Phys. B13, 2099 (1980)

[5] O. Heil, R.D. DuBois, R. Maier, M. Kuzel, K.-O. Groeneveld, Phys. Rev. A (1990, submitted to be printed)

[6] O. Heil, R. Maier, K.-O. Groeneveld (1990, to be published)

[7] G. Bernardi, S. Suarez, P. Focke, W. Meckbach, Nucl. Instr. Meth. B33 (1988) 321

[8] Rinn, Müller, Eichenauer, Salzborn, Rev. Sci. Instr. 53, 829 (1982)

[9] Atomic Data and Nuclear Data Tables Vol. 18, no. 5, November 1976

[10] N.F. Mott, H.S.W. Massey, The Theory of Atomic Collisions, 3. ed. (Clarendon Press, Oxford, 1965), p. 489

[11] M.H.Day, J. Phys. B14, 231 (1981)

[12] R.D. DuBois, S.T. Manson, Phys. Rev. A35, 2007 (1987)

[13] R.D. DuBois, A. Kövèr, Phys. Rev. A40, 1989, 3605-3612

[14] Y. Nakai, T. Shirai, T. Tabata, R. Ito, Atomic Data and Nuclear Data Tables, Vol. 37, 1987, S. 69

TWO-CENTER EFFECTS IN THE IONIZATION OF Ne by H^+ AND $^3He^{2+}$ IMPACT AT INTERMEDIATE ENERGIES[1]

S. Suarez, C. Garibotti, W. Meckbach, G. Bernardi, P. Focke, R. Pregliasco
Centro Atómico, 8400 S.C. de Bariloche, Argentina
G. Sigaud
Depto. de Física, PUC, Cx. Postal 38071,Rio de Janeiro 22453, Brasil

In a recent paper[(1)] we reported measurements of the double differential cross section for electron emission in the collision of H^+ and He^{++} projectiles on a He-target, at incident energies of 50 and 100keV/u. We now present such measurements performed with a Ne-target, using the same experimental setup and procedure. Spectra have been measured as a function of the electron energy E_e in the range from 5 to 600 eV at selected emission angles between -180 and +180 deg. For proton impact, earlier measurements by Rudd and collaborators[(2)] are available, covering 1.5 to 1057 eV in energy and 10 to 160 deg. in angle. Furthermore from a large number of experiments[(3)] total cross sections are reported; detached for the different channels contributing to electron emission, i.e. single and multiple ionization as well as transfer ionization.

We first fix our attention on the emission of electrons near the forward direction with energies E_e smaller than the electron equivalent ion energy E_{eq}, that is velocities V_e smaller than V_p, the electron transfer to the continuum (ETC)-peak velocity. For the He-target the angular electron distributions at fixed electron energies showed a maximum that became steeper at lower impact energies. An additional narrow ridge, about 2 deg wide, was found before in this laboratory[(4)] but was later shown[(5)] to be due to an experimental artifice, that is an extended gas target traversed by the projectile beam.

In Fig. 1 we present "transverse distributions" as measured, represented as a function of $v_T = v_e \sin\theta$ for two selected electron velocities $v_e = 1.43$ and 1.0 au. The ordinate scale is arbitrary, but relatively normalized with reference to the data shown. The mentioned high broad maximum, centered at the forward direction, is confirmed. We had noted that this feature is an evidence of the two center force, and is not described by current theoretical approaches[(6)].

1 Supported by the Consejo Nacional de Invest. Cientificas y Tecnicas, Argentina.

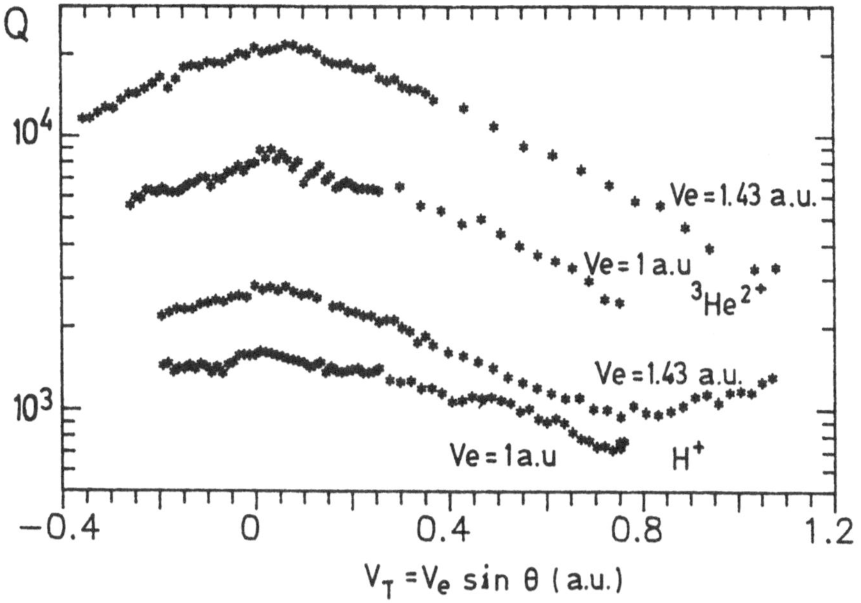

Fig. 1: Double differential cross sections for $^3He^{2+}$ and H^+ on He at 100 keV/u, represented for two fixed electron energies as a function of the transverse velocity $v_T = v_e \sin \theta$. The ordinate scale is arbitrary. Data are relatively normalized.

Meanwhile the above mentioned ridge attracted interest as a possible evidence for electrons stranded in the saddle of the potential between the two Coulomb attractive centers. A remarkable maximum, which quite naturally becomes evident upon representing the cross section $d\sigma/dv_e\, d\Omega$ as a function of v_e, has been considered as due to this effect.

Calculations using the classical trajectory Monte Carlo (CTMC) approach(7) predicted a large fraction of electrons moving with the velocity of that saddle point (v_s). A simple interpretation led to a shift of the saddle towards the target and hence to a decrease of v_s, when the projectile charge Z_p is increased. This effect would result in a shift in the maximum of $d\sigma/dvd\Omega$. Otherwise, one(8) and three dimensional(9) models for two colliding zero range potentials give a small probability to have an electron in the saddle region.

That shift was not confirmed neither by recent measurements performed in this laboratory with H^+ and $^3He^{2+}$ colliding on He(10), nor by the present data for Ne targets.

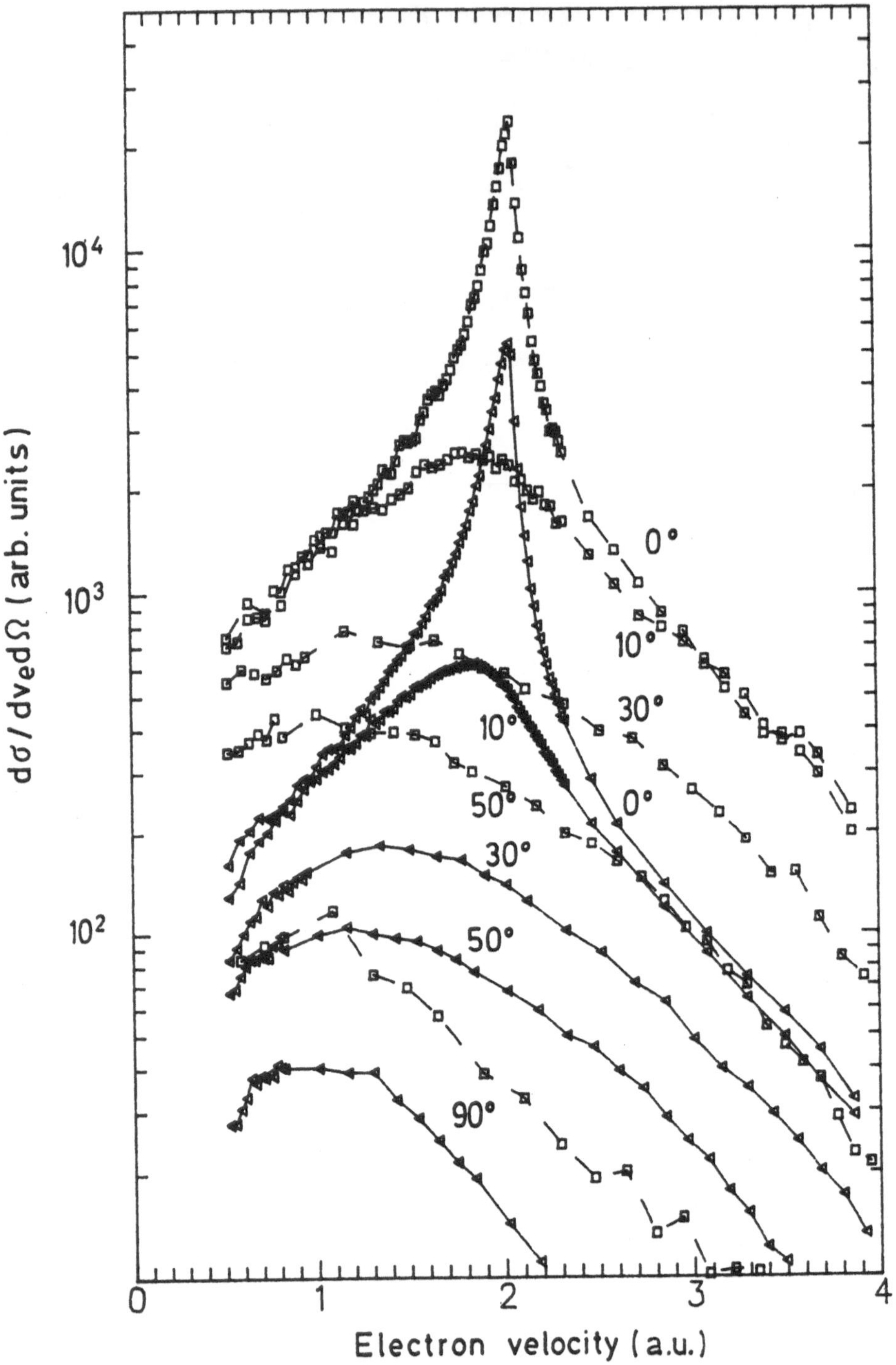

Fig. 2: Velocity cross sections (dσ/dvdΩ) of electrons ejected in collisions of H^+ and $^3He^{2+}$ with Ne at 100keV/u, at different emission angles as a function of the electron velocity. □ $^3He^{2+}$ and Δ: H^+ projectiles.

In Fig. 2 we show the cross section $d^2\sigma/dv_e d\Omega$ for H^+ and He^{2+} projectiles on Ne plotted as a function of V_e at different emission angles. No dependence of the position of the maximum of that cross section, on the projectile charge, is observed.

As a main feature we finally analyze, as before in the case of the He-target[1], the projectile charge (Z_p) dependence of the electron energy distributions measured at a fixed emission angle θ. As in ref. 11. We define the ratio:

$$R(E_e,\theta) = (d\sigma/dE_e d\Omega)_{He^{++}}/(d\sigma/dE_e d\Omega)_{H^+}$$

of the cross section for He^{++} to that for H^+ projectiles. For the He-target we had reported an upwards step of R in the forward direction localized at E_{eq}[1] In the lower energy range, $E_e < E_{eq}$, R depends only weakly on E_e. For 100 keV/u and θ = 0 deg. its value was R = 4, as predicted by the Z_p^2 scaling that results from first order perturbative calculations. However, for increasing θ, the ratio R was found to decrease or, "saturate", and tend to 1 for large emission angles.

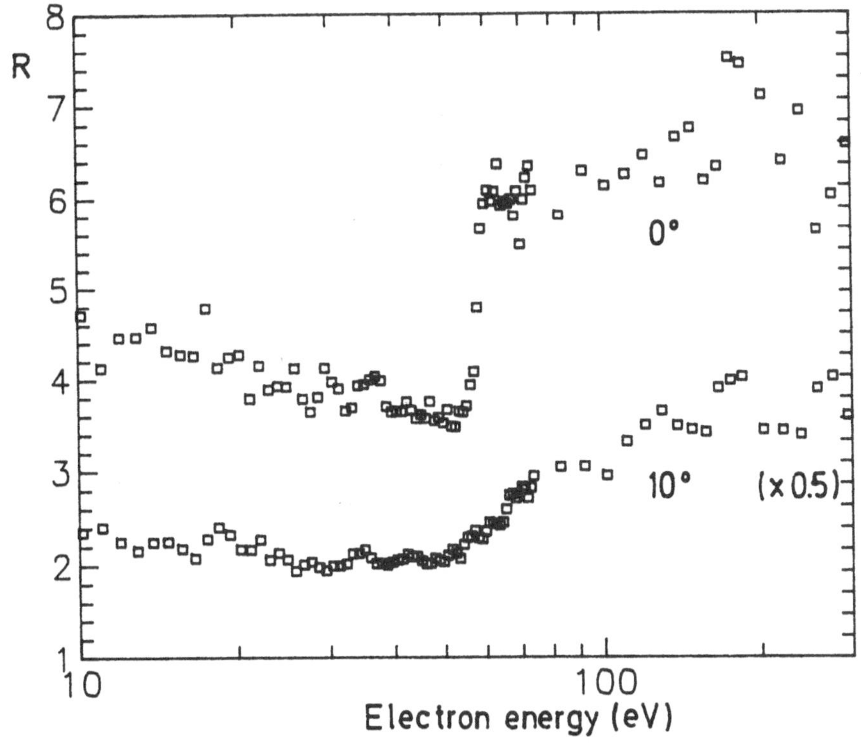

Fig. 3: Ratio between the double differential cross sections for $^3He^{2+}$ and H^+ on Ne at 100 keV/u as a function of the energy, for the fixed ejection angles indicate. The ratio for θ = 10° is multiplied by 0.5.

In Fig. 3 we show $R(E_e)$ as it results for 100keV/u He^{++} and H^+ on Ne at $\theta = 0$ and 10 degs. In the forward direction we observe again the characteristic stepwise increase at $E_e = E_{eq}$ that becomes smooth as θ increases. As in the case of the He-target[1] the value of the ratio is R = 4 at the low-energy side of the step. A different feature in the present data is that for Ne-target there is a gradual increase of R for lower energies, where for He this ratio remained constant.

We note that, according to the data of Du Bois[3], in the case of the Ne-target the transfer ionization (TI) channel contributes a fraction of 6.5% and up to 28% respectively for 100keV/u H^+ and He^{++}, whereas these values were only about 1% and 10% for the He target. In this connection it is significant to mention the conjecture[12,13] that, in the process of TI, first two target electrons are transfered to a highly correlated projectile state, of which one is subsequently lost into the continuum. Then, as electron loss into the continuum (ELC) is known to furnish a more symmetrical cusp as compared to ECC, the larger contribution from TI in the case of He^{++} might eventually contribute to the stepwise increase of R at $E_e = E_{eq}$. However, the fact that the height of this step (from R = 4 to 6) for Ne is the same as that found for the He-target, (whereas the contribution from TI is lower) may be an indication against this interpretation. Furthermore, it remains significant that the CDW-EIS approximation, which describes the process of single ionization in terms of a two center Coulomb interaction of the emitted electron, has been found to be able to describe a jump in the ratio R[1]. On the other hand the CTMC computation of collisional electron emission[14] that includes all the electron mentioned emission channels also leads to the characteristic stepwise increase of $R(E_e,\theta)$ at $\theta = 0$, $E_e = E_{eq}$.

We conclude that electron emission for the Ne target shows essentially the main features observed by us for He.

References:

1.- G. Bernardi, S. Suárez, P. Fainstein, C. Garibotti, W. Meckbach and P. Focke, Phys. Rev. A40, 6863, 1989.

2.- M.E. Rudd and T. Jorgensen, Phys. Rev. 131, 666, 1963; J.B. Crooks and M.E. Rudd, Phys. Rev. A3, 1628, 1971, M.E. Rudd and D.H. Madison, Phys. Rev. A14, 128, 1977.

3.- M.B. Shah and M.B. Gilbody, J. Phys. B18, 899, 1985; R.D. Du Bois, Phys. Rev. Lett. 52, 2348, 1984; Phys. Rev. A33, 1595, 1986.

4.- W. Meckbach P. Focke, A. Goñi, S. Suárez, J. Macek and M. Menendez, Phys. Rev. Lett. 57, 1587, 1986.

5.- G. Bernardi, S. Suárez, P. Focke and W. Meckbach in Lecture Notes in Physics, Vol. 294, Ed. D. Berenyi and G. Hock, pp. 286 and 295, 1987.

6.- C. Garibotti, M. Martiarena and D. Zanette, Lect. Notes in Phys., Vol. 294, Ed. D. Berenyi and G. Hock, p. 262, 1987.

7.- R. Olson, Phys. Rev. A33, 4397, 1986.

8.- J. Burgdorfer, J. Wang and A. Barany, Phys. Rev. A38, 4919, 1988.

9.- M. Rakovic and E. Solovev, Phys. Rev. A41, 3635, 1990.

10.- G. Bernardi, P. Fainstein, C. Garibotti and S. Suárez, J. Phys. B23, L139, 1990.

11.- G. Bernardi, P. Fainstein, C. Garibotti and S. Suárez, J. de Physique, C1, 50, 189, 1990.

12.- S. Datz et al., Phys. REv. A41, 3559, 1990.

13.- L. Andersen, M. Frost, P. Hvelplund and M. Knudsen, Phys. Rev. Lett. 52, 518, 1984.

14.- C. Reinhold and R. Olson, Phys. Rev. A39, 3861, 1989.

C Double Electron and Resonance Processes

Chairman: A.K. EDWARDS

C.1 Invited Surveys

J.A. TANIS
Resonant Recombination in Ion–Atom Collisions

SH.D. KUNIKEEV, V.S. SENASHENKO and V.A. SIDOROVICH
Production of Autoionizing States of Fast Charged Particle by Double Electron Capture

C.2 Invited Contributions

TH. STÖHLKER, C. KOZHUHAROV, E.A. LIVINGSTON, P.H. MOKLER, Z. STACHURA, A. SZYMANSKI and A. WARCZAK
Evidence for Resonant Double Capture and Single Excitation in Ge^{31+} + Ne Collisions

R. GAYET and A. SALIN
Capture with Ionisation in Proton–Helium Collisions

A.K. EDWARDS, R.M. WOOD, J.L. DAVIS, M.W. DITTMANN, J.F. BROWNING, M.A. MANGAN and R.L. EZELL
Two-Electron Processes in Projectile-H_2 Collisions

C.3 Contributions (Posters)

V.A. SIDOROVICH
The Energy Dependence of the Double Ionization Cross Section of Helium at High Energies

G.V. AVAKOV, L.D. BLOKHINTSEV, S.P. KREKOTEN, D.A. SAVIN, L.A. STOTLAND and A.M. SHIROKOV
Doubly Differential Detachment Cross Section for fast H^- on He Including Projectile Excitation to H

RESONANT RECOMBINATION IN ION-ATOM COLLISIONS

John A. Tanis
Department of Physics, Western Michigan University
Kalamazoo, Michigan 49008 USA

Abstract

Resonant recombination in atomic collisions is reviewed. This process, often referred to as resonant transfer excitation (RTE), occurs when electron capture (transfer) is accompanied by projectile excitation in a single-collision with the electron-electron interaction mediating the combined capture and excitation. The close relationship of RTE to dielectronic recombination (DR) can be exploited to provide significant tests of theoretical calculations of this latter fundamental ion-electron recombination process. An overview of resonant recombination results obtained to date along isoelectronic and isonuclear sequences is presented and compared with theory. Certain discrepancies with theory exist and explanations which have been advanced are noted. Apart from its connection with DR, the RTE process also demonstrates the importance of the electron-electron interaction in the dynamics of fast ion-atom collisions. In this regard, results are discussed showing the role of resonant recombination in ion channeling and in relativistic heavy ion-atom collisions.

I. INTRODUCTION

The recombination of ions is a subject of considerable interest as it involves fundamental interaction mechanisms, and, when mediated by the electron-electron interaction, the recombination takes place resonantly. For free-electron-ion collisions, the electron-electron interaction gives rise to dielectronic recombination[1] (DR) which occurs when electron capture is accompanied by simultaneous excitation of the ion resulting in the formation of a doubly-excited intermediate state. Subsequent deexcitation by photon emission completes the recombination of the ion as shown in Fig. 1a. Formation of the intermediate state proceeds via the inverse of an Auger transition, and is resonant for relative velocities corresponding to outgoing Auger-electron energies. Many different electronic configurations are possible in the intermediate state as indicated in Fig. 1b. It is common practice to denote these transitions as KLL, KLM, KLO, KMM... etc., or, in the case of L-shell excitation, LMM, LMN,..., etc. Since the intermediate state can be formed by excitation of the K or L (or higher) shells, deexcitation can occur by the emission of K or L x rays.

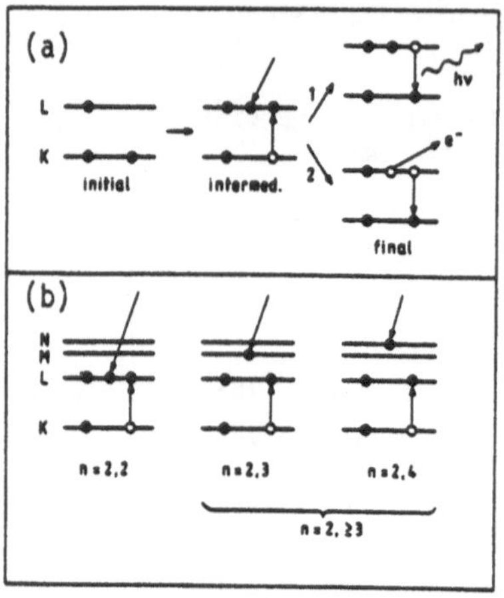

Fig. 1. (a) Schematic of the DR process for a Li-like ion. The intermediate doubly-excited state is formed via electron capture accompanied by projectile excitation. This intermediate state subsequently decays by either photon emission (1) or electron emission (2). (b) Schematic showing how DR takes place through different intermediate states. The notation n=2,2, n=2,3, etc., denotes the principal qunatum numbers of the two active electrons in the DR process.

In an ion-atom collision, the electron-electron interaction manifests itself in a manner completely analogous to DR when the ion velocity is much greater than the bound target electron velocity, i.e., the bound electrons are considered "quasi-free." This process, referred to as resonant transfer excitation[2] (RTE), can be treated in the impulse approximation[3] and is related to DR with a cross section given by

$$\sigma_{RTE} \sim \sigma_{DR} \, \Sigma J_i(p_{iz}) \qquad (1)$$

where σ_{DR} is the dielectronic recombination cross section, and $J_i(p_{iz})$ is the probability of finding a particular target electron with momentum component p_{iz} along the beam axis. Hence, the resonant DR maxima are broadened by the momentum distribution of the target electrons. A number of experimental[2] and theoretical[3,4] studies have clearly demonstrated that RTE closely approximates DR, and this close relationship permits RTE to be used as a test of theoretical DR cross sections. A typical example illustrating this close connection between DR and RTE is shown in Fig. 2 for Ca^{17+} + He collisions.[5]

Much of the interest in RTE stems from this close relationship to DR since, over a broad energy range, this latter mechanism is the principal mechanism by which free electrons recombine with ions. Apart from its intrinsic fundamental interest, DR is important in the study of astrophysical plasmas,[1] and in the development of nuclear fusion plasmas.[6] Cross sections for DR have been calculated[4,7-9] extensively, and recent measurements utilizing ion traps, electron coolers, and storage rings[10-13] are providing the first stringent test of DR theory.

Apart from its connection to DR, RTE also demonstrates the importance of the electron-electron interaction in the dynamics of fast ion-atom collisions. Since it can result in recombination of an ion, RTE competes with direct capture (radiative and nonradiative) resulting from independent-particle interactions between the charged ion and a target electron. (In RTE electron capture is the result of the

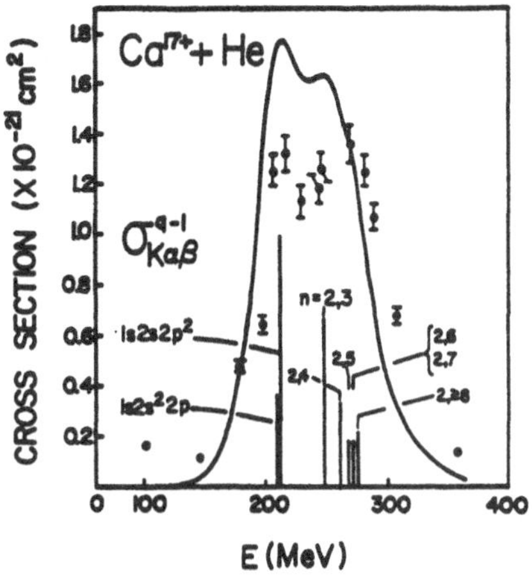

Fig. 2. RTE cross sections for Ca^{17+} + He. Data (Ref. 5) are for calcium K x rays coincident with single capture. The vertical bars give the calculated positions and relative intensities of the intermediate states for dielectronic recombination (see Fig. 1b). The solid curve is the calculated RTE cross section.

electron-electron interaction.) As an example, for Ca^{q+} + H_e collisions[14], it was found that RTE accounts for nearly half of the total single-electron capture in the energy region where RTE is important. This result suggested that electron capture by relativistic heavy ions might be almost entirely due to RTE. Subsequent calculations[15] of total electron-capture cross sections, obtained by adding together the expected cross sections for direct single-capture[16] and RTE, for U^{89+} in H_e, Be, and C indicated that RTE should be readily observable above the monotonically decreasing direct capture cross section.

To investigate RTE experimentally, this particular reaction channel needs to be isolated from competing channels. From Fig. 1a it can be seen that the signature for RTE is x-ray or Auger emission associated with electron capture. Two basic techniques exist to investigate RTE: (1) low-resolution coincidence methods in which electron capture is associated with the appropriate decay process (photon or Auger emission), and (2) high-resolution x-ray or Auger measurements in which specific intermediate states are identified. To date, RTE has been identified and investigated by measuring the energy dependence of (a) coincidences between photons associated with excited projectiles which have captured an electron,[2] (b) Auger-electron emission associated with transfer plus excitation events,[17] and (c) total single-electron capture cross sections to which RTE contributes substantially enough so that resonant behavior is observed.[18]

Ion-x-ray coincidence measurements of RTE have been conducted for ions with atomic numbers ranging from 9-32 and incident charge states ranging from H-like to Al-like, high-resolution Auger measurements have been used for C, O, and F ions in H-like and Li-like charge states, and total single-electron capture measurements have been used to study RTE for He-like and Li-like U ions. Thus, in general, Auger measurements have been used only for low Z ($\leq$ 9) ions, while x-ray and "singles" measurements have been used for higher Z ($\geq$ 9) ions.

Because this conference is devoted mainly to the study of ion-atom collisions, the present discussion will be limited to RTE, and, in particular, those collisions which result in recombination of the ion. Results obtained by several investigators will be reviewed and discussed from the standpoint of providing tests of DR cross sections. Additionally, discrepancies with theory will be noted along with explanations which have been advanced. Finally, three significant RTE studies conducted recently will be discussed in some detail.

II. ATOMIC NUMBER AND CHARGE-STATE DEPENDENCE OF RTE

The close relationship between RTE and DR permits the former to be used as a test of theoretical DR cross sections. Until very recently, most studies[2,4] of RTE involving K-shell excitation have been conducted for ions with $9 \leq Z \leq 32$, while the only extensive study for L-shell excitation[19,20] is for Nb^{q+} ions. Important tests of DR come from studies along isoelectronic or isonuclear sequences. Extensive results along the Li-like isoelectronic sequence for $9 \leq Z \leq 92$ and along the isonuclear sequence for H-like to Ne-like charge states of Ca^{q+} ions and the isonuclear sequence for Ne-like to Al-like Nb^{q+} ions are available. The substantial agreement between theory and experiment for KLL and KLM transitions of the Li-like isoelectronic sequence (U results are for He-like ions) is displayed in Fig. 3. This figure shows the ratio of the theoretical-to-experimental maximum RTE cross section values as a function of projectile atomic number. For Z values $\geq$ 20 the KLL and KLn ($n \geq M$) transitions can be separately identified. The U results, which will be discussed in more detail below, are significant because they provide a test of relativistic DR theory.

From Fig. 3 it is seen that there is good agreement between theory and experiment for all ions investigated except for F^{6+} where the discrepancy is a factor of about two, and possibly for Si^{11+} where the discrepancy is about 1.25.

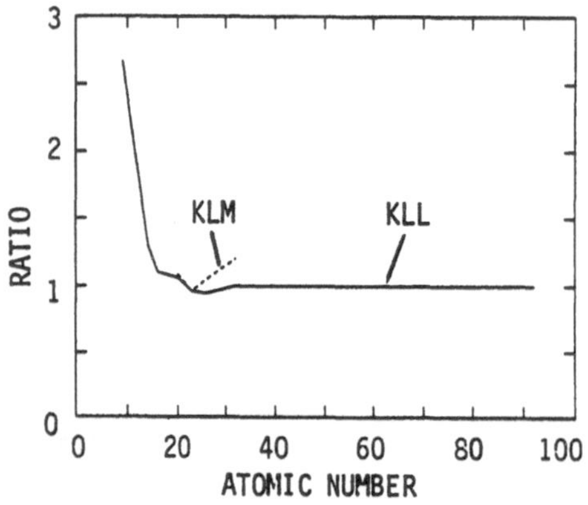

Fig. 3. Ratio of theoretical-to-experimental RTE cross section maxima as a function of atomic number Z. Results for KLL and KLM transitions are separately identified. In calculating the ratios, data from Refs. 5, 18, 21, 22, 24, 29, 45, and 46 were used, and calculations from Refs. 3, 7, 22, 44, and 47 were used.

For all higher Z ions theory and experiment agree to better than ± 10%. The reason for the large discrepancy in the case of F^{6+} is unknown at this time.

Another important test of DR theory is its accuracy in predicting the charge-state dependence of DR cross sections. Because of the close relationship between DR and RTE, the latter has been used to probe this charge-state dependence as summarized in Fig. 4 where the maximum value of the measured and calculated RTE cross sections are plotted against the number of electrons on the incident ion. Results are shown for incident Ca^{q+} (Ref. 21) ions, Fe^{q+} (Ref. 22) ions, and Nb^{q+} (Ref. 20) ions. The results for Ca and Fe involve K-shell excitation while those for Nb ions involve L-shell excitation. The near factor-of-two decrease between two electrons and one electron on the incident ion reflects the fact that, in the former case, there are two K electrons which can participate in the RTE process, whereas, for the latter there is only one.

The good agreeement between theory and experimant in Fig. 4 for KLL and KLM transitions provides further evidence of the accuracy with which DR involving K-shell excitation may be calculated for these very highly-charged ions. In view of the K-shell results, the large discrepancy between theory [9,23] and experiment[20] for RTE involving L-shell excitation as seen in Fig. 4 is difficult to understand, and no explanation has been forthcoming to date.

Another unresolved discrepancy is the difference between RTE theory and experiment at high energies where the theory falls off faster than the data as seen in Fig. 2. This same general trend has been observed for Li-like sulfur ions[24] and

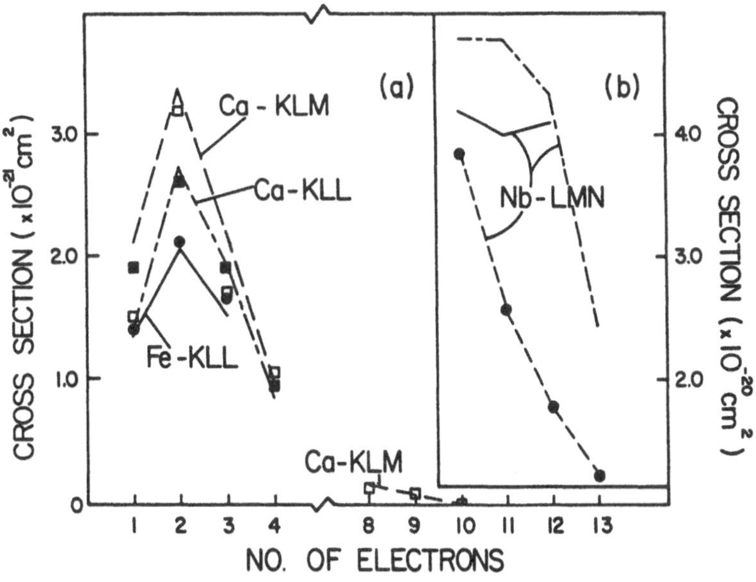

Fig. 4. Charge-state dependence of RTE cross-section maxima. The data for Ca, Fe, and Nb are from Refs. 21, 22, and 20, respectively. The theory for Ca and Fe are from Refs. 4 and 22, respectively, while the theory for Nb is from Ref. 23 (solid line) and Ref. 9 (broken line).

Ne-like niobium ions,[19] and has been a prevailing feature of nearly all RTE investigations to date.[2]

As a means of explaining this discrepancy between theory and experiment, the uncorrelated transfer excitation [25,26] mechanism has been invoked. For UTE (also called 2eTE) the excitation is due to a direct electron-electron interaction (similar to electron impact excitation) while the capture process is due to an electron-nucleus interaction, and these excitation and capture interactions are independent. The UTE mechanism was first proposed[27] in 1985 as a possible expalanation for resolving the high-energy discrepancy between theory and experiment in the RTE cross sections after attempts to explain it through the contribution of 1s --> nln'l' (n,n' $\geq$ 3) transitions or through a core field effect proved unsuccessful.[28] Some experimental evidence for UTE in x-ray measurements of RTE was reported by Schulz et al.[29] in F^{6+} + H_2 collisions, while more definitive evidence[30] was obtained in high-resolution Auger-electron measurements of RTE for F^{8+} + H_2 collisions.

To evalute quantitatively the effect of UTE, Hahn and Ramadan[26] have carried out a calculation of its contribution in Nb^{31+} + H_2 collisions and compared the result with the experimental measurements[19] of L-x-ray emission coincident with electron capture in these same collisions. The results are shown in Fig. 5a in which it is seen that the calculated UTE contribution accurately accounts for the

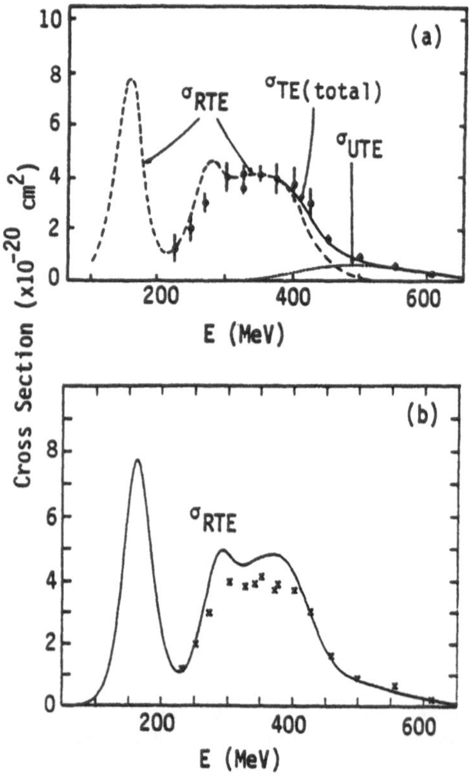

Fig. 5. Cross sections for L x-ray emission coincident with single-electron capture in Nb^{31+} + H_2 collisions. Data in (a) and (b) are from Ref. 19. (a) Comaprison of data with calculated RTE and UTE cross sections from Ref. 26. (b) Comparison of data with calculated RTE cross sections from Ref. 9. These latter RTE calculations include contributions from LNn, n$\geq$N, while the former do not.

discrepancy between experiment and the predicted RTE cross section in the high-energy region.

However, by including RTE contributions from higher transitions LNn, LOn,..., n≥ N, Badnell[9] has found the data to be nearly fully accounted for by the RTE mechanism alone for these same Nb^{31+} + H_2 collisions as seen in Fig. 5b. From this latter result it is concluded that any contribution from UTE must be small compared to the RTE contribution in the high-energy region for this collision system. In other calculations of RTE involving K-shell excitation in Ca^{q+} ions, Badnell[8] was able also to account for much of the discrepancy between theory and experiment at high energies by including the contribution of KMn, n≥ M transitions.

From these apparently conflicting results of Refs. 9 and 26, it appears that the importance of UTE in the transfer-excitation process is not fully understood. It should also be noted that UTE calculations are difficult and the uncertainties are such that the calculated values are reliable only to about a factor of two,[25,26] thereby making it difficult to unambiguously determine the origin of the high-energy discrepancy between theory and expeiment in the RTE cross sections. Thus, additional systematic measurements coupled with more reliable calculations of its magnitude will be required to determine the effect of UTE on RTE cross sections in the high-energy region.

III. SPECIFIC RECENT INVESTIGATIONS OF RESONANT RECOMBINATION

1. Resonant Capture in Ion Channeling

When fast heavy ions are channeled between the ordered rows of a crystal lattice, small-impact parameter collisions with the target atoms are reduced considerably. As a result these channeled ions can have an anomalously high probability for maintaining their initail charge state in moving through crystals many times thicker than amorphous targets of equilibrium thickness.[31] The so-called "frozen" charge state represents that fraction of the beam which is well-channeled and thus avoids close collisions with the target atoms of the crystal. Recently, radiative electron capture (REC) was found[32] to account for nearly all of the electron capture by 25 MeV/u Xe^{53+} ions channeled in silicon crystals. (REC is a process in which electron capture is accompanied by the simultaneous emission of a stabilizing photon,[33] i.e., the inverse of the photoelectric effect.)

These considerations suggest the possibility that resonant capture may be important in ion channeling. This process, again involving capture accompanied by projectile excitation via the electron-electron interaction and resulting in the formation of a doubly-excited intermediate state, is very closely related to RTE in ion-atom collisions and to DR in free-electron-ion collisions. If the "size" of the incident ions is small enough, the channeled ions interact mainly with the valence electrons in the crystal. These electrons, which are nearly free, have a small distribution of momenta and would be expected to give rise to narrow resonance widths. Hence, resonant capture in channeled ions should be essentially identical with the free-electron-ion process of DR.

In recent experiments with H-like S^{15+} and Ca^{19+} and He-like Ti^{20+} ions channeled in silicon crystals, it was found[34] that the exit charge-state distributions and emitted projectile K x-ray intensities were strongly influenced by resonant capture. These experiments were carried out with comparatively thick crystals in which the ions had a very high probability of undergoing charge changing following capture. Hence, the resonant capture had to be discerned indirectly through its effect on these subsequent charge-changing collisions. Additionally, it was not possible in this work to separate those ions which were well channeled from those which were poorly channeled in the crystal, so that the fate of these two very different classes of channeled ions could not be separately followed.

In another study, resonant capture in ion channeling was investigated using thin crystals so that the well-channeled ions could be identified through high-resolution measurements of their energy losses.[35] In this work, beams of 267-320 MeV $Ti^{19,20+}$ ions were directed along the <110> axis in Au crystals of thickness ~1200A. Results of this work show that electron capture by the well-channeled ions is significantly enhanced when the projectile velocity corresponds to ejected KLL Auger-electron energies from doubly-excited states in the rest frame of the ion as shown in Fig. 6. Furthermore, the capture maxima exhibit very narrow (nearly an order of magnitude sharper than those observed in similar RTE experiments[36] with H_2 targets) resonance widths which can be understood in terms of the small distribution

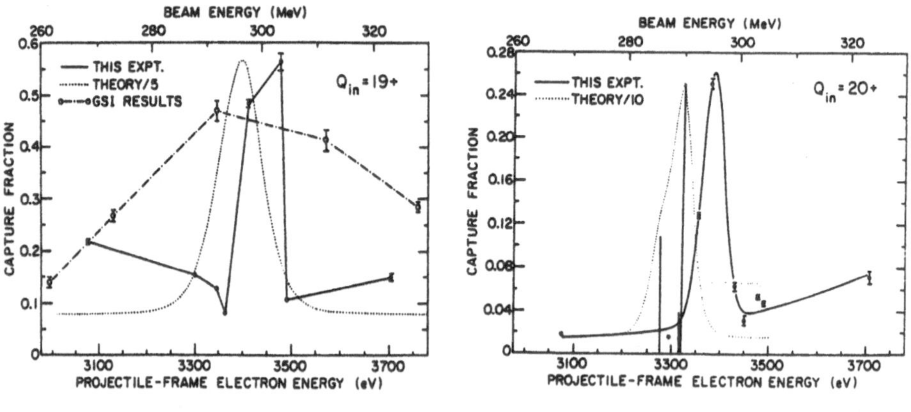

<u>Fig. 6</u>. Measured fraction of well-channeled ions emerging as Ti^{18+} or Ti^{19+} when Ti^{19+} and Ti^{20+}, respectively, were incident along the <110> axis of a 1200 A Au crystal. The solid lines are drawn to guide the eye. Results for Ti^{19+} on H_2 from Ref. 36 are shown for comparison. The dotted curves are theoretical estimates using the Compton profile of H_2 divided by factors of 5 and 10, respectively, for incident 19+ and 20+ ions.

of electron momenta "seen" by the well-channeled ions. These resonance widths are, in fact, narrower than those observed for DR maxima with cooled ions in an electron beam ion trap (EBIT).[10]

From Fig. 6 is is seen that the resonance energy for Ti^{19+} occurs about 40 eV higher in energy than for Ti^{20+} in agreement with theoretical calculations of the relative Auger energies for these ions. However, both resonances are also shifted by about 40 eV on an absolute scale to energies higher than anticipated from theory. Presently, it is not known whether this discrepancy is a consequence of electronic screening while in the solid, a conseqence of the resonant capture process itself, or if there is an error in the theoretical calculation of the Auger transition energies.

2. Resonant transfer excitation for $Fe^{23,24,25+}$ ions.

An ion of particular importance in the study of DR is iron, largely because of its importance in the understanding of astrophysical plasmas. Detailed theoretical calculations[37] of DR cross sections for Fe^{15+}, Fe^{23+}, and Fe^{25+} have been reported, and the cross sections are found to decrease strongly with increasing projectile charge state. As mentioned earlier, the close relationship between DR and RTE permits the latter to be used as a test of theoretical DR cross sections. Additionally, it is noteworthy that, in spite of the close relationship between RTE and DR, no direct experimental comparison has ever been made between these two processes. Presently, there is work in progress[38] to measure DR cross sections for Fe^{23+} and, possibly, for Fe^{25+}. Thus, RTE measurements for iron not only provide information on DR, but may also allow the first direct comparison between these two fundamental and related recombination processes.

An experimental and theoretical study of RTE for Fe^{23+}, Fe^{24+}, and Fe^{25+} ions colliding with H_2 has recently been reported.[39] The measurements were made by observing coincidences between electron capture events and Fe K x-ray emisssion resulting from decay of the intermediate doubly excited states created in the RTE process, e.g., $Fe^{23+}(1s^22s) + e^- \rightarrow Fe^{22+}(1s2s2p^2) \rightarrow Fe^{22+}(1s^22s2p) + h\nu$. The theoretical calculations were based on an isolated resonance approximation using Thomas-Fermi-Dirac-Amaldi radial wavefunctions and the (Coulomb) Breit-Pauli Hamiltonian.[40-42]

The cross sections obtained in Ref. 40 are shown in Fig. 7. The measurements extend only to 525 MeV which was the maximum attainable energy for the iron ions used in this work. Thus, the measurements cover the region corresponding to intermediate resonance states involving KLL transitions only. For Fe^{25+} the large cross section value near 325 MeV is due to electron capture to $n \geq 2$ followed by K x-ray emission when this singly-excited state decays. The smooth curves are absolute calculations of the RTE cross sections and have not been normalized to the data. The maximum at higher energies in the theoretical curves is due to KLM, KLN,

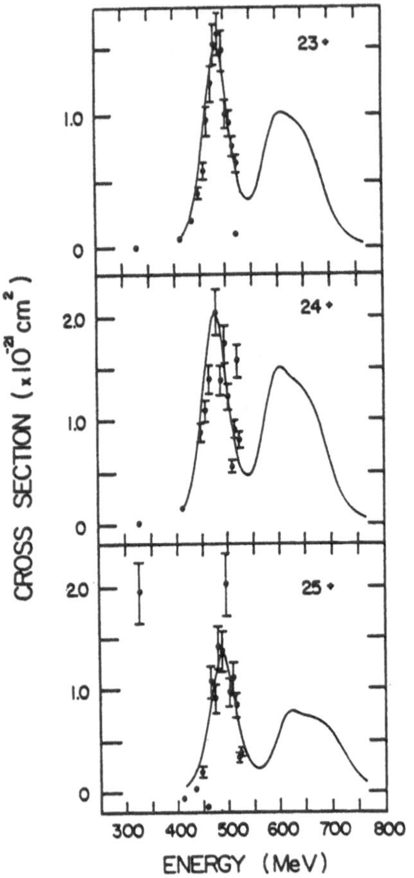

Fig. 7. Cross sections for Fe K x-ray emission in $Fe^{23,24,25+}$ + H_e collisions from Ref. 22. The calculated RTE cross sections shown by the smooth curves were obtained from theoretical DR cross sections averaged over the electron momentum distribution of the target. The lower-energy maximum corresponds to resonant KLL transitions, while the higher energy maximum is due to KLM, KLN, KLO,... transitions.

KLO,... transitions. In general, the agreement between theory and experiment is observed to be very good for each charge state. Earlier in this paper the charge-state dependence of RTE was discussed, and the cross section maxima for the present Fe data are shown in Fig. 4, again displaying the good agreement between theory and experiment.

Thus, these results demonstrate that DR involving K-shell excitation can be accurately predicted for these highly-charged ions. Earlier RTE results[21] for K-shell excitation in H-like to Ne-like Ca ions also showed good agreement with theory.[4,8] However, this good agreement is contrary to results[20] obtained for RTE with L-shell excitation in Na-, Mg-, and Al-like Nb ions for which theory[7,23] and experiment[20] disagree by a factor of two or more (see Fig. 4).

3. Resonant transfer excitation in U^{90+} + H_2 collisions.

There has been considerable recent theoretical interest in RTE for highly-stripped U ions incident on various targets.[15,43,44] In the first of these studies,[15] predicted RTE cross sections were obtained by extrapolating to Z=92 the theoretical nonrelativistic DR cross sections[4] for ions in the range Z=14-26. In the newest of these studies,[43,44] the energy levels, Auger rates, and radiative rates required to determine the DR cross sections were calculated using the multiconfiguration Dirac-Fock model, and the inclusion of relativistic effects is found to change significantly the Auger rates and energies. For example, the lowest energy transition, arising from the 1s2l2l' intermediate state splits into three peaks, and these peaks occur at higher energies than in the nonrelativistic calculation. The DR and RTE cross sections calculated by Chen[43] and by Pindzola and Badnell[44] are in very good agreement with each other for U^{89+} projectiles. The latter authors have also calculated the RTE cross section for U^{90+} in H_2 and these cross sections are predicted to be significantly larger than those for U^{89+} in H_2.

Recent measurements[18] of the total single-electron capture cross sections for U^{90+} + H_2 collisions are shown in Fig. 8a. Two prominent peaks and an indication of a third superimposed on a montonically decreasing background (due to direct capture) are observed. The maxima are attributed to the contribution of RTE to single capture. The three peaks arise from the formation, with increasing energy, of the $1s2s^2$ + $1s2s2p_{1/2}$ + $1s2p^2{}_{1/2}$, $1s2s2p_{3/2}$ + $1s2p_{1/2}2p_{3/2}$, and $1s2p^2{}_{3/2}$ intermediate states. Three more peaks are expected, starting at about 151 MeV/u due to the formation of the 1s2l3l' intermediate state.

By fitting a smooth function to the direct capture background as shown in Fig. 8a, the RTE cross section for U^{90+} in H_2 can be found by subtracting this "background" from the measured total single-electron capture cross section. The resulting RTE cross section is shown in Fig. 8b. It is seen that the magnitude of the measured RTE cross section is in excellent agreement with the calculation of Pindzola and Badnell.[44] It is emphasized that the magnitudes of both the measured and calculated cross sections are absolute and have not been normalized to one another.

The positions of the maxima in the measured cross sections are noted to be slightly lower in energy than the calculated values. The difference of about 2 MeV/u is considerably larger than the estimated absolute uncertainty (± 0.5 MeV/u) in the determination of the experimental energies. The origin of this discrepancy between the measured and calculated energies is presently unknown.

These results have thus extended the study of RTE to very high atomic number, and provide a quantitative test of DR theory in a region where relativistic effects are expected to be important. Apart from testing DR cross sections, however, these results also show the importance of the electron-electron interaction, as manifested by the fact that RTE contributes substantially to the total single-electron capture cross section in the energy region where RTE occurs.

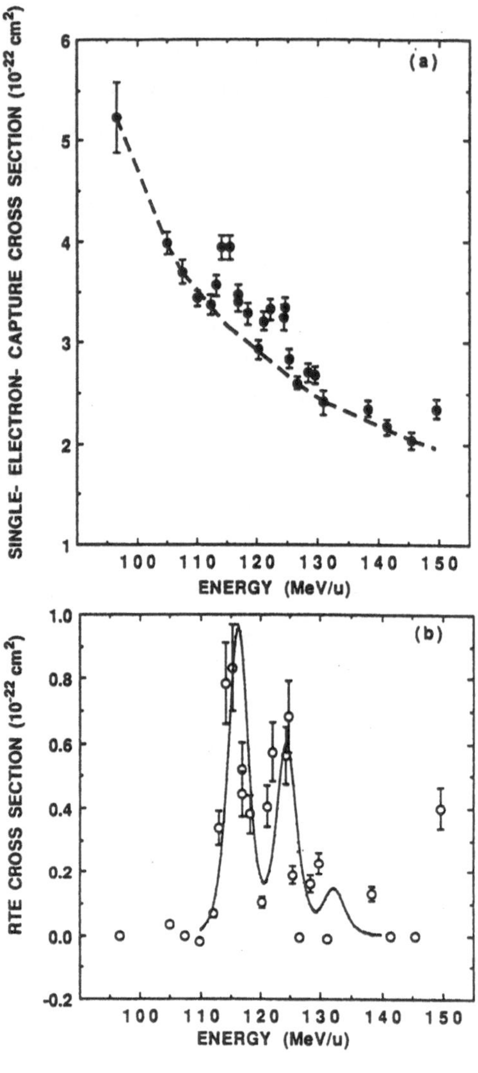

Fig. 8. Cross sections for U^{90+} in H_2. (a) Data points are the measured total single-electron capture cross sections. Error bars show the relative uncertainty in these measured cross sections. The dashed curve is an empirical fit to the direct capture background. (b) Data points show the experimentally derived RTE cross sections (total capture cross section minus the empirical fit). Error bars give the absolute uncertainty in these experimental cross sections. The solid curve is the predicted[44] RTE cross section.

IV. SUMMARY AND CONCLUSIONS

Resonant recombination in atomic collisions has been reviewed. It is shown that the close relationship of resonant transfer excitation (RTE) to dielectronic recombination (DR) can be exploited to provide important tests of theoretical calculations for the latter fundamental ion-electron recombination process. Apart from its connection with DR, however, RTE demonstrates the importance of the electron-electron interaction in the dynamics of fast ion-atom collisions.

As a test of DR, measurements of RTE have been used to provide information along isoelectronic and isonuclear sequences. For the isoelectronic sequence, RTE has been investigated for ions with atomic number ranging from 9-92, and the agreement with calculations based on DR cross sections is very good except for the case of F ions where theory is a factor of two bigger than experiment. The RTE measurements for uranium ions provide the first test of relativistic DR theory. Measurements of RTE have also been conducted along isonuclear sequences for H-like to Ne-like charge states of Ca^{q+} ions and for Ne-like to Al-like charge states of Nb^{q+} ions. In the case of Ca ions (involving K-shell excitation), the agreement between theory and experiment is good while for Nb ions (involving L-shell excitation) theory and experiment disagree substantially in both magnitude and energy dependence and the reason for this discrepancy is not understood.

On the high-energy side of the RTE maximum the theory has been found to consistently and significantly overestimate the data. Although this discrepancy has been attributed to uncorrelated transfer excitation (UTE), it is not yet clear whether high-energy RTE transitions, previously negelcted in calculations, might in fact account for most of the discrepancy.

With regard to the effect of RTE in the dynamics of ion-atom collisions, this resonant process has been shown to play a significant role in ion channeling and in relativistic electron capture. In channeling, RTE can alter significantly the emerging charge-state distributions and x-ray emission of well-channeled ions compared to poorly-channeled ions or ions moving in amorphous solids. For relativistic heavy (uranium) ions, RTE has been found to contribute substantially to the total single-electron capture thereby providing an important recombination channel at very high velocites.

In conclusion, resonant recombination in ion-atom collisions has provided considerable insight into the electron-electron interaction and the few-body problem. Although much has been learned, furthur studies are likely to provide more stringent tests of theory and thereby lead to a fuller understanding of fundamental atomic interactions and processes.

Acknowledgements

The author is deeply indebted to the many colleagues with whom he has worked closely for several years. This work was supported in part by the Division of Chemical Sciences, Office of Basic Energy Sciences, Office of Energy Research, U. S. Department of Energy.

REFERENCES

1. A. Burgess, Astrophys. J. 139, 776 (1964); and 141, 1588 (1965).
2. See J. A. Tanis, Nucl. Instrum. Meth. Phys. Res. A262, 52 (1987), and references therein.
3. D. Brandt, Phys. Rev. A27, 1314 (1983).
4. See Y. Hahn and K. J. LaGattuta, Phys. Rep. 166, 196 (1988), and references therein.
5. J. A. Tanis, E. M. Bernstein, W. G. Graham, M. P. Stockli, M. Clark, R. H. McFarland, T. J. Morgan, K. H. Berkner, A. S. Schlachter, and J. W. Stearns, Phys. Rev. Lett. 53, 2551 (1984).
6. M. Bitter, S. von Goeler, S. Cohen, K. W. Hill, S. Sesnic, F. Tenney, J. Timberlake, U. I. Safronova, L. A. Vainshtein, J. Dubau, M. Loulergue, F. Bely-Dubau, and L. Steenman-Clark, Phys. Rev. A29, 661 (1984).
7. N. R. Badnell, Phys. Rev. A40, 3579 (1989).
8. N. R. Badnell, Phys. Rev. A42, 209 (1990).
9. N. R. Badnell, Phys. Rev. A42, 204 (1990).
10. D. A. Knapp, R. E. Marrs, M. A. Levine, C. L. Bennett, M. H. Chen, J. R. Henderson, M. B. Schneider, and J. H. Scofield, Phys. Rev. Lett. 62, 2104 (1989).
11. L. H. Andersen, P. Hvelplund, H. Knudsen, and P. Kvistgaard, Phys. Rev. Lett. 62, 2656 (1989).
12. R. Ali, C. P. Bhalla, C. L. Cocke, and M. Stockli, Phys. Rev. Lett. 64, 633 (1990).
13. G. Kilgus, J. Berger, P. Blatt, M. Grieser, D. Habs, B. Hochadel, E. Jaeschke, D. Krämer, R. Neumann, G. Neureither, W. Ott, D. Schwalm, M. Steck, R. Stokstad, E. Szmola, A. Wolf, R. Schuch, A. Müller, M. Wagner, Phys. Rev. Lett. 64, 737 (1990).
14. W. G. Graham, E. M. Bernstein, M. W. Clark, J. A. Tanis, K. H. Berkner, P. Gohil, R. J. McDonald, A. S. Schlachter, J. W. Stearns, R. H. McFarland, T. J. Morgan, and A. Müller, Phys. Rev. A33, 3591 (1986).
15. W. G. Graham, E. M. Bernstein, M. W. Clark, and J. A. Tanis, Phys. Lett. A125, 134 (1987).
16. W. E. Meyerhof, R. Anholt, J. Eichler, H. Gould, Ch. Munger, J. Alonso, P. Thieberger, and H. E. Wegner, Phys. Rev. A32, 3291 (1985).
17. J. K. Swenson, Y. Yamazaki, P. D. Miller, H. F. Krause, P. F. Dittner, P. L. Pepmiller, S. Datz, and N. Stolterfoht, Phys. Rev. Lett. 57, 3042 (1986).
18. W. G. Graham, K. H. Berkner, E. M. Bernstein, M. W. Clark, B. Feinberg, M. A. McMahan, T. J. Morgan, W. Rathbun, A. S. Schlachter, and J. A. Tanis, submitted to Phys. Rev. Lett.
19. E. M. Bernstein, M. W. Clark, J. A. Tanis, K. H. Berkner, R. J. McDonald, A. S. Schlachter, J. W. Stearns, W. G. Graham, R. H. McFarland, T. J. Morgan, J. R. Mowat, D. W. Mueller, and M. P. Stockli, J. Phys. B20, L505 (1987).
20. E. M. Bernstein, M. W. Clark, J. A. Tanis, W. T. Woodland, K. H. Berkner, A. S. Schlachter, J. W. Stearns, R. D. DuBois, W. G. Graham, T. J. Morgan, D. W. Mueller, and M. P. Stockli, Phys. Rev. A40, 4085 (1989).
21. J. A. Tanis, E. M. Bernstein, M. W. Clark, W. G. Graham, R. H. McFarland, T. J. Morgan, J. R. Mowat, D. W. Mueller, A. Müller, M. P. Stockli, K. H. Berkner, P. Gohil, R. J. McDonald, A. S. Schlachter, and J. W. Stearns, Phys. Rev. A34, 2543 (1986).
22. M. W. Clark, J. A. Tanis, E. M. Bernstein, N. R. Badnell, R. D. DuBois, W. G. Graham, T. J. Morgan, V. L. Plano, A. S. Schlachter, and M. P. Stockli, submitted to Phys. Rev. A.
23. Y. Hahn, J. N. Gau, G. Omar, and M. P. Dube, Phys. Rev. A36, 576 (1987).
24. J. A. Tanis, E. M. Bernstein, M. W. Clark, W. G. Graham, R. H. McFarland, T. J. Morgan, B. M. Johnson, K. W. Jones, and M. Meron, Phys. Rev. A31, 4040 (1985).
25. Y. Hahn and H. Ramadan, Nucl. Instrum. Meth. Phys. Res. B43, 285 (1989).
26. Y. Hahn and H. Ramadan, Phys. Rev. A40, 6206 (1989).
27. Y. Hahn, 1985, unpublished.

28. Y. Hahn, Second US-Mexico Symposium on Atomic and Molecular Physics: Two Electron Phenomena, Cocoyoc, Mexico (1986); published in Notas de Fisica, eds. I. Alvarez, C. Cisneros, J. de Urquijo, and T. J. Morgan (Instituto de Fisica, UNAM, Cuernavaca, Mor., Mexicao, 1987), vol. 10, no. 2, pp. 91-104.
29. M. Schulz, R. Schuch, S. Datz, E. L. B. Justiniano, P. D. Miller, and H. Schöne, Phys. Rev. A38, 5454 (1988).
30. M. Schulz, J. P. Giese, J. K. Swenson, S. Datz, P. F. Dittner, H. F. Krause, H. Schöne, C. R. Vane, M. Benhenni, and S. M. Shafroth, Phys. Rev. Lett. 62, 1738 (1989).
31. S. Datz, F. W. Martin, C. D. Moak, B. R. Appleton, and L. B. Bridwell, Radiat. Eff. 12, 163 (1972).
32. S. Andriamonje, M. Chevallier, C. Cohen, J. Dural, M. J. Gaillard, R. Genre, M. Hage-Ali, R. Kirsch, A. L'Hoir, B. Mazuy, J. Mory, J. Moulin, J. C. Poizat, J. Remillieux, D. Schmaus, and M. Toulemonde, Phys. Rev. Lett. 59, 2271 (1987).
33. H. W. Schnopper, H. D. Betz, J. P. Delvaille, K. Kalata, A. R. Sohval, K. W. Jones, and H. E. Wegner, Phys. Rev. Lett. 29, 898 (1972).
34. S. Datz, C. R. Vane, P. F. Dittner, J. P. Giese, J. Gomez del Campo, N. L. Jones, H. F. Krause, P. D. Miller, M. Schulz, H. Schöne, and T. M. Rosseel, Phys. Rev. Lett. 63, 742 (1989).
35. A. Belkacem, E. P. Kanter, K. E. Rehm, E. M. Bernstein, M. W. Clark, S. M. Ferguson, J. A. Tanis, K. H. Berkner, and D. Schneider, Phys. Rev. Lett. 64, 380 (1990).
36. S. Reusch, P. H. Mokler, R. Schuch, E. Justiniano, M, Schulz, A. Müller, and Z. Stachura, Nucl. Instrum. Meth. Phys. Res. B23, 137 (1987); P. H. Mokler and S. Reusch, Z. Phys. D8, 393 (1988).
37. D. C. Griffin and M. S. Pindzola, Phys. Rev. A35, 2821 (1987).
38. A. Müller, private communication.
39. M. W. Clark, J. A. Tanis, E. M. Bernstein, N. R. Badnell, R. D. DuBois, W. G. Graham, T. J. Morgan, V. L. Plano, A. S. Schlachter, and M. P. Stockli, submitted to Phys. Rev. A.
40. W. Eissner, M. Jones, and H. Nussbaumer, Comput. Phys. Commun. 8, 270 (1974).
41. N. R. Badnell, J. Phys. B19, 3827 (1986).
42. N. R. Badnell and M. S. Pindzola, Phys. Rev. A39, 1685 (1989).
43. M. H. Chen, Phys. Rev. A41, 4102 (1990).
44. M. S. Pindzola and N. R. Badnell, Phys. Rev. A42, in press.
45. M. Clark, D. Brandt, J. K. Swenson, and S. M. Shafroth, Phys. Rev. Lett. 54, 544 (1985).
46. S. Reusch, Y. Awaya, T. Kambara, P. H. Mokler, D. J. McLaughlin, A. Müller, R. Schuch, and M. Schulz, Fifteenth International Conference on the Physics of Electronic and Atomic Collisions, Brighton, United Kingdom, 1987, Abstracts of Contributed Papers, edited by J. Geddes, H. B. Gilbody, A. E. Kingston, C. J. Latimer, and H. J. R. Walters, p. 642.
47. C. P. Bhalla and K. R. Karim, Phys. Rev. A39, 6060 (1989).

Production of Autoionizing States of Fast Charged Particles by Double Electron Capture

Sh.D. Kunikeev, V.S. Senashenko and V.A. Sidorovich

Institute of Nuclear Physics, Moscow State University, Moscow, USSR

1 Introduction

In the work of Zouros *et al.* [1] the absolute cross sections for the production of the lowest autoionizing resonances of the projectile excited in collisions of He^{2+}+He as a result of the state-selective double-electron capture (SSDEC) were measured. Salin *et al.* [2] made the first calculations of the production cross sections of the $(2s^2)^1S$, $(2s2p)^1P$ and $(2p^2)^1D$ autoionizing states. In their calculations they used the independent electron approximation and the impact parameter method. The transition amplitude for each electron was calculated in the continuum distorted wave approximation (CDW). The results of these calculations significantly exceed (from 6 to 75 times) the cross sections observed. More complicated theoretical treatment of double electron processes within the close-coupling method (AO^+) was carried out in Ref. [3]. The SSDEC cross sections were calculated in two variants of close-coupling approximation: (i) the amplitude of double electron capture was determined via the single electron capture amplitudes according to the independent electron model; (ii) the amplitude of double electron capture was determined on the basis of a double electron version of the close coupling model. The results of the calculations made in [3] are in a rather good agreement with the experiment for intermediate energies.

The present work suggests the quantum mechanical method of calculation of the SSDEC cross sections in the framework of the independent electron model. In this method the autoionizing states (AIS) are populated as a result of successive one-electron transitions in the field of screened nuclear charges of the target and the projectile. The electron–electron correlations are allowed for in the final state of the system.

2 Basic formulae

Let us consider SSDEC for the system

$$^3\mathrm{He}^{2+} + ^4\mathrm{He}(1s^2) \rightarrow \mathrm{He}^{**}(n_1 l_1 n_2 l_2) + \mathrm{He}^{2+}$$

$$\hookrightarrow \mathrm{He}^+(1s) + e^-. \tag{2.1}$$

The total Hamiltonian of the system in neglecting of the internuclear interaction is

$$\mathrm{H} = \hat{\mathrm{K}}_{pt} + \sum_{i=1}^{2}(\hat{\mathrm{K}}_{ti} - \mathrm{Z}_p/\mathrm{r}_{pi} - \mathrm{Z}_t/\mathrm{r}_{ti}) + 1/\mathrm{r}_{12}, \tag{2.2}$$

where $\hat{\mathrm{K}}_{pt}$ and $\hat{\mathrm{K}}_{ti}$ are the kinetic energy operators of relative motion of heavy particles and the i-th electron, respectively; r_{pi} and r_{ti} are the distances between the i-th electron and the projectile and the target atom nucleus, respectively; r_{12} is the interelectron distance; Z_p and Z_t are the charges of the nuclei of the ion and the target atom.

The corresponding SSDEC amplitude is

$$\tau_{\mathrm{ex}}^{\mathrm{LM}} = < \chi_f^{\mathrm{LM}} \mid \mathrm{V}_i \mid \chi_i > + < \chi_f^{\mathrm{LM}} \mid \mathrm{V}_f(\mathrm{E}_f^+ - \mathrm{H})^{-1}\mathrm{V}_i \mid \chi_i >, \tag{2.3}$$

here χ_i and χ_f^{LM} are the asymptotic initial and final state wave functions; V_i and V_f are the parts of the total interaction potential nondiagonalized in χ_i and χ_f^{LM}, respectively; E_f is the total energy of the system in the final state. In the intermediate energy region under consideration the Born term in the amplitude (2.3) provides a negligible contribution [4] and can be omitted. The second term in (2.3) accounts for multistep transitions. We consider here only the two-step mechanism of the AIS production of a fast particle. We shall determine the asymptotic wave function χ_i in the Hartree–Fock approximation [5]. χ_f^{LM} will be described through a configuration interaction treatment [6] using antisymmetrized hydrogenic orbitals. As the intermediate state wave function $\chi_\alpha^\pm$ we shall use an approximation expression for the wave function of the bound electron in the field of two Coulomb centers (see, e.g. [7])

$$\chi_\alpha^\pm = (2\pi)^{-3/2}\exp(i\boldsymbol{K}_\alpha\boldsymbol{R}_\alpha)\varphi_{n_1 l_1 m_1}(\boldsymbol{r}_{p1})\psi_\alpha(\boldsymbol{r}_{t2}) \times \phi_{\boldsymbol{v}_i}^\pm(\nu_t', \boldsymbol{r}_{t1})\phi_{\boldsymbol{v}_i}^\pm(\nu_p', \boldsymbol{r}_{p2}), \tag{2.4}$$

$$\phi_{\boldsymbol{v}}^\pm(\nu, \boldsymbol{r}) = exp(-\pi\nu/2)\Gamma(1 \pm i\nu)_1F_1(\mp i\nu, 1, \pm ivr - i\boldsymbol{vr}),$$

$$\nu_t' = -\mathrm{Z}_t'/v_i, \qquad \nu_p' = -\mathrm{Z}_p'/v_i,$$

where $\boldsymbol{R}_\alpha$ and $\boldsymbol{K}_\alpha$ are the relative distance radius-vector and the momentum of relative motion of heavy particles; $\boldsymbol{v}_i$ is the projectile velocity; $\mathrm{Z}_{p(t)}'$ is the effective charge of the projectile (of the target) in the intermediate state $\mid \alpha >$. The cdistorting factors $\phi_{\boldsymbol{v}_i}^\pm$ in (2.4) account for the effects of the Coulomb electron scattering on charges Z_p' and Z_t' in the continuum. We also assume that (i) the passive (second) electron state ψ_α in the transition $\mid i > \rightarrow \mid \alpha >$ is not distorted by the projectile field ($\nu_p' = 0$) and (ii) the passive (first) electron state $\varphi_{n_1 l_1 m_1}$ in the transition $| \alpha > \rightarrow | f >$ is not distorted by the field of the residual ion of the target ($\nu_t' = 0$).

In accordance with these approximations the amplitude (2.3) can be transformed as

$$\tau_{\text{ex}}^{\text{LM}} = -\frac{\pi i}{v_i} \sum_{\substack{n_1 l_1 m_1 \\ n_2 l_2 m_2}} \alpha_{\text{LM}}(n_1 l_1 m_1, n_2 l_2 m_2)$$

$$\times \int d\boldsymbol{Q}_{i\alpha} \tau^{\pm}_{n_1 l_1 m_1}(i \rightarrow \alpha) \tau^{\pm}_{n_2 l_2 m_2}(\alpha \rightarrow f), \tag{2.5}$$

where the weight factors α_{LM} characterize the configuration interaction. The one-electron transition amplitudes are of the form

$$\tau^{\pm}_{n_1 l_1 m_1}(i \rightarrow \alpha) = \text{N}_{\alpha i} < \varphi_{n_1 l_1 m_1} \mid -\frac{\text{Z}_p}{r_{p1}} \mid exp(i\boldsymbol{K}_1 \boldsymbol{r}_{p1}) >$$

$$\times < exp(i(\boldsymbol{K}_1 + \boldsymbol{v}_i)\boldsymbol{r}_{t1}) \phi^{\pm}_{\boldsymbol{v}_i}(\nu_t', \boldsymbol{r}_{t1}) \mid \psi_i >, \tag{2.6}$$

$$\tau^{\pm}_{n_2 l_2 m_2}(\alpha \rightarrow f) =< \varphi_{n_2 l_2 m_2} \mid -\frac{\text{Z}_p'}{r_{p2}} \mid exp(i\boldsymbol{K}_2 \boldsymbol{r}_{p2} >$$

$$\times < exp(i(\boldsymbol{K}_2 + \boldsymbol{v}_i)\boldsymbol{r}_{t2}) \phi^{\pm}_{\boldsymbol{v}_i}(\nu_t, \boldsymbol{r}_{t2}) \mid \psi_\alpha >$$

$$+(\text{E}_r - \epsilon_{n1} - \epsilon_{n2}) < \varphi_{n_2 l_2 m_2} \mid exp(i\boldsymbol{K}_2 \boldsymbol{r}_{p2}) >< exp(i(\boldsymbol{K}_2 + \boldsymbol{v}_i)\boldsymbol{r}_{t2}) \mid \psi_\alpha >, \tag{2.7}$$

$$\boldsymbol{K}_1 = K_{i\alpha} \boldsymbol{n}_i + \boldsymbol{Q}_{i\alpha}, \quad \boldsymbol{K}_2 = K_{\alpha f} \boldsymbol{n}_i - \boldsymbol{Q}_{i\alpha} - \boldsymbol{Q}_{if},$$

$$K_{i\alpha} = -v_i/2 + \Delta\epsilon_{i\alpha}/v_i, \quad K_{\alpha f} = -v_i/2 + \Delta\epsilon_{\alpha f}/v_i, \quad \nu_t = -\text{Z}_t/v_i,$$

where $\text{N}_{\alpha i} =< \psi_\alpha \mid \psi_i >$ is the overlap integral of the wave functions of the second (passive) electron initial and intermediate states; $\Delta\epsilon_{i\alpha}(\Delta\epsilon_{\alpha f})$ is the defect of the binding energy of the initial (intermediate) and intermediate (final) states; $\boldsymbol{n}_i$ is the unit vector in the incident beam direction; $\boldsymbol{Q}_{i\alpha(f)}$ is the transverse component of the momentum transferred by the projectile in the transition $| i > \rightarrow | \alpha > (| f >)$; E_r is the energy of the resonance; $\epsilon_n = -\text{Z}_p^2/(2n^2)$. The second term in the relation (2.7) is of a correlative nature, since the allowance for the electron correlations in the AIS wave function [6] has the result that $\text{E}_r \neq \epsilon_{n_1} + \epsilon_{n_2}$.

3 Calculation results and discussion

On the basis of the present method we have calculated the production cross sections of the $(2s^2)^1\text{S}$, $(2s2p)^1\text{P}$ and $(2p^2)^1\text{D}$ autoionizing states in the He^{2+}+He collisions at projectile energies E_p=150–500 keV. The calculated results along with the available experimental data [1] and the calculations of other authors [2, 3] are presented in Fig. 1.

The comparison of calculations shows that electron–electron correlations lead to significant decrease of cross sections, substantially improving the agreement between theory and experiment. Figure 1 shows that the calculations in the high-energy CDW-approximation [2] significantly exceed the experimental data [1]. The

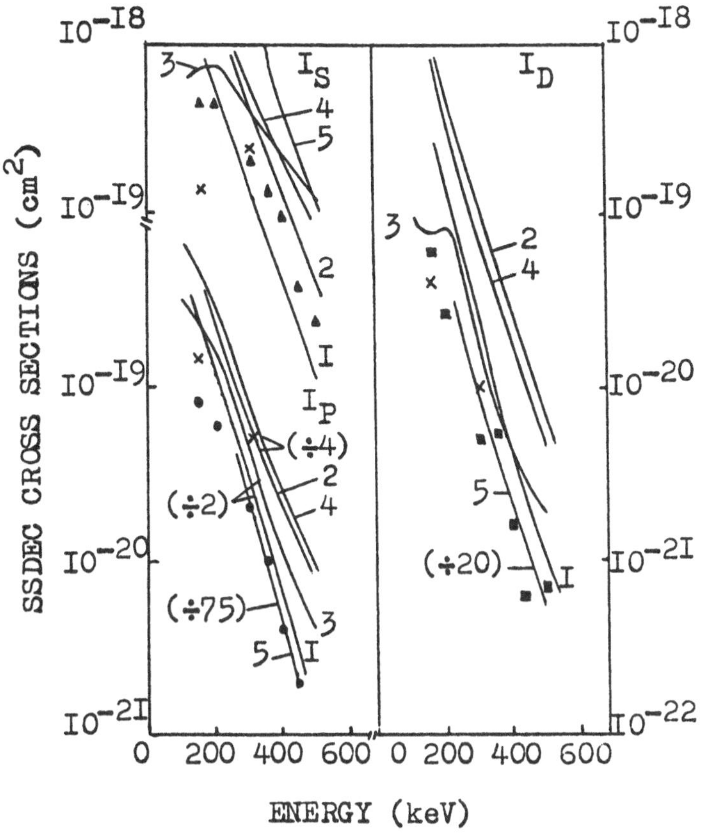

Fig. 1. The production cross sections of the $(2s^2)^1$S, $(2s2p)^1$P and $(2p^2)^1$D autoionizing states in the He^{2+}+He collisions versus the projectile energy. Curves represent the theoretical calculations: 1 and 2 – present calculations with and without allowance for electron–electron correlations, respectively; 3 and 4 – the calculations in the independent electron version of the close-coupling model [3] with and without allowance for the relaxation of the second electron after removal of the first electron from the target atom, respectively; 5 – the results of the CDW calculation [2], taken from Ref. [1]. Points represent the experimental data [1]. Crosses denote the calculation in the double electron version of the close-coupling model [3].

calculation in the close-coupling approximation [3] gives a much better agreement between theory and experiment at the intermediate energies. However, this agreement becomes worse with increasing energy. Since at $v_i > 1$ the role of the continuum states turns out to increase, the allowance for the latter by means of including the united atom pseudostates into total wave function expansion of colliding particle system is not sufficient.

In our consideration the electron capture through the continuum is approximately taken into account by distorting factors $\phi^{\pm}_{v_i}$. As a whole, our calculations (curves 1 and 2) at intermediate collision energies are in good agreement with the corresponding calculations in the close-coupling approximation (curves 3 and 4). With increasing energy ($E_p > 400$ keV) the agreement of our results with the experimental data becomes slightly better.

References

1. T.J.M. Zouros, D. Schneider and N. Stolterfoht: Phys. Rev. **A35** 1963 (1987)
2. A. Salin, H. Bachau and R. Gayet: as quoted in Ref. [1] *(unpublished)*
3. A. Jain, C.D. Lin and W. Fritsch: Phys. Rev. **A39** 1741 (1989)
4. J.H. McGuire, E. Salzborn and A. Müller: Phys. Rev. **A35** 3265 (1987)
5. L.C. Green, M.M. Mulder, M.N. Levis and J.W. Woll: Phys. Rev. **93** 757 (1954)
6. V.V. Balashov, S.I. Grishanova, I.M. Kruglova and V.S. Senashenko: Optika i Spektroskopiya **28** 859 (1970) *(in Russian)*
7. Dž. Belkić, R. Gayet and A. Salin: Phys. Rep. **56** 279 (1979)

Evidence for Resonant Double Electron Capture and Single Excitation in $Ge^{31+} \rightarrow$ Ne collisions

Th. Stöhlker [1], C. Kozhuharov, E.A. Livingston[2], P.H. Mokler[1]

GSI-Darmstadt, FRG

Z. Stachura, A. Szymanski, A. Warczak

University and Inst. of Nuclear Physics, Krakow, Poland

Evidence for resonant two-electron capture and excitation, i.e. the time-reversed double-Auger process with radiative stabilization, was found in collisions between hydrogenic $_{32}Ge^{31+}$ projectiles and Ne gas atoms. The multiply excited projectile states produced in the collision decay via x-ray cascades. The measurement of coincidences between two-projectile K-x rays associated with double electron capture at collision energies between 4.5 MeV/u and 11.5 MeV/u provides us with its excitation function for this resonant process which shows a strong enhancement at beam energies below 7.5 MeV/u in accordance with the expected resonance location. The cross sections found for this process are surprisingly large compared to the measured cross sections for resonant single-electron capture and excitation in $Ge^{31+} \rightarrow H_2$ collisions. Considering also the non-correlated two-electron capture and excitation process we estimate for the resonant region a contribution of the non-resonant process to be about 1/3 of the measured cross section.

1 Introduction

In collisions between highly charged few-electron projectiles and target atoms, resonant electron capture and excitation (RTE) is one important projectile charge exchange process. It has been studied intensively during the last years for different projectiles up to Z_{PROJ}=92 and various gas and solid targets [1], [2], [3], [4]. Considering the impulse approximation, RTE is the time reversed Auger process where the energy gained by the capture of one target electron is used to excite one projectile electron leading to a doubly excited projectile state. This excited state may then decay via photon emission. The different resonances for this process are located at adiabaticity parameters of $0.5 \leq \eta \leq 0.75$ ($\eta = \frac{v_{KIN}^2}{v_K^2}$, where v_K denotes the K-shell electron velocity and v_{KIN} the velocity of the projectile).

Now, considering the double Auger process and applying again the principle of detailed balance leads to a new process, the two or multiple-RTE, i.e. the resonant capture of two target electrons by simultaneous excitation of one projectile electron. For hydrogenic projectiles this process forms a triply exited intermediate state with two K-shell vacancies which can stabilize radiatively via the emission of two K x-rays. Therefore, the fingerprint of double RTE is the simultaneous emission of two projectile K-photons associated with the capture of two electrons into the projectile. Due to the capture of two electrons the different double RTE resonances are located at low-intermediate

[1]GSI-Darmstadt and University Gießen, Germany

[2]Notre Dame University, IN, USA

projectile energies. Using for the binding energies a hydrogenic scaling the resonances are at an adiabaticity parameter of $\eta = \frac{1}{8}$ for the KL-LL resonances and of $\eta = \frac{1}{3}$ for the KL-MM resonances, respectively.

In this paper we report on an experiment identifying the double RTE process in collisions between hygrogenic Ge projectiles and Ne atoms at beam energies between 4.5 MeV/u and 11.5 MeV/u [5].

2 Experimental Method

In order to provide hydrogenic Ge projectiles at low-intermediate energies the acceleration-stripping-deceleration technique was applied [6]. Ge ions with a charge state 17+ were accelerated up to an energy of 11.5 MeV/u. After passing through a carbon stripper foil the projectiles were decelerated by using the single cavities of the UNILCAC accelerator down to the final collision energy. The charge state analysed and well collimated beam was then directed onto a three stage, differential pumped Ne gas target. After passing the target cell the ejectiles were charge state analysed by a dipole magnet and the projectiles associated with one and two captured electrons (Ge^{30+}, Ge^{29+}) were detected by fast scintillator counters with an efficency of practically 100 %. In order to allow x-ray/x-ray/particle coincidence measurements the target area was viewed by two Si(Li) detectors with a solid angle of 1% of 4π each. The x-ray detection was also used to check the charge state of the incident beam by observing the characteristic K-REC photons (Radiative Electron Capture), a process only possible for projectiles having at least one vacancy in the K-shell. The pressure of the gas target was measured by a capacitance manometer and varied in such a way that for each beam energy single collision condition was guaranteed (between 200 and 9.3 mbar corresponding to beam energies of 11.5 MeV/u and 4.56 MeV/u).

3 Results

In Fig. 1 the measured yield of triple coincidences σ_{KKQ2}, i.e. the simultaneous emission of two Ge K x-rays (KK) associated with projectiles having captured two electrons (Q2), normalized to the Faraday-cup and the gas pressure is plotted. We estimate the systematic error of the triple coincidence measurement to be of the order of 50%. The vertical arrows shown at the top of the figure mark the resonance energies for double RTE whereas the solid lines reflect the estimated full width at half maximum of the Compton-profile (estimated according to [7]) associated with the different RTE resonances. Due to the Compton-profiles of the Ne target electrons the resonance regions overlap. Nevertheless (compare Fig. 1), the measured excitation function shows an enhancement for triple coincidences at energies below 7.5 MeV/u, the energy region where the resonances for double RTE are located.

Because of the modest statistics (the error bars plotted in Fig. 1 are only of statistical origin) the measured data were divided into two parts corresponding to the low energy regime where the resonances are located (4.5 MeV/u $\leq E_{KIN} \leq$ 7.0 MeV/u) and the high energy regime (8.5 MeV/u

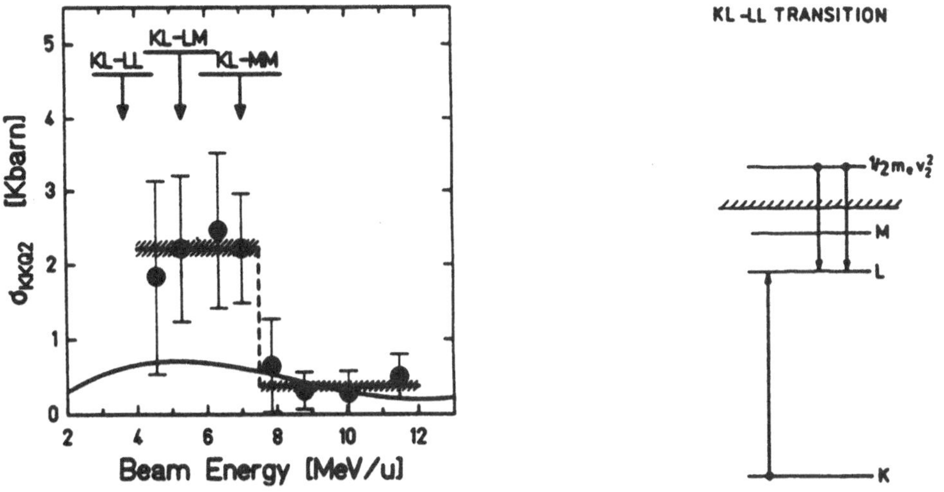

Figure 1: Measured excitation function σ_{KKQ2} for the simultaneous emission of two K-x-rays associated with projectiles having captured two electrons. The arrows mark the resonance energies for double RTE and the horizontal bars (top) reflect the calculated widths of the resonances. The shaded areas represent the errors of the fitted constants. The estimated double NTE contribution on the measured excitation function is given by the solid line. In addition, a schematic presentation of KL–LL double RTE transition in highly ionized atoms is plotted.

$\leq E_{KIN} \leq 11.5$ MeV/u). Fitting a simple step-like function to the data, the result shows an increase of cross section for the low energy regime compared to the high energy regime by about a factor of 5.77 ± 1.02 corresponding to a statistical significance of 4.86 σ [6] (the shaded areas in Fig. 1 represent the fit errors).

In order to estimate the influence of the competing non-resonant electron capture and projectile excitation process (NTE) [8] on the measured cross section for triple coincidences, we calculated the excitation function for NTE associated with two electron capture.

Assuming that the probability for projectile K-shell excitation (P_{EXC}) and the probability to capture n target electrons (P^n_{CAP}) are totally indepedent, the cross section for this non-resonant process is given by $\sigma^n_{NTE} = 2\pi \int P_{EXC}(b) \times P^n_{CAP}(b) b db$ (where b denotes the impact parameter). We calculated the excitation probability by using the SCA formalism [9]. Because it was found that the measured total electron capture cross sections (single and double capture) in $Ge^{31+} \rightarrow$ Ne collisons are very well reproduced by the nCTMC-code from Olson [10], this code was also used to calculate the impact parameter dependent electron capture probabilities [11]. Figure 2 presents the calculated cross sections for single NTE (associated with the capture of one electron), double NTE (associated with the capture of two electrons) and for total NTE in $Ge^{31+} \rightarrow$ Ne collisions.

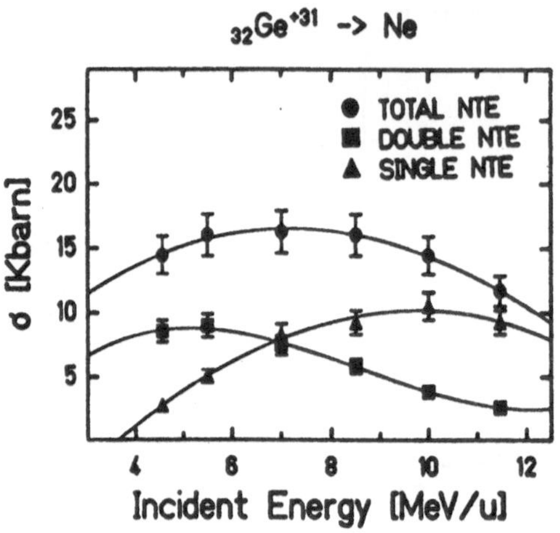

Figure 2: Calculated excitation functon for single NTE (associated with the capture of one electron into the projectile), double NTE (associated with the capture of two electrons into the projectile), and total NTE.

The predicted excitation function for double NTE shows a flat maximum at low collision energies and a decreasing cross section with increasing beam energy. This result for double NTE has to be multiplied by a factor of 1/7 in order to take into account electron screening, metastable states, and the fluorescence yields not included in the calculation. Based on this estimation the averaged cross section at the low energy regime for double NTE amounts to 0.91 ± 0.44 kb and for double RTE to 1.32 ± 0.65 kb.

Considering the detailed balance the cross section for RTE (single and double) is proportional to the corresponding Auger-rates (Γ_A^{1e}, Γ_A^{2e}). For medium Z ions Γ_A^{2e} is roughly one order of magnitude smaller than Γ_A^{1e} [12]. Therefore, the extracted cross section for double RTE seems to be surprisingly high compared to the measured cross section for the single KLL-RTE resonance in $Ge^{31+} \rightarrow H_2$ collisions of $\sigma_{RTE} \sim 1$ Kb. However, in the case of the double RTE experiment with a Ne target 8 quasifree electrons are available for the resonant two electron capture process. Therefore Γ_A^{2e} has to be multiplied by the binominal factor $\frac{8!}{2!(8-2)!} = 28$ in order to compare the results for double RTE with the experimental results for single RTE in $Ge^{31+} \rightarrow H_2$ collisions. Hence, taking into account the 8 quasifree electrons of Ne, the extracted cross section of the double RTE process seems not to contradict the already measured cross section for the KLL-RTE resonance of Ge^{31+} [4].

4 Conclusion

The experimental results reported here provide evidence for resonant two electron capture and simultaneous projectile K-shell excitation in heavy ion atom collisions. Under consideration of the competing non-resonant capture and excitation process the measured cross sections for double RTE

Capture with ionisation in proton-helium collisions

R Gayet and A Salin

Laboratoire des Collisions Atomiques, Université Bordeaux I,

351 Cours de la Libération, 33405 Talence Cedex, France

There has been in the last few years an increasing tendency to identify atomic processes with a (classical) billard-ball game. The reason for this can be traced to a generalisation of the coincidence between the classical and quantum-mechanical capture processes at high energies for the three-body Coulomb problem (Thomas process). We consider in the present contribution the reaction:

$$H^{+} + He \rightarrow H + He^{++} + e \qquad (1)$$

A billard-like mechanism has also been proposed for this reaction (Thomas, 1927, Shakeshaft and Spruch, 1979, Briggs and Taulbjerg, 1979). The incoming proton would collide with the first electron. Subsequently, this electron would collide with the second one. If, as a result of this second reaction, one electron has the same velocity as the projectile, then one considers that capture has occured. The corresponding kinematical conditions determines for which scattering angle of the projectile and ejected electron such a reaction is likely to occur. The corresponding angle is named *critical angle*. Also, as the second collision is an electron-electron collision, this mechanism has been considered as a good case where *dynamical correlation* is important. Recently, Horsdal *et al* (1986) have measured the reaction (1) as a function of the projectile scattering angle. When one looks at the relative yield of capture with and without ionisation as a function of the projectile scattering angle, a peak is observed around 0.5 mrad, very close to the position of the critical angle defined above (0.55 mrad). The latter experiment has been considered as a confirmation that the billard-like process had indeed been observed.

In the present work we calculate the differential cross-section for the reaction (1) in an independent electron model. One considers that the projectile interacts independently with both electrons so that the transition amplitude can be written as the product of an ionisation and capture amplitudes.

For the impact energies and projectile scattering angles considered here, the eikonal approximation is fully justified (Mc Carroll and Salin, 1978, and references therein).

in collision between hydrogenic Ge projectiles and Ne gas atoms are found to be consistent with the already measured cross section for the single KLL-resonance of Ge^{31+}. Therefore, we point out that multiple resonant capture and excitation processes must be taken into account when dealing with inner shell excitation processes.

References

[1] J.A. Tanis, E.M. Bernstein, W.G. Graham, M. Clark, S.M. Shafroth B.M. Johnson, K.W. Jones, M. Meron, Phys. Rev. Lett. 49 (1982) 1325.

[2] J.A. Tanis, Nucl. Instr. Methods A 262 (1987) 52.

[3] W.G. Graham, K.H. Berkner, E.M. Bernstein, M.W. Clark, B. Feinberg, M.A. McMahan, T.J. Morgan, W. Rathbun, A.S. Schlachter and J.A. Tanis, submitted to Phys. Rev. Lett.

[4] P.H. Mokler, Th. Stöhlker, Ch. Kozhuharov, J. Ullrich, S. Reusch, Z. Stachura, A. Warczak, A. Müller, R. Schuch, E.A. Livingston, M. Schulz, Y. Awaya, T. Kambara, invited paper to X-90 (Conference on X-Ray and Inner Shell Processes, Knoxville, 1990)

[5] A. Warczak, Z. Stachura, A. Syzmanski, Th. Stöhlker, Ch. Kozhuharov, A.E. Livingston, P.H. Mokler, S. Reusch, Phys. Lett. A 146 (1990) 122.

[6] P.H. Mokler, D.H.H. Hoffmann, W.A. Schönfeldt, D. Maor, W.E. Meyerhof, Z. Stachura, Nucl. Instr. Meth. (B4) (1984) 34

[7] F. Biggs, L.B. Mendelsohn and J.B. Mann, ADANDT 3 (1975) 201.

[8] M. Clark, D. Brandt, J.K. Swenson and S.M. Shafroth, Phys. Rev. Lett. 54 (1985) 544.

[9] J. van den Bos, F.J. de Heer, Physika 34 (1967) 333.

[10] R.E. Olson in Electronic and Atomic Collisions, edited by H.B. Gilbody, W.R. Newell, F.H. Read, and A.C.H. Smith (Elsevier Science, New York, 1988), pp. 271 – 285

[11] Th. Stöhlker, Ch. Kozhuharov, E.A. Livingston, P.H. Mokler, R.E. Olson, A. Szymanski, Z. Stachura, A. Warczak, scientific report 1989, GSI90-1 (ISSN0174-0814 1990) 135.

[12] T. Aberg, in: Atomic inner-shell processes, Vol.1, ed. B. Crasemann (Academic Press, New York, 1975) p. 353

Let us call $A_{if}(\rho; \kappa)$ the amplitude for the production of reaction (1) when the impact parameter is ρ and the ejected electron has a momentum κ. Then, the differential cross-section for scattering of the projectile into the solid angle $d\Omega_p$ around the direction (θ_p, ϕ_p) is given by (in the laboratory frame):

$$\frac{d\sigma_{if}}{d\Omega_p} = \frac{M_p^2 v^2}{4\pi^2} \int d\kappa \left| \int d\rho \exp(i\eta.\rho) A_{if}(\rho; \kappa) \right|^2 \tag{2}$$

Here, η is the transverse momentum transfer:

$$\eta = M_p v \theta_p$$

where M_p is the proton mass and v its velocity. As the integration over ρ involves the amplitude and not the probability, we must use an independent electron model for the amplitudes (Lüdde and Dreizler, 1985, Gayet, 1989):

$$A_{if}(\rho; \kappa) = a_c(\rho) a_i(\rho; \kappa) \rho^{2i\nu} \tag{3}$$

where a_c is a one electron capture amplitude, a_i is a one electron ionisation amplitude and $\nu = Z_p Z_T / v$ where $Z_{p,T}$ are the projectile and target nuclear charge. The term $\rho^{2i\nu}$ arises from the internuclear interaction. We have evaluated a_c and a_i from well established approximations. The capture amplitude has been calculated with the CDW program of Belkić *et al* (1984). In the latter program, one calculates the amplitude as a function of transverse momentum transfer η. However a transformation similar to (2) produces readily the transition amplitude as a function of ρ as explained by Belkić and Salin (1978). The initial $1s^2$ state of helium is described by the Hartree-Fock function of Clementi and Roetti (1974) which is consistent with an independent electron model. Similarly, the ionisation amplitudes have been calculated with the Born-Hartree-Fock-Slater model of Madison (1973). Full details on the method which allows the determination of these amplitudes as a function of impact parameter will be given elsewhere (Salin, 1989).

For comparison purposes, we have also calculated the one electron capture amplitude with the Oppenheimer-Brinkman-Kramers (OBK) approximation. In the latter case, the initial wave function was chosen as a hydrogenic $1s$ function with effective charge 1.69.

The theoretical results for the differential capture cross-section are of course those of Belkić and Salin (1978). Comparison with the experimental results is made in figure 1.

The agreement is very good, except for a slight oscillation for angles larger than 0.4 mrad that is not seen in the experiment. This oscillation is always observed in the CDW approximation around the position of the Thomas peak (for a detailed comparison see Belkić *et al* , 1979).

On figure 2a we have plotted the branching ratio for capture with and without ionisation together with the experimental results of Horsdal *et al* (1986). We first note that the absolute value of this ratio is very close to the experimental one. We do observe a maximum as in the experiment. Superposed on this maximum, we get an extra structure which can be correlated to the oscillation in the CDW theory. If we replace the CDW cross section for capture without ionisation by the experimental values, we obtain excellent agreement with experiment at 300 keV. This results shows that the experimental peak can be reproduced with an independent electron model. It is not due to the billard-like mechanism described above since the electron-electron interaction is not included in the dynamics of our calculations.

To understand the origin of the peak, it is of interest to do the same calculations with the OBK amplitude instead of the CDW one. As shown in figures 2a and 2b, no peak appears. Calculations on single ionisation (Salin, 1989) show that scattering around 0.5 mrad corresponds to a maximum in the average energy of the electrons as for small impact parameters. One could therefore think that the sharp decrease of the capture amplitude with impact parameter induces through the product (3) a relatively larger scattering around 0.5 mrad. We have therefore replaced the capture amplitude in (3) by $\exp(-\lambda\rho)$ with various values of λ. Results are always very close to the OBK results.

In fact, the interpretation is more complex. The OBK amplitude is purely imaginary. On the other hand CDW amplitudes, as all second order amplitudes, are complex. Rivarola (1984) has shown that the differential capture cross section is very sensitive to the phase of the capture amplitude. This is what we observe also here. We have checked that variations of the amplitude phase changes drastically the pattern of figure 2. It is understandable that the differential cross section should be very sensitive to the phase of the amplitude. When one goes to the classical limit of the two dimensional Fourier transform in (2) through a stationary phase approximation, then the relation between scattering angle and impact parameter is governed by the phase of the transition amplitude (Mc Carroll and Salin, 1968). In other terms, in our approach, it is the

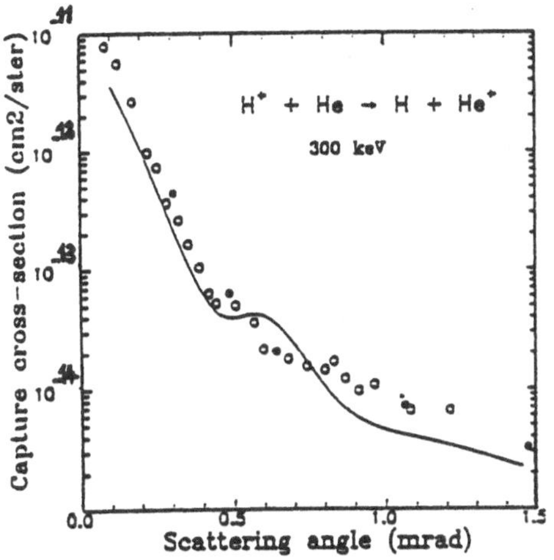

Figure 1: Electron capture cross-section for 300 keV proton impact on Helium as a function of projectile scattering angle.
Full curve: present CDW calculations. Experiments: ○ Bratton et al (1977), • Horsdal et al (1986).

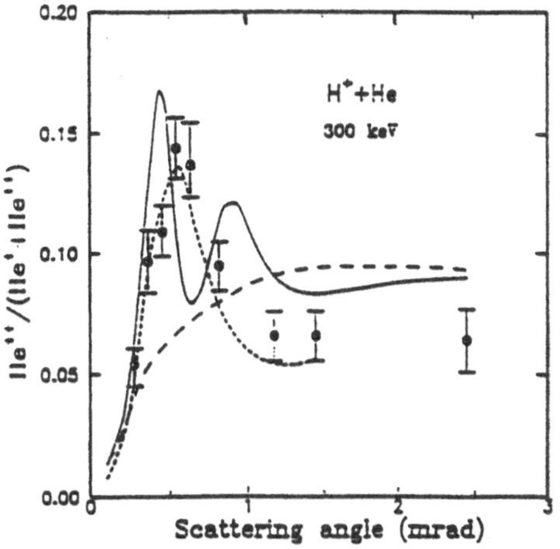

Figure 2a: Relative yield of He^{++} in electron capture by protons from Helium as a function of projectile scattering angle for a collision energy of 300 keV.
Experimental points are from Horsdal et al (1986). Full curve: present results; short dash: experimental data of figure 1 are used for the global capture; long dash: results with the OBK amplitude for capture.

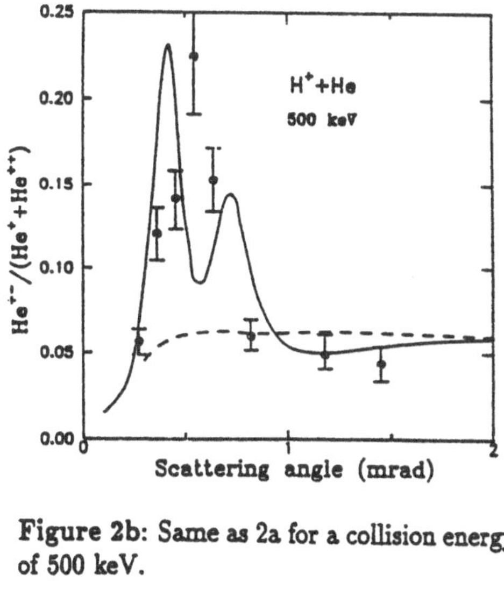

Figure 2b: Same as 2a for a collision energy of 500 keV.

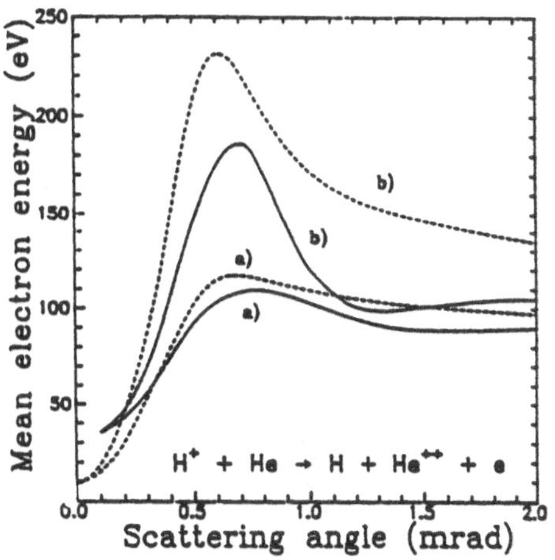

Figure 3: Mean energy of electrons ejected from Helium by proton impact as a function of projectile scattering angle. Impact energy is: a) 300 keV; b) 500 keV.
Full curve: ionisation with capture; dashed curve: ionisation without capture.

combined phase of the ionisation and capture amplitudes which creates the structure observed in the experiment.

We conclude that the experiment of Horsdal *et al* cannot be invoked as a proof of the billard-like mechanism (with which no quantitative comparison is available for this particular experiment) since an alternative mechanism can reproduce the experimental structure. We give in figure 3 the average electron energy as a function of projectile scattering angle for ionisation with and without capture. Capture does not change much the average electron energy. Measurements of the electron distribution could therefore discriminate among the two interpretations. In our theory, we find no structure in the energy distribution of the electrons whereas the billard-like mechanism would produce electrons with a velocity whose absolute value is close to that of the projectile.

References

Belkić Dž, Gayet R and Salin A 1979 *Phys. Rep.* **56** 279–369

Belkić Dž, Gayet R and Salin A 1984 *Comp. Phys. Comm.* **32** 385–97

Belkić Dž and Salin A 1978 *J. Phys. B: At. Mol. Phys.* **11** 3905–11

Bratton J, Cocke C L and Macdonald J R 1977 *J. Phys. B: At. Mol. Phys.* **10** L517–20

Briggs J S and Taulbjerg K 1979 *J. Phys. B: At. Mol. Phys.* **12** 2565–73

Clementi E and Roetti C 1974 *Atomic Data and Nuclear Data Tables* **14** 177–478

Gayet R 1989 *J. Physique C1* **50** 53–70

Horsdal E, Jensen B and Nielsen K O 1986 *Phys. Rev. Lett.* **57** 1414–6

Lüdde H J and Dreizler R M 1985 *J. Phys. B: At. Mol. Phys.* **18** 107–12

McCarroll R and Salin A 1978 *J. Phys. B: At. Mol. Phys.* **11** L693–5

McCarroll R and Salin A 1968 *J. Phys. B: At. Mol. Phys.* **1** 163–71

Madison D H 1973 *Phys. Rev. A* **8** 2449–55

Rivarola R D 1984 *Phys. Rev. A* **30** 1122–4

Salin A 1989 *J. Phys. B: At. Mol. Opt. Phys.* **22** 3901–14

Shakeshaft R and Spruch L 1979 *Rev. Mod. Phys.* **51** 369–405

Thomas L H 1927 *Proc. Roy. Soc. London* **114** 561–76

TWO–ELECTRON PROCESSES IN PROJECTILE–H_2 COLLISIONS*

A. K. Edwards, R. M. Wood, J. L. Davis, M. W. Dittmann,
J. F. Browning and M. A. Mangan
Department of Physics and Astronomy
University of Georgia
Athens, GA, 30602, USA

R. L. Ezell
Department of Chemistry and Physics
Augusta College, Augusta, GA, 30910, USA

1. Introduction

Molecular hydrogen is a two–electron system that has been studied extensively such that its excited states, its potential curves and its ionization potential are well known.[1] For the doubly excited states of H_2, a number of potential curves have been calculated,[2] but they have not been subjected to the same level of experimental scrutiny as have the other H_2 potential curves. The properties of its molecular ion H_2^+ are also well known, and its potential cruves have been tabulated for use in numerical calculations.[3,4]

Molecular hydrogen has the unique property that for any process involving the excitation or ionization of both electrons, a final state is formed that will dissociate into two energetic fragments. One of these fragments is charged and can be easily detected. It is this property that enables the study of two–electron processes in projectile–H_2 collisions.

Figure 1 illustrates the interactions included in the model used to describe the experimental results. The Goldstone diagrams of Fig. 1(a–c) depict those events which occur by a single projectile interaction and an electron–electron or electron–hole interaction. The direction of time is from the bottom of each figure to the top. Particles travel forward in time and holes are shown as propagating backwards. The interactions drawn with an X and a dashed line signify interactions between the projectile and the target electron. Figure 1(d) corresponds to an uncorrelated double collision event.

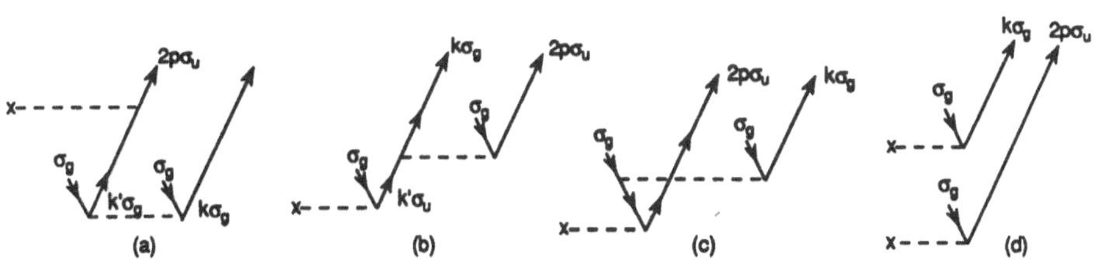

Figure 1 Goldstone diagrams for interactions involved in excitation of the $2p\sigma_u$ state of H_2^+. (a) Ground state correlation followed by projectile–interaction. (b) Projectile interaction followed by electron–electron interaction, and (c) electron–hole interaction. (d) Double collision.

The enhancement of the double ionization cross sections of helium by negatively charged projectiles, relative to those of positively charged projectiles, has been attributed to an interference between the single collision scattering amplitudes and the double collision amplitudes.[5,6] This interference effect in the double ionization process has been observed[7] in H_2. It also occurs in H_2 for the two–electron processes of ionization plus excitation and double excitation.[8] The cross section measurements for projectile–H_2 collisions that we have reported[7,8,9] have been made with the detectors placed at 90° relative to the beam direction. Because of the rapid dissociation of the target H_2, the detector records only those events occurring with the molecular axis aligned with the detector. With the detector being placed at other angles, cross sections are now being measured as a function of the angle between the projectile beam and the molecular axis.

In discussing the effects of molecular orientation on the cross sections, one investigates the single particle matrix elements of the projectile–H_2 interaction shown in the Goldstone diagrams. The expected angular dependencies have been reported elsewhere[10] and are listed in Table I. These results assumed a dipole interaction with the projectile where the ionized electron was either s–wave or p–wave. New results for angular measurements of 1 MeV/amu p+ on H_2 collisions are reported here and compared to similar results for equivelocity electron bombardment.

2. Experimental Procedure

The charged–fragment ions from the H_2 target are energy analyzed by an electrostatic, hemispherical analyzer placed at a specified angle relative to the beam direction. A kinetic–energy spectrum (0 to 15 eV) is recorded of all H^+ fragments from all dissociative states. A fitting routine[8] is used to determine the contribution from individual states to the total spectrum. The kinetic–energy spectrum of each individual state is generated from the known potential curves and used in the fitting program. Details of the apparatus and the fitting procedure are given elsewhere.[8]

3. Results

The angular dependencies of the cross sections for the total yield of H^+ fragments and for the excitation of selected states are shown in Figs. 2 and 3, respectively. The results are for 1 MeV/amu electron and proton projectiles, and they have been normalized at 90° in order to better demonstrate the effect of the orientation of the molecular axis on the cross sections. The smooth lines are fits to the data of the form $(1 + a \cos^2\theta_r + b \cos^4\theta_r)$ where θ_r is the angle of the internuclear axis relative to the beam direction. The $\cos^4\theta_r$ term arises from nondipole contributions to the first Born term and from d–wave contributions to the interaction. It is included in this work in order to improve the fit to the data over earlier work,[10] which only included a term in $\cos^2\theta_r$, and to demonstrate its relative importance. The error bars on the data illustrate a relative error of 5% rather than the absolute error of 15%. The relative errors are included in the polynomial fit[11] and contribute to the uncertainties in the values of a and b listed in Table II.

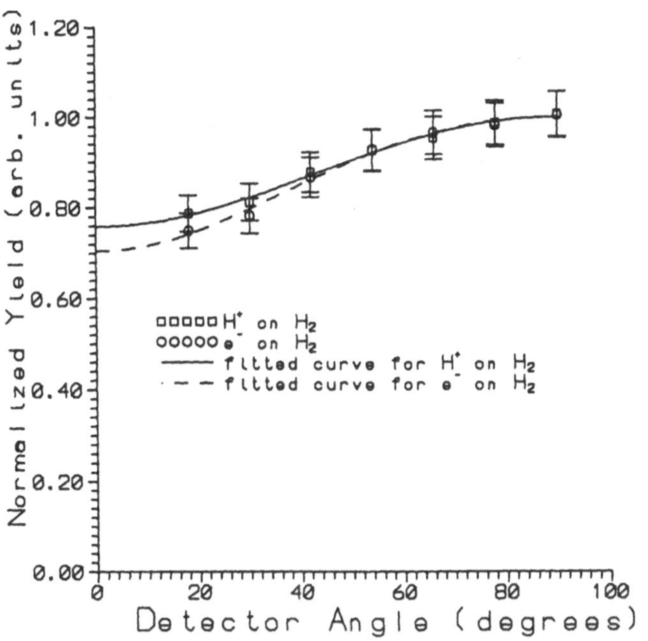

Figure 2 Dependence of the relative cross sections for all two–electron processes on the orientation of the molecular axis relative to the projectile direction.

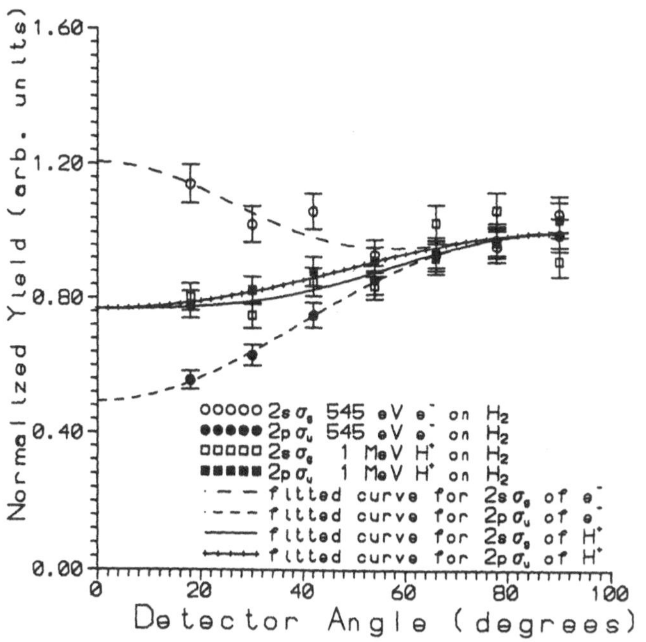

Figure 3 Dependence of the relative cross sections for ionization plus excitation to the $2p\sigma_u$ and $2s\sigma_g$ states of H_2^+ on the orientation of the molecular axis relative to the projectile direction.

Figure 2 shows the angular dependence of the total cross sections for all two–electron processes for both e^- and p^+ collisions. Very little difference is observed in the angular dependence for the two projectiles. However, there is a difference if individual states are measured. Figure 3 shows the orientation effects for the cross sections for ionization plus excitation of the $2p\sigma_u$ and $2s\sigma_g$ states by electrons and protons. The proton data show very little difference between these two states, whereas the electron data show a significant difference. Efforts are underway to explain these differences in terms of dipole and nondipole contributions to the collisions.

Table I. Angular dependence of the cross section on the orientation of the molecular axis (θ_r) and momentum transfer (θ_K) relative to the projectile direction. A dipole interaction with the projectile was assumed.

Final State	Interaction	Angular Dependence
$2p\sigma_u ks\sigma_g$	σ_u	$\frac{1}{2}\sin^2\theta_K[1+(2\cot^2\theta_K-1)\cos^2\theta_r]$
$2p\pi_u ks\sigma_g$	π_u	$(\frac{1}{2}\sin^2\theta_K+\cos^2\theta_K)[1+\left[\frac{\tan^2\theta_K-2}{\tan^2\theta_K+2}\right]\cos^2\theta_r]$
$\left.\begin{matrix}2s\sigma_g kp\sigma_u\\2s\sigma_g kp\pi_u\end{matrix}\right\}$	σ_u, π_u	$\left[\frac{A}{2}\sin^2\theta_K+B(\cos^2\theta_K+\frac{1}{2}\sin^2\theta_K)\right]\left[1+\frac{(A-B)(2\cot^2\theta_K-1)}{(A+B)(2\cot^2\theta_K+1)}\cos^2\theta_r\right]$ where A~ the $kp\sigma_u$ contribution and B~ the $kp\pi_u$ contribution.

Table II. Values of a and b derived from the fit of $1 + a\cos^2\theta_r + b\cos^4\theta_r$ to the measured angular dependencies for 1 MeV/μ electron and proton collisions.

	Electrons		Protons	
Final State	a	b	a	b
$2s\sigma\sigma_g$	−0.309±0.220	0.514±0.254	−0.416±0.195	0.184±0.210
$2p\sigma_u$	−0.387±0.182	−0.121±0.186	−0.270±0.204	0.039±0.218
$2p\pi_u$	0.286±0.219	−0.481±0.237	0.348±0.231	−0.500±0.249
Double Ionization	0.829±0.270	−0.437±0.311	0.430±0.237	−0.397±0.266
Double Excitation	−0.543±0.183	0.126±0.191	−0.759±0.171	0.289±0.175
Total	−0.198±0.203	−0.097±0.216	−0.226±0.204	−0.015±0.218

References

1. G. Herzberg, "Diatomic Molecules," (Van Nostrand, New York, 1950).

2. S. L. Guberman, J. Chem. Phys. 78, 1404 (1983).

3. D. R. Bates, K. Ledsham and A. L. Stewart, Phil. Trans. Roy. Soc. (London), A 246, 215 (1953).

4. T. E. Sharp, At. Data 2, 119 (1971).

5. J. H. McGuire, Phys. Rev. Lett. 49, 1153 (1982).

6. J. F. Reading and A. L. Ford, J. Phys. B 20 3747 (1987).

7. A. K. Edwards, R. M. Wood, A. S. Beard and R. L. Ezell, Phys. Rev. A 37, 3697 (1988).

8. A. K. Edwards, R. M. Wood, J. L. Davis and R. L. Ezell, Phys. Rev. A 42, 1367 (1990).

9. A. K. Edwards, R. M. Wood and R. L. Ezell, Phys. Rev. A 42, 1799 (1990).

10. A. K. Edwards, R. M. Wood, M. W. Dittmann, J. F. Browning, M. A. Mangan and R. L. Ezell, Nucl. Instr. and Meth. (to be published).

11. P. R. Bevington, "Data Reduction and Error Analysis for the Physical Sciences," (McGraw Hill Book Co., New York, 1969).

---

*Supported by the National Science Foundation under grant number PHY9005112.

The Energy Dependence of the Double Ionization Cross Section of Helium at High Energies

V.A. Sidorovich

Institute of Nuclear Physics, Moscow State University, Moscow 119899, USSR

1 Introduction

The absolute magnitudes of the double ionization cross section of helium by protons at proton energies E=2.31–40 MeV, calculated by Ford and Reading [1] in the first Born approximation, are in rather good agreement with the experimental data [2]. At E>5 MeV the experimental data are lacking and the calculated results [1] show that the double ionization cross section σ^{2+} follows closely $\ln E/E$ or $1/E$ dependences. The $\ln E/E$ dependence of σ^{2+} upon the projectile energy is obtained in the "shake-off" calculations of the double ionization cross section [3]. However, as it will be demonstrated in what follows, in the double ionization of helium by charged particles at high collision energies the accounting of the correlation interaction of atomic electrons is not restricted to the "shake-off" mechanism only.

The aim of the present report is to find out the helium double ionization mechanisms and to determine asymptotic behaviour of the helium double ionization cross section by charged particles at high collision energies.

2 Mechanisms of the double ionization

Consider the double ionization of helium in collision

$$\mathrm{p}^{\pm} + \mathrm{He}(1s^2) \rightarrow \mathrm{p}^{\pm} + \mathrm{He}^{2+} + 2e^- \tag{2.1}$$

The total hamiltonian of the system has the form

$$\mathrm{H} = \frac{\hat{\boldsymbol{P}}^2}{2\mathrm{M}} + \sum_{j=1}^{2}\left(\frac{\hat{\boldsymbol{p}}_j^2}{2} - \frac{\mathrm{Z}_2}{r_j}\right) + \frac{1}{|\boldsymbol{r}_1 - \boldsymbol{r}_2|} - \sum_{j=1}^{2}\frac{\mathrm{Z}_1}{|\boldsymbol{R} - \boldsymbol{r}_j|} + \frac{\mathrm{Z}_1\mathrm{Z}_2}{R} \tag{2.2}$$

Here $\hat{\boldsymbol{P}}$ and $\hat{\boldsymbol{p}}_j$ are the momentum operators of the relative motion and the j-th electron of the He atom; $\boldsymbol{R}$ and $\boldsymbol{r}_j$ are the radius vectors of the ion and the j-th electron of the He atom with respect to the atomic nucleus; Z_1 and Z_2 are projectile and nuclear target charges, respectively; M is the reduced mass of the colliding particles.

Write the atomic Hamiltonian H_a in the form

$$H_a = H_a^o + W^c, \tag{2.3}$$

where H_a^o is the atomic Hamiltonian in the independent particle model

$$H_a^o = \sum_{j=1}^{2} \left(\frac{\hat{\boldsymbol{p}}_j^2}{2} - \frac{Z_2}{r_j} + w_j \right), \tag{2.4}$$

W^c is the correlation part of the atomic Hamiltonian

$$W^c = V^c - \sum_{j=1}^{2} w_j, \tag{2.5}$$

$V^c = 1/|\boldsymbol{r}_1 - \boldsymbol{r}_2|$, w_j is an average potential produced by the second electron of helium which the j-th electron feels.

We shall consider the scattering problem in the impact parameter method. We assume that the projectile moves along a straight line $\boldsymbol{R}(t) = \boldsymbol{b} + \boldsymbol{v}t$ (where $\boldsymbol{b}$ is the impact parameter vector and $\boldsymbol{v}$ is the projectile velocity), while the atomic electron wavefunction is independent of the relative distance between the colliding particles and determined from the equation

$$i\frac{\partial \psi(t)}{\partial t} = \tilde{V}(t)\psi(t); \qquad \psi(t)|_{t\to\pm\infty} = | \phi_{I(F)}(\boldsymbol{r}_1, \boldsymbol{r}_2) >, \tag{2.6}$$

where

$$\tilde{V}(t) = \exp(iH_o t) \sum_{j=1}^{2} \frac{Z_1}{|\boldsymbol{R} - \boldsymbol{r}_j|} \exp(-iH_o t) \tag{2.7}$$

is the interaction operator in the interaction representation; H_o is the free motion Hamiltonian. The functions $| \phi_{I(F)}(\boldsymbol{r}_1, \boldsymbol{r}_2) >$ are eigenfunction of the atomic Hamiltonian. In determining $| \phi_K >$, we shall regard the correlation interaction operator W^c, entering into the Hamiltonian H_a, as perturbations. If we restrict ourselves to the first order of the potential W^c, we shall get

$$| \phi_K > = | \phi_K^o > + \sum_{P \neq K} \frac{| \phi_P^o >< \phi_P^o | W^c | \phi_K^o >}{E_K^o - E_P^o + i\delta} \tag{2.8}$$

where

$$| \phi_K^o > = \frac{1 + \mathcal{P}_{12}}{\sqrt{2}} \prod_{j=1}^{2} | \varphi_{p_j}(\boldsymbol{r}_j) > \tag{2.9}$$

are eigenfunctions of the Hamiltonian H_a^o; E_p^o are eigenvalues for $|\phi_P^o>$; $P = (p_1, p_2)$; $p_j = \{n_j l_j m_j\}$ is the complete set of quantum numbers for the j-th electron of the He atom; $|\varphi_p(\boldsymbol{r})>$ are the Coulomb wavefunctions. $P \neq K$ only if K describes the state of both the electrons in the discrete spectrum; otherwise, it is necessary to introduce in the energy denominator the small imaginary quantity $i\delta$ which has the effect of moving the pole off the real axis [4]. $\mathcal{P}_{12}$ is the permutation operator.

The double ionization amplitude $A(\boldsymbol{b})$ as a function of the impact parameter $\boldsymbol{b}$ it is convenient to write as a sum of two terms

$$A(\boldsymbol{b}) = A_1(\boldsymbol{b}) + A_2(\boldsymbol{b}), \tag{2.10}$$

which correspond to two different ionization mechanisms. Using the Fourier transform for the potential $-Z_1/|\boldsymbol{R}(t) - \boldsymbol{r}_j|$, in the second order in $\tilde{V}(t)$ and W^c we shall have

$$A_1(\boldsymbol{b}) = -\frac{Z_1^2\sqrt{2}}{\pi^2}\sum_{j=1}^{2}\sum_{l\neq j}^{2}(1+\mathcal{P}_{i_j i_l})\int\frac{d^3q}{q^2}\exp(-i\boldsymbol{q}_\perp\boldsymbol{b})$$
$$\times <\varphi_{f_j}|\exp(i\boldsymbol{q}\boldsymbol{r})|\varphi_{i_l}>\delta(\epsilon_{f_j}-\epsilon_{i_l}-q_\parallel v)\int\frac{d^3p}{p^2}\exp(-i\boldsymbol{p}_\perp\boldsymbol{b}) \tag{2.11}$$
$$\times <\varphi_{f_l}|\exp(i\boldsymbol{p}\boldsymbol{r})|\varphi_{i_j}>\delta(\epsilon_{f_l}-\epsilon_{i_j}-p_\parallel v)$$

and

$$A_2(\boldsymbol{b}) = \frac{iZ_1}{\pi\sqrt{2}}\int\frac{d^3q}{q^2}\exp(-i\boldsymbol{q}_\perp\boldsymbol{b})\delta(E_F - E_I - q_\parallel v)\sum_{j,m=1}^{2}\sum_{l\neq j}^{2}\sum\!\!\!\!\!\!\int_{k_m}$$
$$\{(1+\mathcal{P}_{f_1 f_2})\frac{<\varphi_{f_1}\varphi_{f_2}|W^c|\varphi_{k_m}\varphi_{i_l}>}{E_F^o - E_K^o - i\delta}<\varphi_{k_m}|\exp(i\boldsymbol{q}\boldsymbol{r})|\varphi_{i_j}> + \tag{2.12}$$
$$+(1+\mathcal{P}_{i_1 i_2})<\varphi_{f_j}|\exp(i\boldsymbol{q}\boldsymbol{r})|\varphi_{k_m}>\frac{<\varphi_{k_m}\varphi_{f_l}|W^c|\varphi_{i_1}\varphi_{i_2}>}{E_I^o - E_K^o}\}.$$

Here E_L and E_L^o are the electron energies of the target atom in state L with the inclusion and with the neglect of the correlations, respectively; ϵ_{i_j} and ϵ_{f_j} are the energies of the j-th electron in the initial and final states, respectively; $\boldsymbol{q}$ is the momentum transfer, $q_\perp$ and $q_\parallel$ are its orthogonal and parallel components relative to the velocity $\boldsymbol{v}$. The amplitudes $A_1(\boldsymbol{b})$ and $A_2(\boldsymbol{b})$ describe the transitions of two atomic electrons into the continuum as a result of the interaction of each electron with the projectile and as a result of the interaction of one electron with the projectile with the subsequent (previous) correlation interaction between the atomic electrons. Both the amplitudes contain electron correlations through the interaction operator $\tilde{V}(t)$. Correlations of this kind were called by McGuire the scattering correlations [3b]. The allowance for the scattering correlations ensures the correct (correlated) value of the electron energy (the eigenvalues of the atomic Hamiltonian H_a enter in the arguments of the δ-functions). In addition to the scattering correlations the second amplitude $A_2(\boldsymbol{b})$ contains also the correlations arising in the wavefunctions. The allowance for the scattering correlations in the amplitude $A_1(\boldsymbol{b})$ does not violate the independence of one-electron events. The amplitude $A_1(\boldsymbol{b})$ coincides with the amplitude determined in the independent electron approximation. The only difference is that in the former approximation the energies of electron states are determined within the self-consistent field model and in the present case the energies have correct (correlated) values. For the more detailed studying of the correlation amplitude $A_2(\boldsymbol{b})$ we shall transform it to the form

$$A_2(\boldsymbol{b}) = \frac{iZ_1}{\pi\sqrt{2}} \int \frac{d^3q}{q^2} \exp(-i\boldsymbol{q}_\perp \boldsymbol{b}) \delta(E_F - E_I - q_\parallel v) \sum_{j,m=1}^{2} \sum_{l \neq j}^{2}$$

$$\{\pi i(1+\mathcal{P}_{f_1f_2}) \int \kappa_m d\Omega_m < \varphi_{f_1}\varphi_{f_2}|V^c|\varphi_{\kappa_m}\varphi_{i_l} >< \varphi_{\kappa_m}|\exp(i\boldsymbol{q}\boldsymbol{r})|\varphi_{i_j} > + \quad (2.13)$$

$$+(1+\mathcal{P}_{f_1f_2}) \sum\!\!\!\!\!\!\int_{k_m} \frac{P}{E_F^o - E_K^o} < \varphi_{f_1}\varphi_{f_2}|V^c|\varphi_{k_m}\varphi_{i_l} >< \varphi_{k_m}|\exp(i\boldsymbol{q}\boldsymbol{r})|\varphi_{i_j} > +$$

$$+(1+\mathcal{P}_{i_1i_2}) \sum\!\!\!\!\!\!\int_{k_m} \frac{P}{E_I^o - E_K^o} < \varphi_{f_j}|\exp(i\boldsymbol{q}\boldsymbol{r})|\varphi_{k_m} >< \varphi_{k_m}\varphi_{f_l}|V^c|\varphi_{i_1}\varphi_{i_2} >\}.$$

Here P means that the integral over dE_K is taken in the sense of the principal value; $\boldsymbol{\kappa}_1$ and $\boldsymbol{\kappa}_2$ are the electron momenta in the states f_1 and f_2, respectively; $\boldsymbol{\kappa}_m$ is the electron momentum acquired after its interaction with the projectile. The value of $\boldsymbol{\kappa}_m$ is determined from the relation

$$\kappa_m^2/2 = \kappa_1^2/2 + \kappa_2^2/2 - \epsilon_{i_l} \quad (2.14)$$

The first term in (2.13) describes the process when one electron, after the interaction with the projectile, acquires the momentum $\boldsymbol{\kappa}_m$ and then scatters on the second electron and, as a result, both the electrons get in the continuum. The second and the third terms describe the transitions of the two electrons in the continuum as a result of two interactions $\tilde{V}(t)$ and V^c. The expression (2.13) contains no terms with the matrix element of operator w_j. Thus, the "shake-off" mechanism should not contribute to the double ionization amplitude when only the first correlation correction to the wavefunction is made.

3 Asymptotics of the helium double ionization cross section

The calculations of the double ionization cross section of helium by protons [5] showed that the correlation mechanism provides the main contribution to the double ionization cross section at high collision energies. The dependence of the amplitude $A_2(\boldsymbol{b})$ upon the projectile energy E is contained only in the factor

$$N = \int \frac{d^3q}{q^2} \exp(-i\boldsymbol{q}_\perp \boldsymbol{b}) < \varphi_l|\exp(i\boldsymbol{q}\boldsymbol{r})|\varphi_m > \delta(E_F^o - E_I^o - q_\parallel v) \quad (3.1)$$

Expanding the operator $\exp(i\boldsymbol{qr})$ in a power series and next using only the first two terms, which provide the main contribution at high collision energies, we shall have the next asymptotics for the helium double ionization cross section at high collision energies

$$\sigma^{2+} = \mathrm{B}(\mathrm{C} + \ln \mathrm{E})/\mathrm{E}, \tag{3.2}$$

where B, C are some constants. Calculations shown that in the double ionization of helium by protons the parameter C is equal to -5.54 when E is determined in keV. So, the term proportional to 1/E should effect strongly the behaviour of the cross section σ^{2+} even at very high proton energies. This is in agreement with the results of measurements of the double ionization cross section of helium by electrons [6].

References

1. A.L. Ford and J.F. Reading: J. Phys. **B21** L685 (1988)
2. L.H. Andersen, P. Hvelplund, H. Knudsen, S.P. Møller, K. Elsener, K.-G. Rensfelt and E. Uggerhøj: Phys. Rev. Lett. **57** 2147 (1986)
3. J.H. McGuire: **a)** Phys. Rev. Lett. **49** 1153 (1982); **b)** Nucl. Instrum. Methods Phys. Res. Sect. **B10/11** 17 (1985)
4. L.D. Landau and E.M. Lifshits: Quantum Mechanics (Nauka, M., 1989).
5. V.A. Sidorovich: Proc. of Contr. Papers XVI ICPEAC (N.-Y., 1989), p. 467
6. B.L. Schram, A.J.H. Boerboom and J. Kistemaker: Physica **32** 185 (1966)

D Electron Correlation and Post Collision Interaction Effects

Chairman: V.V. BALASHOV

D.1 Invited Surveys

N. STOLTERFOHT
Interference and Correlation in Energetic Ion–Atom Collisions: The Role of Dielectronic Processes

A.L. FORD and J.F. READING
Electron Correlation in High Energy Ion–Atom Collisions

D.2 Invited Contributions

T. VAJNAI, L. SARKADI, Á. KÖVÉR and J. VÉGH
Study of the Line Shape of Auger Electrons Excited by Direct Ionization and Electron Capture

M. BARAT, P. RONCIN, C. ADJOURI, N. ANDERSEN, M.N. GABORIAUD and L. GUILLEMOT
Orientation Propensity in Electron Capture by Multiply Charged Ions

D.3 Contributions (Posters)

S. RICZ, J. VÉGH, B. SULIK, D. VARGA, E. TAKÁCS, I. KÁDÁR and D. BERÉNYI
Anomaly in Auger Spectra of Fast Heavy Ion Impact on Ne in Forward and Backward Direction

K. TŐKÉSI, S. RICZ, E. TAKÁCS and B. SULIK
e^- Impact Energy Dependence of the L_3 Subshell Alignment of Ar

M. BARAT, P. RONCIN, M.N. GABORIAUD, L. GILLEMOT and H. LAURENT
Multiple Electron Capture in Highly Charged Ion–Atom Collisions at keV Energies. Correlated and Incorrelated Mechanisms

E.M. BERNSTEIN, A. KAMAL, S.M. FERGUSON and J.A. TANIS
Resonant Transfer and Excitation in ${}_{12}M^{9+}+H_2$

Interference and Correlation in Energetic Ion-Atom Collisions: The Role of Dielectronic Processes

N. Stolterfoht[+]

Laboratoire de Spectroscopie Atomique, I.S.M.R.a,
F-14050 Caen Cedex, France

Abstract

The influence of electron correlation on interference effects in energetic ion-atom collisions is analyzed. Electron-correlation phenomena are discussed in terms of dielectronic processes which are due to two-electron transitions produced by the electron-electron interaction. The collision systems are treated using the semi-classical approximation up to second order. It is shown that first-order amplitudes are either real or imaginary depending on their symmetries whereas the second-order amplitudes are in general complex. In a multi-electron system, two-step processes produced by the nucleus-electron interaction are shown to proceed via two associated paths canceling important parts of the transition amplitude by interference effects. As an example for the process of single excitation interferences between first- and second-order amplitudes are shown to be influenced by the exchange interaction due to the Pauli principle. Furthermore, double excitation of He and Li is discussed in view of interferences between the nucleus-electron and electron-electron interaction. For Li this interference is expected to be influenced by configuration interaction involved in the states $1s2s^2\,{}^2S$ and $1s2p^2\,{}^2S$.

1 Introduction

In the past few years, energetic ion-atom collisions have received a great deal of attention with respect to electron-correlation effects which are due to the mutual interaction of two electrons occurring in the Coulomb field of the nucleus involved[1-3]. Detailed information about the collision mechanisms may be obtained when electron-correlation effects interfere with two-electron processes produced by the nucleus-electron interaction. Particular attention has been devoted to transfer-excitation[4-6], as well as to double ionization[7-9] and excitation[10–13] to study interferences between events produced by electron-electron and nucleus-electron interactions. A remarkable feature of ion impact is that anti-matter particles can be used to investigate such interference effects for projectiles of opposite charge but equal mass. Studies of He double ionization[9] involving antiproton impact have been performed at the Low Energy Antiproton Ring (LEAR) facility at CERN (Geneva). These measurements have created a lively discussion about electron-correlation effects in ion-atom collisions[1,2,14-16].

The formal analysis of electron correlation is based on the concept that the nucleus-electron interaction is associated with one-body operators whereas the electron-electron interaction is attributed to two-body operators[17]. The electron-electron interaction is produced by the Coulomb force which is also responsible for the nucleus-electron interaction. However, the nucleus is much heavier than the electron so that in the mutual nucleus-electron interaction, it appears as if the electron is the only one which reacts. In this case it is understandable that the nucleus-electron interaction exhibits a single-particle character. In contrast to this feature the electron-electron interaction has a two-particle character. When the nucleus is replaced by an electron, the scattering center is accelerated producing a receding Coulomb field. However, part of the electron-electron interaction may also be represented by a single-particle potential. It is emphasized that this part is associated with the presence of an attractive nucleus. Due to the coupling with the nucleus, the state of the bound electron may remain unaltered, although it participates in the interaction with the incident electron. The electron remains in the ground state when the energy transferred during the collision does not exceed the energy of the first excited state.

If the energy transfer to the bound electron is not sufficient, this energy will be "absorbed" by the nucleus. Such a bound electron seems to be part of the nucleus and, thus, it appears as a "heavy" particle whose interaction exhibits a single-particle character, i.e., whose potential is monoelectronic. The two-particle aspect of the electron-electron interaction becomes important when the bound electron is removed from its ground state. The removed electron acts as a "light" particle which produces the receding Coulomb field, i.e., whose potential is dielectronic. Hence, two-electron events produced by the electron-electron interaction are referred to as **dielectronic processes**[18,19] which are considered as the essential phenomena associated with electron correlation in ion-atom collisions. It is noted that the term **dielectronic** has been introduced by Massey and Bates[20].

In this work, the dielectronic processes are discussed in conjunction with the interplay between the electron-electron and nucleus-electron interactions. Specific emphasis is placed on the derivation of the phases of the transition matrix elements which are evaluated in first and second order using time-dependent perturbation theory. The characteristics of two-electron transitions are first treated separately for the nucleus-electron interaction and the electron-electron interaction. Then, interference effects in collision processes are considered when both interactions are involved. The analysis is restricted to excitation processes occurring at one atomic center. For more details about electron correlation in ion-atom collisions the reader is referred to previous review articles[1-3,18-22]. It is noted that, in this work, atomic units are used if not otherwise stated.

2 Time-Dependent Treatment of the Collision

2.1 Partition of the Hamiltonian

The present analysis of energetic ion-atom collisions is based on a time-dependent treatment where the nuclei are treated classically[23]. Within this semiclassical approximation (SCA) the internuclear distance is obtained as a function of time, i.e., $R = R(t)$, and the Hamiltonian which governs the collision system becomes time dependent. This (electronic) Hamiltonian consists of different terms:

$$H = T^{e} + V^{n} + V^{e} \tag{1}$$

where T^{e} represents the kinetic energy of the electrons, V^{n} the electron-nucleus interaction and V^{e} the electron-electron interaction. The latter quantities are obtained as

$$V^n = -\sum_j \left[\frac{Z_a}{|\vec{r}_j - \vec{R}_a|} + \frac{Z_b}{|\vec{r}_j - \vec{R}_b|} \right] \tag{2}$$

and

$$V^e = \sum_{k>j} \frac{1}{|\vec{r}_k - \vec{r}_j|} \tag{3}$$

where $\vec{r}_j$ is the position of the jth electron, $\vec{R}_a$ and $\vec{R}_b$ are the positions of the nuclei, and Z_a and Z_b are their respective charges.

As discussed previously[3] the concept of electron correlation is based on the partition of the total Hamiltonian into terms of one- and two-body operators. An introduction to the method of partitioning the electron-electron interaction is given in the textbook by Condon and Shortley[17] from which additional information may be obtained. The one-body operators are associated with the interactions postulated within the framework of the independent-particle model whereas the two-body operators represent interactions leading beyond that model. Thus, the electron-electron interaction is partitioned into

$$V^e = W^e + V^c \tag{4}$$

$$\text{where} \quad W^e = \sum_j w_j \quad \text{and} \quad V^c = \sum_{j<k} q_{jk} \tag{5}$$

The effective potential W^e is given by one-electron operators w_j representing the mean field postulated within the independent-particle model and the residual interaction V^c consists of two-electron operators q_{jk}. In the following, W^e and V^c will also be referred to as **monoelectronic** and **dielectronic potentials**, respectively. The dielectronic potential may be used to partition the full Hamiltonian

$$H = H^o + V^c \tag{6}$$

where the IPM Hamiltonian $H^o = T^e + V^n + W^e$ incorporates the interactions which are represented by one-body operators only.

To study scattering correlation it is instructive to consider the time-dependent interactions relevant during the collision[1]. One may set

$$V^n = V^{n(i)} + V^{n(sc)} \quad \text{and} \quad V^e = V^{e(i)} + V^{e(sc)} \tag{7}$$

where the initial atomic potentials $V^{n(i)} = V^n(t=-\infty)$ and $V^{e(i)} = V^e(t=-\infty)$ are constant in time. The nuclear scattering potential $V^{n(sc)}(t)$ is obtained by summing over the terms $Z_a/|\vec{r}_j^{\,b} + \vec{R}|$ and $Z_b/|\vec{r}_j^{\,a} + \vec{R}|$ where $\vec{r}_j^{\,a} = \vec{r}_j - \vec{R}_a$ and $\vec{r}_j^{\,b} = \vec{r}_j - \vec{R}_b$ are the electronic positions relative to the nucleus at which the electron is initially centered. These terms are time dependent since the internuclear separation $\vec{R} = \vec{R}_b - \vec{R}_a$ varies with time. Analogously, the operator $V^{e(sc)}(t)$ is obtained by summing over the terms $1/|\vec{r}_j^{\,a} - \vec{r}_k^{\,b} - \vec{R}|$ which are again time dependent through the internuclear separation $\vec{R}$. Moreover, the mono-electronic and dielectronic potentials are split into partitions relevant before and during the collision, i.e., $W^e = W^{e(i)} + W^{e(sc)}$ and $V^c = V^{c(i)} + V^{c(sc)}$. More details about the individual interactions have been discussed previously[3].

2.2 Coupling Matrix Elements

The foregoing discussion shows that there are various possibilities for partitioning the Hamiltonian. For high-energy collisions it is favorable to split the full Hamiltonian into an asymptotic (or atomic) Hamiltonian $H^{(a)}$ and a time-dependent perturbation V(t):

$$H = H^{(a)} + V(t) \tag{8}$$

At this point it is left open whether $H^{(a)}$ accounts for correlation. Generally, the unperturbed Hamiltonian $H^{(a)}$ is identified with the asymptotic IPM Hamiltonian $H^o(t=-\infty)$ from eq. (6) and electron correlation represented by V^c is treated as a perturbation. Then, the perturbation potential V splits into one- and two-body operators according to

$$V = V^{sn} + V^c \tag{9}$$

where the screened nuclear potential $V^{sn} = V^{n(sc)} + W^{e(sc)}$ summarizes the one-body operators. When electron correlation is incorporated into the asymptotic states, $H^{(a)}$ is to be identified with the full asymptotic Hamiltonian $H(t=-\infty)$ and the dielectronic interaction V^c is to be replaced by $V^{c(sc)}$ relevant during the scattering only.

The equation $H^{(a)}|\Phi_j^a\rangle = E_j^a|\Phi_j^a\rangle$ yields the eigenenergies and eigenstates of $H^{(a)}$. The eigenstates are used to evaluate the coupling matrix elements

$$V_{jj'}(t) = \langle\Phi_{j'}^a|V(t)|\Phi_j^a\rangle \tag{10}$$

which may further be developed by means of the partition (9) yielding $V_{jj'} = V_{jj'}^{sn} + V_{jj'}^c$. The matrix elements $V_{jj'}^{sn}$ and $V_{jj'}^c$ are formed analogously to $V_{jj'}$ using the interactions V^{sn} and V^c, respectively. It is important to note that these interactions couple different categories of states. The screened nucleus interaction V^{sn} composed of one-body operators couples states which differ in **one** spin orbital whereas the dielectronic interaction V^c composed of two-body operators couple states which

Table I: *Symmetry of the matrix elements produced by the screened nuclear interaction V^{sn} and the dielectronic interaction V^c. The quantity $\Delta\Pi$ denotes the change of parity and ΔM denotes the change of magnetic quantum number in the transition from the initial to the final state.*

Interaction	Transition	$\Delta\Pi$	$\Delta M=0$	$\Delta M=1$
V^{sn}	Monopole, Quadrupole, ...	No	even	odd
V^{sn}	Dipole, ...	Yes	odd	even
V^c	Monopole	No	even	—

differ in **two** spin orbitals. Therefore, for a given category of states, only one of the matrix elements $V_{jj'}^{sn}$ and $V_{jj'}^{c}$ is important.

Furthermore, it should be emphasized that the coupling matrix elements considered here have specific symmetries. It is assumed that the initial, intermediate, and final states have parity as a good quantum number and they are located at one center only. Then, the matrix elements $V_{jj'}^{sn}(t)$, and $V_{jj'}^{c}(t)$ are **even** or **odd** with respect to the time variation depending on the change of parity and the magnetic quantum number. The symmetries of the matrix elements are shown in Table I.

2.3 Transition Amplitudes in First and Second Order

In the following analysis of interference effects in energetic ion-atom collisions we consider the transition amplitude up to second order:

$$A_{if}^{(II)} = A_{if}^{(1)} + A_{if}^{(2)} \tag{11}$$

where $A_{if}^{(1)}$ and $A_{if}^{(2)}$ are the first- and second-order terms of the time-dependent perturbation theory expansion. They are given by[24]

$$A_{if}^{(1)} = -i\int_{-\infty}^{\infty} V_{if}(\tau)e^{i\omega_{if}\tau}d\tau \tag{12}$$

and

$$A_{if}^{(2)} = -\sum_{k}\int_{-\infty}^{\infty} V_{kf}(\tau)e^{i\omega_{kf}\tau}\int_{-\infty}^{\tau} V_{ik}(\tau')e^{i\omega_{ik}\tau'}d\tau'd\tau \tag{13}$$

where $\omega_{jj'} = E_{j'}^{a} - E_{j}^{a}$ are transition energies and $V_{jj'}$ are coupling matrix elements from eq. (10). The sum in eq. (13) is extended over the eigenfunction spectrum of $H^{(a)}$. Hereafter, an individual term of the perturbation expansion will be denoted A_{if}^{k} where the specification of the additional label k shall indicate that the amplitude is of second order.

Interference effects are governed by the phases of the matrix elements. To obtain information about these phases it is useful to consider separately their real and imaginary parts. It is assumed that the initial and final wave functions are real as is the case for bound states. Hence, in the following, excitation of the electrons is considered. From the relation $e^{i\omega\tau} = \cos\omega\tau + i\sin\omega\tau$ it follows that

$$A_{if}^{(1)} = F_{if}^{(1)} + iG_{if}^{(1)} \tag{14}$$

where

$$F_{if}^{(1)} = \int_{-\infty}^{\infty} V_{if}(\tau)\sin\omega_{if}\tau\, d\tau \tag{15}$$

and

$$G_{if}^{(1)} = -\int_{-\infty}^{\infty} V_{if}(\tau)\cos\omega_{if}\tau\, d\tau \tag{16}$$

Similarly for the second-order terms it follows that

$$A_{if}^{(2)} = F_{if}^{(2)} + iG_{if}^{(2)} \tag{17}$$

where

$$\begin{aligned} F_{if}^{(2)} = & -\sum_k \int_{-\infty}^{\infty} V_{kf}(\tau)\cos\omega_{kf}\tau \int_{-\infty}^{\tau} V_{ik}(\tau')\cos\omega_{ik}\tau'\, d\tau'\, d\tau \\ & + \sum_k \int_{-\infty}^{\infty} V_{kf}(\tau)\sin\omega_{kf}\tau \int_{-\infty}^{\tau} V_{ik}(\tau')\sin\omega_{ik}\tau'\, d\tau'\, d\tau \end{aligned} \tag{18}$$

and

$$\begin{aligned} G_{if}^{(2)} = & -\sum_k \int_{-\infty}^{\infty} V_{kf}(\tau)\cos\omega_{kf}\tau \int_{-\infty}^{\tau} V_{ik}(\tau')\sin\omega_{ik}\tau'\, d\tau'\, d\tau \\ & - \sum_k \int_{-\infty}^{\infty} V_{kf}(\tau)\sin\omega_{kf}\tau \int_{-\infty}^{\tau} V_{ik}(\tau')\cos\omega_{ik}\tau'\, d\tau'\, d\tau \end{aligned} \tag{19}$$

It is noted that the quantities $F_{if}^{(1)}$, $G_{if}^{(1)}$, $F_{if}^{(2)}$, and $G_{if}^{(2)}$ are real. The phases of the transition amplitudes $A_{if}^{(1)}$ and $A_{if}^{(2)}$ follow from their properties as complex numbers. The first-order amplitudes are either real or imaginary depending on whether the coupling matrix element V_{if} is **odd** or **even**, respectively. The rules for the first-order amplitudes are summarized in Table II.

Table II: *Complex properties of the first-order transition amplitude representing the screened nuclear interaction V^{sn} and the dielectronic interaction V^c. Notations as in Table I.*

Interaction	Transition	ΔΠ	ΔM=0	ΔM=1
V^{sn}	Monopole, Quadrupole, ...	No	imaginary	real
V^{sc}	Dipole, ...	Yes	real	imaginary
V^c	Monopole	No	imaginary	—

For the second-order transition amplitudes such a simple rule does not apply. In this case it is important to note that both a real and an imaginary part contributes to the transition amplitude.

A priori, it is difficult to predict whether the real or the imaginary part dominates. However, we are primarily interested in fast collisions. For high incident energies it follows that the terms including the cosine function are expected to be significantly larger than those including the sine function. This is due to the fact that $\sin\omega_{if}\tau \ll 1$ for small collision times. Hence, the terms in eqs. (18) and (19) containing sine functions are generally small. Consequently, the first term in eq. (18) is most important since it contains only cosine functions. This term contributes to $F_{if}^{(2)}$ which represents the real part of the transition amplitude.

3 Two-Electron Transitions Produced by V^{sn} and V^{c}

3.1 Associated Paths within the IPM

In this section we consider two-electron processes induced by the screened nucleus-electron interaction V^{sn}. These uncorrelated processes are treated within the framework of the independent particle model in which the transition probability for a two-electron transition is given by the product of the related single-particle transition probabilities[1,3]. In the following, it will be shown that this product form is based on a pair of second-order transitions involving significant interference effects.

Within the independent-particle model, the initial, intermediate, and final states are obtained as single configuration states. Hence, as shown in Figure 1, the initial and final state are given by

$$\Phi_i^o = \det(\varphi_a\varphi_b\ldots) \qquad \text{and} \qquad \Phi_f^o = \det(\varphi_c\varphi_d\ldots) \tag{20}$$

where the symbol det stands for the (normalized) Slater determinant and $\varphi_a, \varphi_b, \varphi_c$, and φ_d are spin orbitals specifying the active electrons in the initial and final state. In this case two intermediate states are important where one electron did undergo a transition and the other one did not:

$$\Phi_r^o = \det(\varphi_c\varphi_b\ldots) \qquad \text{and} \qquad \Phi_s^o = \det(\varphi_a\varphi_d\ldots) \tag{21}$$

The paths attributed to these intermediate states labelled r and s will be referred to as **associated paths** (Figure 1). The related amplitudes are given by

$$A_{if}^{r} = -\int_{-\infty}^{\infty} V_{rf}(\tau)e^{i\omega_{rf}\tau}\int_{-\infty}^{\tau} V_{ir}(\tau')e^{i\omega_{ir}\tau'}d\tau' d\tau \tag{22}$$

and

$$A_{if}^{s} = -\int_{-\infty}^{\infty} V_{sf}(\tau)e^{i\omega_{sf}\tau}\int_{-\infty}^{\tau} V_{is}(\tau')e^{i\omega_{is}\tau'}d\tau' d\tau \tag{23}$$

In the following it will be assumed that the electron orbitals are frozen, i.e., that they are not influenced by changes of the mean field postulated within the framework of the independent particle model. From the frozen energies it follows that

$$\omega_{ir} = \omega_{sf} = \omega_{ac} \quad \text{and} \quad \omega_{is} = \omega_{rf} = \omega_{bd} \tag{24}$$

where ω_{ac} and ω_{bd} are the energy differences for the single-electron orbitals involved. Moreover, from the frozen wave functions it follows that

$$V_{ir} = V_{sf} = V_{ac} \quad \text{and} \quad V_{is} = V_{rf} = V_{bd} \tag{25}$$

where $V_{ac}=s_{ac}<\varphi_c|v|\varphi_a>$ and $V_{bd}=s_{bd}<\varphi_d|v|\varphi_b>$ are single-electron matrix elements including the statistical factors s_{ac} and s_{bd}, respectively. The single-particle operator v is the part of V which is relevant for the active electron. It is seen that within the frozen-orbital approach the multi-electron matrix elements reduce to single-electron matrix elements.

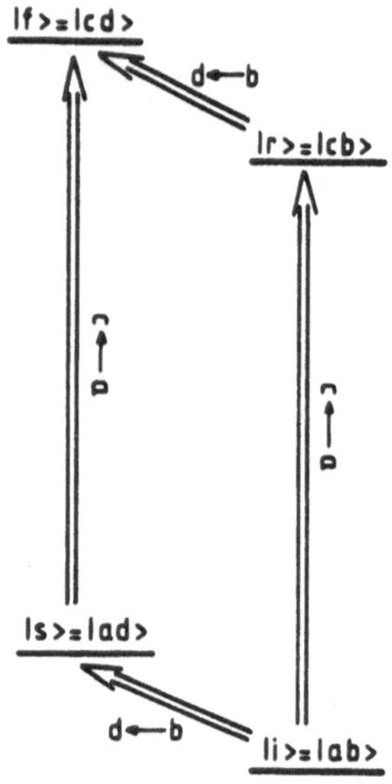

Figure 1: *Associated paths leading from the initial state |i> to the final state |f> via the intermediate states |r> and |s> for which the order of the one-electron transitions is interchanged.*

Using these identities and the well-known integration rule $\int f dx \int g dx = \int (f\int g dx + g\int f dx)dx'$ one obtains for the sum of the associated amplitudes

$$A_{if}^{r} + A_{if}^{s} = -\int_{-\infty}^{\infty} V_{bd}(\tau) e^{i\omega_{bd}\tau} d\tau \int_{-\infty}^{\infty} V_{ac}(\tau') e^{i\omega_{ac}\tau'} d\tau' \tag{26}$$

Hence, by introducing the single-electron amplitudes

$$a_{ac} = -i\int_{-\infty}^{\infty} V_{ac}(\tau) e^{i\omega_{ac}\tau} d\tau \quad \text{and} \quad a_{bd} = -i\int_{-\infty}^{\infty} V_{bd}(\tau) e^{i\omega_{bd}\tau} d\tau \tag{27}$$

one obtains the usual product form of the amplitude for the two-electron transition

$$A_{if}^{r} + A_{if}^{s} = a_{ac}\, a_{bd} \tag{28}$$

Thus, the product form is obtained by adding coherently the amplitudes for the associated paths relevant within the framework of the independent particle model.

The important point to be made here is that, in general, the amplitudes A_{if}^{r} and A_{if}^{s} involve real and imaginary parts as it is usual for the second-order amplitudes. Hence, one may set

$$A_{if}^{r} = F_{if}^{r} + iG_{if}^{r} \qquad \text{and} \qquad A_{if}^{s} = F_{if}^{s} + iG_{if}^{s} \tag{29}$$

When these amplitudes are added coherently, significant interference effects occur. This follows from the fact that the coherent sum of the associated amplitudes A_{if}^{r} and A_{if}^{s} is either real or imaginary. This, in turn, follows from eq. (28) and the fact that the first-order amplitudes a_{ac} and a_{bd} are either real or imaginary (Table II). Therefore, when the amplitudes from eq. (29) are added, either the pair F_{if}^{r} and F_{if}^{s} or the pair G_{if}^{r} and G_{if}^{s} cancel by means of interference effects.

These interferences may affect the dominant part of the transition amplitude. As noted before, in fast collisions, the real parts F_{if}^{r} and F_{if}^{s} are most important. They are canceled, however, if the associated two-electron process involves, e.g., a dipole and a monopole transition. It should be emphasized that the conclusion of the pure real or imaginary amplitude holds only within the independent-particle model involving frozen orbitals (IPM-FO). It is expected that deviations from this complete cancellation of the amplitudes F_{if}^{r} and F_{if}^{s} or G_{if}^{r} and G_{if}^{s} could be taken as an indication for the breakdown of the IPM-FO.

3.2 Dielectronic Processes

In this section we consider dielectronic processes which are due to mutual two-electron transitions produced by the electron-electron interaction. These processes occur in the collision system as well as in the separated atoms after the collision, when the system is prepared in a state which is non-stationary with regard to the interaction V^c, e.g., when it is prepared in a single-configuration state. The occurrence of two-electron processes is associated with the fact that the dielectronic interaction V^c couples states which differ in precisely **two** spin orbitals describing the active electrons. It is anticipated that the dielectronic processes play a central role in the field of dynamic electron correlation. As the dielectronic processes have been discussed before[18,19] only a brief description is given here.

In the following, channel diagrams are presented to discuss the dielectronic processes for an atomic three-body system. Moreover, the three-body phenomena are applied by analogy to four-body systems relevant for ion-atom collisions. It should be mentioned also that channel diagrams have already been used by Fano[25] many years ago. Since then, these diagrams have been applied in numerous studies to clarify configuration interaction in conjunction with autoionization [see, e.g., 26,27]. In accordance with that previous work it is shown that electron-correlation effects are associated with configuration interaction (CI).

Configuration interaction can be visualized in the channel diagram of single-configuration states for He given in Figure 2 indicating energy levels of the eigenstates Φ_j^o of the IPM Hamiltonian H^o from eq. (6). Couplings occur between states which differ in two spin orbitals involving bound and continuum electrons. The term "closed channel" here refers to doubly excited states whereas the term "open channel" refers to single- and double-ionization states. Since the single-configuration states Φ_j^o are not eigenstates of the full Hamiltonian, the interaction between the different states is possible. From Figure 2 it is seen that interactions occur within the channels and across the channel borders. **Bound-state** CI occurs within the closed channel. **Free-bound state** CI occurs between the closed channel and single ionization channels. **Interchannel** CI occurs within the single ionization channels[26]. The coupling between the single and double ionization channel is referred to as **transchannel** CI. Finally, coupling within the double ionization (DI) channel is referred to as **DI interchannel** CI.

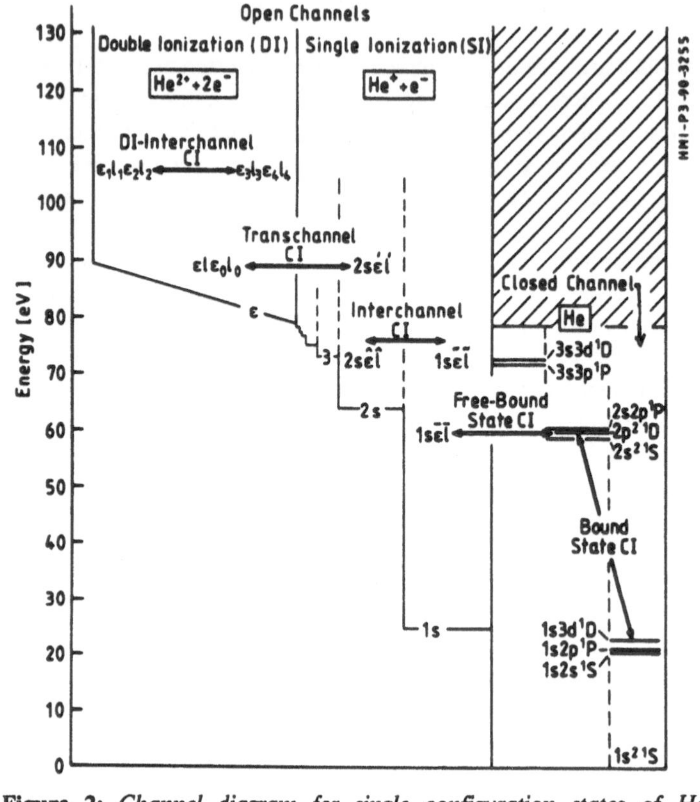

Figure 2: *Channel diagram for single configuration states of He indicating various types of configuration interaction (CI).*

Due to the energy conservation in the unperturbed system, the interactions along horizontal arrows are favored. However, interactions along nearly vertical arrows are also possible. The selection rule for the CI coupling is that the two single-configuration states have the same total angular momentum and parity. If the spin-orbit coupling scheme is valid, as it is for the case of He, then the orbital angular momentum and the spin must also be equal.

4 Examples for Interference Effects

4.1 Single Excitation Involving Electron Exchange

In this section we consider the process of one-electron excitation produced by the screened nucleus-electron interaction. At high energies, dipole transitions are dominant in the single-step process. It is seen from Table II that the related first-order amplitude is real for $\Delta M=0$. Furthermore, it is recalled that the corresponding second-order amplitude is essentially real in fast collisions. Therefore, for high energies and $\Delta M=0$, favorable conditions are created for interference effects between first- and second-order amplitudes in the single excitation process. On the other hand, interference effects are small for $\Delta M=1$ since the first-order amplitude is imaginary (Table II) whereas the second-order amplitude remains essentially real. In the following we shall focus on the $\Delta M=0$ excitation. In this case it is important to note that significant interferences can be concluded for one-electron systems, but this conclusion does not necessarily hold for many-electron systems. Indeed, the situation changes completely when two active electrons participate in the process of single excitation.

In a multi-electron system a two-step process is generally accompanied by an analogous process proceeding via the associated path (Figure 1). For the case of $\Delta M=0$ the **sum** of the associated second-order amplitudes is imaginary as can readily be shown by means of eq. (28) and Table II. Hence, the real parts of the amplitudes due to the associated paths interfere destructively so that they are canceled completely. As a consequence of this cancellation, the pair of second-order amplitudes cannot interfere with the first-order amplitude (which is real). At this point the question arises whether experiments are available yielding evidence for these complex interferences. In the following, it is shown that experimental information about the different interference effects can be obtained from an atomic system in which electron exchange due to the Pauli principle plays an important role.

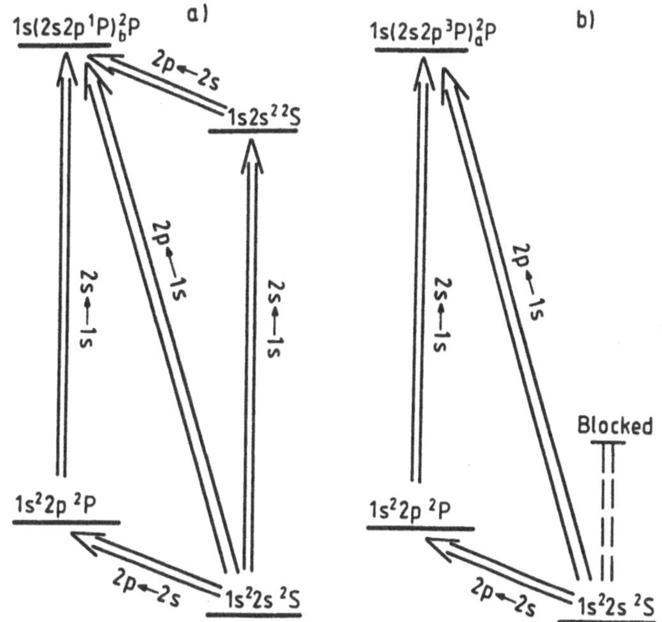

Figure 3: *Diagrams indicating the 2p excitation in a Li-like atomic system via one-step and two-step processes. The right diagram (a) exhibits the phenomenon of Pauli blocking which inhibits the occupation of the 2s orbital by two electrons with parallel spin.*

Figure 3 shows an example for the process of single excitation involving two active electrons. The example refers to the 1s-2p excitation in a Li-like atom such as O^{6+} or Ne^{7+}. These ions have been studied at high incident energies using the method of zero-degree Auger spectroscopy[28-30] which is sensitive primarily to M=0 and, thus, on $\Delta M=0$ as the initial state is an S state. Studies with highly charged ions have the advantage that the 2p orbital is sufficiently far away from the shells with principal quantum numbers $n \geq 3$ so that the interpretation of the system may be based on a few states only. Advanced theoretical work about the 1s-2p excitation process has been performed recently by Gayet et al.[31].

It is seen from Figure 3 that the initial state $1s^2 2s\ ^2S$ and the final state $1s2s2p\ ^2P$ differ in a single electron and that the dominant excitation path is due to a one-step process involving the dipole 1s-2p transition. Furthermore, Figure 3 indicates the associated paths representing the two-step processes via the states $1s^2 2p\ ^2P$ and $1s2s^2\ ^2S$. It is seen that two electrons are active in these processes whereas one of the 1s electrons remains as a passive spectator. Spin flip is unlikely and, thus, the initial, intermediate, and final states are doublets. Two cases are distinguished where the two active electrons form a singlet and a triplet state creating different parent couplings in the final states, i.e., $1s(2s2p\ ^1P)\ ^2P$ and the $1s(2s2p\ ^3P)\ ^2P$. These final states differ sufficiently in energy that they can be distinguished experimentally. The excitation probabilities have been measured separately for each final state as a function of the incident energy[28] and the target nuclear charge[29,30]. The experimental results clearly show that the excitation functions differ in the two final states. This experimental observation can be interpreted as being due to interference effects between the first- and second-order amplitudes in the 1s-2p excitation process.

For the case, in which the final state exhibits a singlet parent term, both associated paths are open (Figure 3a). As discussed for the associated paths in conjunction with Figure 1, the real parts of the second-order amplitudes cancel each other completely and, thus, the interference with the first-order term disappears. However, for the triplet parent term this cancellation does not occur due to the fact that the present multi-electron atom exhibits the phenomenon of **Pauli-blocking**[32]. This mechanism prevents the system from proceeding via the state $1s2s^2\ ^2S$ when the active electrons form a triplet. Two electrons with parallel spin cannot both occupy the 2s orbital and, hence, one of the associated paths is blocked due to the Pauli principle (Figure 3b). For a single path the real part of the second-order amplitudes is maximum so that a significant interference with the first-order amplitude is expected. Experimentally [29] it is found that the excitation probability of the triplet parent term increases in relation to that of the singlet parent term as the interference effect increases. This finding is consistent with the fact that the interference between the first- and second-order amplitudes is constructive as has been found by a theoretical analysis[30].

4.2 Double Excitation Involving Electron Correlation

Recently, considerable interest has been devoted to the process of double ionization of He by fast ions[7-9]. In this case an interference between the two-step process involving two nucleus-electron interactions (TS-2) and the two-step process involving a single nucleus-electron interaction followed by an electron-electron interaction (TS-1) has been predicted[1]. In lowest order the amplitudes for TS−1 is proportional to the projectile charge Z_p whereas the amplitude for TS−2 is proportional to Z_p^2. Therefore, an interference term is produced depending on Z_p^3 and, hence, on the charge sign. The observation of this interference[9] created stimulating controversies [1,2,14,16] about electron correlation effects in ion atom collisions. To obtain more information about these effects McGuire and Deb[33] suggested drawing attention to the process of double excitation.

Double excitation of He has been studied in various laboratories since the mid 1960s (see, e.g., Bordenave-Montesquieu et al.[34] and references therein). Measurements devoted to the charge state-dependence of double excitation of He were carried out more recently[10-13]. To interpret the double excitation data various mechanisms are to be considered. Figure 4a shows different paths leading to double excitation via the closed channel and the single-ionization channel. The successive nucleus-electron interactions are represented by two connected open lines. Also, in Figure 4a, shake-up and resonant free-bound state CI are incorporated and a path is added representing bound state CI, i.e., between the states 1s3d ^1D and $2p^2\,^1D$. Moreover, it was found that initial state CI plays a role[35].

It is expected that the dielectronic processes produced by free-bound state CI should be particularly important since it is resonant. The corresponding path represents the TS-1 mechanism[9] which proceeds via the single ionization channel initiated by nucleus-electron interaction followed by a process similar to dielectronic capture. It should be realized that the interference associated with the TS-1 process produces asymmetric line shapes known as Fano-line profiles[25]. These line profiles make it difficult to extract cross section for double excitation from the experimental data. Hence, there is a conceptual problem in the interpretation of the experimental results in terms of double excitation.

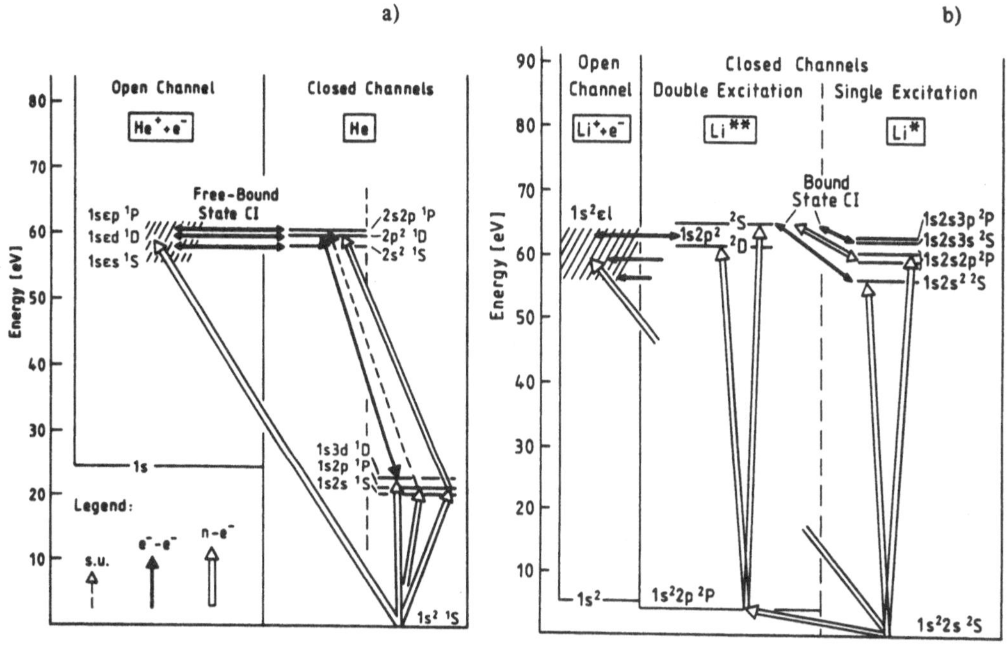

Figure 4: *Channel diagrams for single configuration states of He (a) and Li (b). The transitions induced by a nucleus-electron (n- e⁻) interaction are indicated by open lines and the transitions due to dielectronic processes (e⁻- e⁻) and shake-up (s.u.) are indicated by solid and broken lines, respectively.*

The problems involved in the interpretation of the double excitation are expected to be significantly reduced when the target atom Li is used. It should be emphasized that the double excitation of He affects two equivalent electrons in the 1s orbitals. The double excitation process of Li, however, involves the non-equivalent electrons from the orbitals 1s and 2s. The corresponding channel diagram is given in Figure 4b. The Li spectrum exhibits a relatively small background of continuum electrons due to direct ionization, and the autoionization lines are found to have no asymmetric shapes[36,37]. Hence, the corresponding path involving continuum electrons is probably less important. Rather, shake up and bound-state configuration interaction are expected to gain importance.

Figure 4b shows that the states $1s2s^2\ ^2S$ and $1s2p^2\ ^2S$ are rather close in energy so that CI effects are expected to be significant. It is noted that an interference term depending on Z_p^3 is created as the excitation probabilities for these states depend differently on Z_p. In first order, the amplitudes for the production of the states $1s2s^2\ ^2S$ and $1s2p^2\ ^2S$ are proportional to Z_p and Z_p^2, respectively. It is noted that the amplitudes for the corresponding states $2s^2\ ^1S$ and $2p^2\ ^1S$ of He are both proportional to Z_p^2. Hence, in He the Z_p^3 interference term is probably not produced by bound-state CI. Therefore, the essential difference in the double excitation of He and Li lies in the influence of the bound-state CI on a possible Z_p^3 term.

It was noted previously that the bound-state CI may occur during the collision and/or after the collision partners have separated. In the first case which corresponds to scattering correlation, it is favorable to emphasize the dynamic aspect of the electron correlation effects and, hence, to use the channel diagrams based on the IPM states Φ_j^0 (Figure 4). But these diagrams have the disadvantage that the asymptotic states (in particular, the final state) are not well represented. When electron correlation is important after the collision, we deal with atomic correlation which can be incorporated in the final state[1]. Then, it is favorable to use states which diagonalize (at least within a given channel) the full Hamiltonian H.

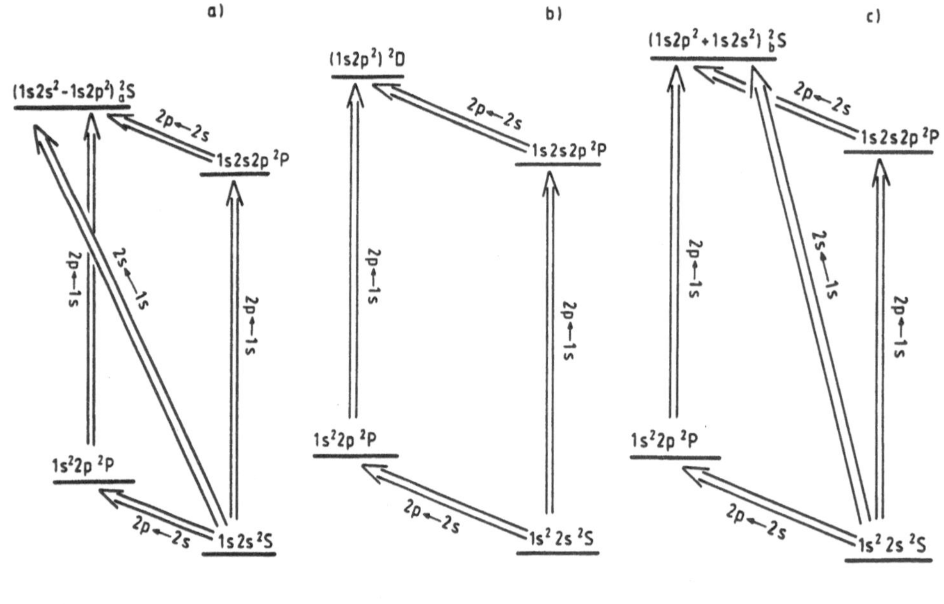

Figure 5: *Diagram involving multi-configuration final states in transitions populating doubly excited states in Li-like atoms. The first component in the final multi-configuration states is supposed to be the dominant component. The diagrams a), b), and c) involve destructive, none, and constructive interferences, respectively.*

Figure 5 shows diagrams for Li-like systems which involve multi configurations for the final states. It is important to point out that a Z_p^3 term may not only be produced when different mechanisms interfer (e.g. TS-1 and TS-2) but also when the corresponding components occurring in the final state depend differently on Z_p. It was noted above that the amplitudes for the production of the components $1s2s^2\ ^2S$ and $1s2p^2\ ^2S$ are in lowest order proportional to Z_p and Z_p^2, respectively. When the transition probability is evaluated the contributions for the different components are added coherently and, hence, the Z_p^3 term is created. The remarkable feature of Li double excitation is that at least three final states exist which exhibit destructive, none, and constructive interference (Figure 5) involving the same Z_p^3 term. Thus, a single Li autoionization spectrum may yield the same information which is otherwise achieved by using projectiles of opposite charges[9].

The analysis of the experimental data[37] is still preliminary but it may be safe to say that interference effects are visible. Further work is needed to interpret the experimental data. As discussed in detail in the previous section we would expect within the IPM-FO that interference effects cancel since both associated paths are open. Similar conclusions have been drawn by Macek[38] and McGuire and Straton[39]. Hence, for future studies it appears as a challenge to verify double excitation by means of interference effects including a breakdown of the frozen-orbital approach and scattering correlation treated in third or higher order.

Acknowledgement

I am grateful to Dr. V. Montemayor, Dr. L. Cocke, Dr. J. McGuire, Dr. J. Macek, Dr. S. Datz, and Dr. W. von Oertzen for helpful communications.

References and Footnotes

+Work performed also at the Hahn-Meitner-Institut Berlin, D-1000 Berlin 39, FRG

1. J.H. McGuire, Phys. Rev. Lett. **49** (1982) 1153, J.H. McGuire, Nucl. Instr. Meth. B10 (1985) 17, J.H. McGuire, Phys. Rev. **A 36** (1987) 1114 and J.H. McGuire, **High-Energy Ion-Atom Collisions**, Lecture Notes in Physics **294**, ed. by D. Berényi and G. Hock (Springer Verlag, Heidelberg, 1988) p. 415
2. J.F. Reading and A.L. Ford, **Electronic and Atomic Collisions**, Invited Papers, edited by H.B. Gilbody, W.R. Newell, F.H. Read and A.C. Smith (North Holland, Amsterdam,1988) p. 693
3. N. Stolterfoht, **Spectroscopy and Collisions of Few-Electron Ions**, edited by M. Ivascu, V. Florescu, and V. Zoran (World Scientific, Singapore, New Jersey, London, 1989) p. 342
4. J.A. Tanis, E.M. Bernstein, M. Clark, W.G. Graham, R.H. McFarland, T.J. Morgan, B.M. Johnson, K.W. Jones, and M. Meron, Phys. Rev. Lett. **31** (1985) 4040 and J.A. Tanis, E.M. Bernstein, M.W. Clark, W.G. Graham, R.H. McFarland, T.J. Morgan, A. Müller, M.P. Stöckli, K.H. Berkner, P. Gohil, A.S. Schlachter, J.W. Stearns, B.M.Johnson, K.W. Jones, M. Meron, and J. Nason, **Electronic and Atomic Collisions**, Invited Papers, edited by D.C. Lorents, W.E. Meyerhof and J.R. Peterson, (North Holland, 1985) p. 425
5. T.J.M. Zouros, D. Schneider, and N. Stolterfoht, Nucl. Instr. Meth. **B31** (1988) 349
6. W. Fritsch and C. D. Lin, Phys. Rev. Lett **61** (1988) 690
7. H. Knudsen, L.H. Andersen, P. Hvelplund, G. Astner, H. Cederquist, H. Danared, L. Liljeby, and K.G. Rensfelt, J. Phys. B **17** (1984) 3545, H. Knudsen, L.H. Andersen, P. Hvelplund, J. Sφrenson, and D. Ciric, J. Phys. B **20** (1987) L253, and references therein
8. A.K.Edwards, R.M. Woods, A.S Beard, and R.L. Ezell, Phys. Rev. **A37** (1988) 3697
9. L.H. Andersen, P. Hvelplund, H. Knudsen, S.P. Mφller, K. Elsener, K.G. Rensfelt, and E. Uggeerhφj, Phys. Rev. Lett. **57** (1986) 2147 and L.H. Andersen, P. Hvelplund, H. Knudsen, S.P. Mφller, A.H. Sφrensen, K. Elsner, K.G. Rensfeld, and E. Uggerhφj, Phys. Rev. A **36** (1987) 3612

10. M. Prost, M. Morgenstern, D. Schneider, and N. Stolterfoht, Proceedings of the Xth Conference of the Physics of Electronic and Atomic Collisions, Abstract of Papers, edited by (Commissariat de Paris, 1977) p 994 and M. Prost, Diplomarbeit, Berlin, 1977, unpublished
11. P.W. Arcuni and D. Schneider, Phys. Rev. A **36** (1987) 3059
12. J.O.P. Pedersen and P. Hvelplund, Phys. Rev. Lett. **62** (1989) 2373
13. J.P. Giese, M. Schultz, J.K. Swenson, H. Schöne, M. Benhenni, S.L. Varghese, C.R. Vane, P.F. Dittner, S.M. Shafroth, and S. Datz, Phys. Rev. A **42** (1990) 1231
14. R.E. Olson, Phys. Rev. A **36**, (1987) 1519
15. L. Végh, Phys. Rev. A 37 (1988) 992
16. J.H. McGuire and J. Burgdörfer, Phys. Rev. **A36** (1987) 4089
17. E.U. Condon and C.H. Shortley, **The Theory of Atomic Spectra** (University Press, Cambridge, 1964)
18. N. Stolterfoht, Physica Scripta **42** (1990) 192
19. N. Stolterfoht, Nucl. Instrum. Methods, to be published
20. H.S.W. Massey and D.R. Bates, Rept. Prog. Phys. **9** (1942) 62
21. J.H. McGuire and J.C. Straton, **The Physics of Electronic and Atomic Collisions**, Invited Papers, edited by A. Dalgarno, R.S. Freund, P.M. Koch, M.S. Lubell, and T.B. Lucatorto (American Institute of Physics, New York,1990) p. 280
22. J.F. Reading and A.L. Ford, **Comments on Atomic and Molecular Physics**, Vol **23** (1990) 301
23. M.R.C. McDowell and J.P. Coleman, **Introduction to the Theory of Ion-Atom Collisions** (North Holland, Amsterdam, 1970) Chap. 4
24. A. Messiah, **Quantum Mechanics**, Vol II (North Holland, Amsterdam, 1970) p. 725
25. U. Fano, Atomic Physics (Plenum Press, New York, 1969) p. 209
26. W. Mehlhorn, **Atomic Inner Shell Processes**, ed. B. Crasemann (Plenum Press, New York, 1985) p. 119
27. N. Stolterfoht, Phys. Reports **146** (1987) 215
28. N. Stolterfoht, P.D. Miller, H.F. Krause, Y. Yamazaki, J.K. Swenson, R. Bruch, P.F. Dittner, P.L. Pepmiller, and S. Datz, Nucl. Instr. Meth. B **24/25** (1987) p. 168
29. D. Schneider, N. Stolterfoht, A. Itoh, G. Schiwietz, W. Zeitz, and U. Wille, **Electronic and Atomic Collisions,** Invited Papers, Eds. D.C. Lorents, W.E. Meyerhof, and J.R. Peterson (North Holland, Amsterdam 1986) p. 671
30. A. Mattis, D. Schneider, G. Schiwietz, T. Schneider, M.G. Menendez, E. Smola, J. Tanis, and N. Stolterfoht, to be published
31. B. Brendlé, R. Gayet, J.P. Rozet, and K. Wohrer, Phys. Rev. Lett. **54** (1985) 2007
32. R.L. Becker, A.L. Ford, and J.F. Reading, J. Phys. B. **13** (1980) 4059 and R.L. Becker, private communication
33. J.H. McGuire and N.C. Deb, in **Atomic Physics with Positrons**, edited by E.A.G. Armour and J.W. Humbentson, NATO Advanced Study Institutes, Ser. B, Vol. **169** (Plenum, New York,1988)
34. A. Bordenave-Montesquieu, A. Gleizes, and P. Benoit-Cattin, Phys. Rev. A **25** (1982) 245
35. T. Ishihara, private communication (1990)
36. P. Ziem, R. Bruch, and N. Stolterfoht, J. Phys. B **8** (1975) L480
37. N. Stolterfoht, X. Husson, D. Lecler, R. Köhrbrück, B. Skogvall, S. Andriamonje, and J.P. Grandin, to be published
38. J. Macek, private communication (1989)
39. J.H. McGuire and J.C. Straton, private communication (1990)

Electron Correlation in High Energy Ion-Atom Collisions

A. L. Ford and J. F. Reading
Physics Department, Texas A&M University
College Station, TX 77843 USA

ABSTRACT

Due to the strong electron-nucleus force in the target atom, many atomic collision processes are well-described within the independent particle model. Even in such cases Pauli-blocking correlations can be important for inclusive cross sections. This talk focuses on ion-atom collisions where dynamical correlations due to interaction between the electrons during the collision play an important role. Such correlations can be important for multielectron transitions in weak coupling cases (high collision energies), since to first order in the projectile interaction and without correlation the cross section is zero. One example to be discussed is the double ionization of helium, where one manifestation of the correlation is the large difference in double ionization produced by protons and antiprotons. For this system both differential and total cross sections will be discussed, as well as the related processes of double excitation and of excitation plus ionization. Another class of cross sections to be discussed are those involving charge transfer: double capture and transfer ionization.

I. INTRODUCTION

Electron correlation effects in ion-atom collisions have received much attention recently. In this paper we will review some of the concepts and results, and point to some of the remaining problems and possible future directions. Several other recent reviews of this subject are available [1-4]. The emphasis given here reflects the directions our own work has taken, but we hope the specific processes we discuss are illustrative of the general features of electron correlation in collisions.

In atomic structure calculations it is straightforward to define correlation as that part of the electron-electron interaction that is not included in Hartree-Fock (HF). For example, the correlation energy in an atom is the difference between the exact total energy and the HF energy. A similar definition for collisions is not so simple, since time-dependent Hartree-Fock (TDHF) has conceptual problems (nonorthogonality of initial and final states and the difficulty of extracting transition amplitudes), is not simple to implement (calculations are tedious), and is not widely used. We will loosely define correlation as being the difference between an exact collision calculation and one done in the independent particle model (IPM). This definition is not precise since many different choices can be made for the single-electron potential used in the IPM. We will also in our definition of correlation use static one-electron potentials in the reference IPM, the same IPM potential in the initial and final channels. In this point of view, shakeoff for example is a correlation effect, even though it arises from a different single particle potential in the final versus the initial state.

Many processes in energetic ion-atom collisions can be described accurately in the IPM, in which dynamical correlation of the electrons is neglected. One class of collision processes where correlation does seem to play an important role is where two or more electrons undergo a transition, particularly in collision systems with small nuclear charges so that the electron-nucleus force does not dominate over the electron-electron force. We will concentrate attention on two-electron systems (helium targets) since here correlation is particularly strong, and there is recent interesting experimental data. There are in fact some large effects (proton/antiproton double ionization differences of a factor of two) that would be entirely absent without correlations.

II. PAULI CORRELATIONS IN THE IPM

In the IPM the collision problem for a bare ion incident on a multi-electron target atom reduces to a set of single-electron problems, the solution of which gives single-electron transition amplitudes. These amplitudes can then be combined to give cross sections for transitions between specific many-electron initial and final states. In the IPM electron correlation plays no role in the collision dynamics, in the sense that only transition amplitudes from uncoupled single-electron equations are calculated. The requirement of antisymmetry of the IPM many-electron wavefunction does introduce "Pauli correlations" among transition amplitudes [5,6]. For example, consider the probability for producing vacancies in orbitals 1 and 2 of the target. If a_{1k} is the transition amplitude for an electron initially in orbital k being scattered into orbital 1, the probability for producing a state with orbital 1 unoccupied is

$$\rho_1 = \sum_{k>N} |a_{k1}|^2 = 1 - \sum_{k=1}^{N} |a_{1k}|^2. \tag{1}$$

The first sum if over all orbitals initially unoccupied and the second sum is over all initially occupied target orbitals. The first expression is simply the probability of scattering the electron initially in orbital 1 to any unoccupied orbital. The second is the probability of *not* scattering an electron from any initially occupied orbital into orbital 1, including elastic scattering. The two expressions are equal by the unitarity of the transition amplitudes. The probability ρ_1 is an inclusive probability. All that is required of the final state is that orbital 1 be unoccupied. All other orbital occupancies are summed over.

The probability ρ_{12} for producing vacancies in orbitals 1 and 2 of the target is given by [5,6]

$$\rho_{12} = \rho_1 \rho_2 - |\sum_{k=1}^{N} a_{1k} a_{2k}^*|^2. \tag{2}$$

The first term is the product of probabilities for independently creating each hole. The second term subtracts from this and may be thought of as the probability of scattering particle 2 into hole 1, or vice versa, through an intermediate state k. The Pauli correlation produced by antisymmetry destroys the independence of probability for this process. Also note that the relative phase between a_{1k} and a_{2k} is important.

The effect of these Pauli correlations has been explored in actual calculations for two cases. One is multiple L-shell vacancy production, where Pauli correlation has the potential for altering the binomial distribution of multiple vacancies [6]. In actual calculations, however, it was found that the Pauli correlation effects were generally small, apparently because of the tendency toward random phases of the amplitudes. The second case is the probability (or cross section) for capture of an electron by the projectile in coincidence with production of a K-shell vacancy in the target [7]. Here cases were found (large projectile charge and low collision energy, so that the transition amplitudes are large) where Pauli correlation corrctions were quite large, but to date there have been no definitive experimental tests of these predicitons.

Note that for spin-independent Coulomb excitation such as produced by ion impact, if the orbitals 1 and 2 are of opposite spin (for example two K-shell orbitals), then when a_{1k} is nonzero a_{2k} must be zero. Thus in this case we recover

$$\rho_{12} = \rho_1 \rho_2, \tag{3}$$

the probability of simultaneously producing both holes is just the product of probabilities for independently producing each hole or vacancy. Thus for the double ionization (or other two-electron transition) of a two-electron target atom or ion initially in the $1s^2$ ground state, the double ionization probability $P_d(b)$ in the IPM is the square of the single ionization probability $P_s(b)$. (We use the language of the semiclassical impact parameter method.) If the single ionization probability is given accurately by first Born, as it is for protons on helium above 1 MeV/u, it scales as Z_p^2, where Z_p is the charge of the bare projectile ion. Then in the IPM $P_d(b)$ scales as Z_p^4 and the ratio R of double to single ionization probabilities, and cross sections, scales as Z_p^2. As will be discussed in more detail in the next section, experiment and subsequent calculations found that the double ionization was markedly different for protons versus electron (and later antiproton) projectiles, in sharp contradiction to the IPM prediction.

III. COLLISION INDUCED TWO-ELECTRON PROCESSES IN HELIUM

We will first consider double ionization of helium by bare projectiles, in the collision energy region from about 0.20 MeV/u to 20 MeV/u. At these collision energies the projectile velocity is much larger than the helium Bohr velocity. Also, the single ionization is given accurately by first Born, in which the projectile interacts only once, with the target electron being ionized. Inclusion of electron correlation in the initial and final state wavefunctions is needed if accurate single ionization cross sections are to be calculated, and this is easily done using configuration-interaction (CI) wavefunctions for the ground state and for the final state continuum pseudostates. But correlation is not essential for single ionization to occur in first Born, and in fact reasonable results can be obtained in the IPM with a judiciously chosen single-electron potential.

The situation for double ionization is quite different. First, experiments [8] with protons, alpha particles, etc. show that above 10 MeV/u the cross section scales as Z_p^2 not as the Z_p^4 of the IPM discussed in the previous section. But this is easy to understand. For a two-electron transition such as double ionization there must be two interactions, one to excite each electron. Two different interactions are available: the projectile-electron interaction and the electron-electron interaction. As the collision energy is increased the projectile-electron interaction becomes less effective. Hence the dominant mechanism at high collision velocities is ionization of one electron by interaction with the projectile and ionization of the other electron through the electron-electron force. This is first Born with correlated wavefunctions, and as it is first Born the cross section scales as Z_p^2. Note that first Born without correlation gives zero for double ionization, or for any other two-electron transition.

As the collision energy is lowered the mechanism where the projectile interacts twice, once with each electron, becomes competitive. This is second Born and the cross section scales as Z_p^4. Correlation is not required for two-electron transitions to occur in second Born, but is needed to calculate the cross section accurately. Experiment [8] shows strong deviation from Z_p^2 as the collision energy is lowered below 10 MeV/u. (In fact, very recent experiments by Watson's group show substantial deviations from Z_p^2 behavior even above 10 MeV/u, for N^{7+} projectiles [9].)

A second, even more interesting feature of the helium double ionization in this energy region is that projectiles with $Z_p = -1$ give about twice the double ionization of projectiles with $Z_p = +1$. This was first noted by McGuire [10] and by Haugen [11] by comparing proton and equivelocity electron projectiles. That it is a projectile charge rather than mass effect has been confirmed by experiments with antiprotons [12]. As first suggested by McGuire [10] and confirmed by calculations we have done [13], the proton/antiproton difference arises from interaction between first and second Born amplitudes, that gives rise to a Z_p^3 term in the cross section. But remember that this Z_p^3 term requires that correlation be present, because without correlation the first Born amplitude is zero and the leading order term in the cross section is then the Z_p^4 of the IPM. The proton/antiproton double ionization difference is a direct consequence of electron correlation during the collision.

There is one subtle point here though [14]. Single ionization is dominated (over 90%) by dipole ($s \rightarrow p$) transitions. That is, the $sp\,^1P$ final state has over 90% of the single ionization cross section in a calculation with $(ss{+}pp)^1S$, $sp\,^1P$, $pp\,^1P$, and $pp\,^1D$ basis states. But if only dipole transitions are included then a given final state cannot be reached in both first and second Born and interference cannot occur. That is, with only dipole transitions the only first Born final state symmetry is $ks\,k'p\,^1P$ and the only second Born final state symmetry is $kp\,k'p\,^1S$ or ^1D. The interference that gives rise to the Z_p^3 term and the proton/antiproton difference requires nondipole transitions.

We have published coupled-states calculations [13] of helium double ionization that include the effects of electron correlation and that give the proton/antiproton difference of the same sign and roughly the same magnitude and energy dependence as observed experimentally. These calculations required the invention and implementation of two new theoretical methods. One is our forced-impulse method (FIM) for including dynamical electron correlation during the collision without the unacceptable computational expense of direct expansion in two-electron pseudostates. In the FIM the electron wavefunctions evolve independently (IPM) for short time steps; at the start of each of these time evolution steps the wavefunction is projected onto fully correlated states. The other new method, required for calculating double ionization, is projection of our continuum pseudostates onto single and double ionization channel states, constructed to satisfy the appropriate boundary conditions. This is necessary for solving the interpolation problem, of separating between the two overlapping single and double ionization continua.

Helium double ionization has also been obtained by Olson using a Monte Carlo classical calculation [15], and he also finds a proton/antiproton difference similar to what is measured. This is a very interesting result, since in our calculations the difference is due to an interference between quantum mechanical amplitudes. It is difficult to see how such an effect can be obtained classically. However, Olson's calculations have been reported only up to a collision energy of 1 MeV; only at somewhat higher energies are our calculations described simply by first Born plus interference between first and second Born. At 1 MeV and below higher Born terms enter as well. It would be of great interest to see what the classical Monte Carlo calculations give above 1 MeV.

The double ionization has also been modeled by Végh [16]. The essence of his model is as follows. Due to their mutual repulsion the two electrons of the helium atom favor positions on opposite sides of the nucleus. During the collision the proton projectile attracts the near-side electron stronger than the far-side one, which, due to the electron-electron interaction, moves away from the projectile. For antiprotons, due to the repulsive interaction, the near-side electron moves away and the far-side electron comes near the projectile. Thus the ejection probability of the far-side electron depends on the sign of the projectile charge. This effect is built into the framework of the IPM by a shift in the impact parameter of the projectile-far-side electron collision. For protons the impact parameter is increased by $\Delta B = 0.8 < r >$, and for antiprotons it is decreased by this amount. This calculation also gives the proton/antiproton difference in qualitative agreement with experiment. Very recently Végh and Burgdörfer [17] have published an analysis of what we have called here the first and second Born terms (in the projectile-electron interaction). Without any actual calculation of matrix elements and transition amplitudes their formal analysis shows that the first Born–second Born interference is constructive for $Z_p < 0$ and destrucitve for $Z_p > 0$, so double ionization is enhanced in the first case and reduced in the second, in agreement with what is seen experimentally.

An analysis of the impact parameter dependent probabilities $P(b)$ for single and double ionization that go into the calculation of our FIM cross sections described above has just been published [18]. We find that the double to single ionization ratio and the proton-antiproton difference is much larger in $P(b)$ at small impact parameter b than observed previously in the cross sections. The mechanisms for producing double ionization and that give rise to the proton-antiproton difference are most effective at small impact parameters. Experiments that probe small b would therefore be of great interest, particularly for antiprotons or bare projectile ions of larger charge, such as He^{2+} or C^{6+}. One such experiment is the double ionization cross section differential in the projectile scattering angle, $d\sigma/d\Omega$, and such measurements have in fact been done using proton projectiles [19,20]. However, our straight-line trajectory, classical path calculations cannot be converted directly to a cross section differential in projectile scattering angle using the classical relation between impact parameter and scattering angle for Coulomb scattering. The difficulty is that scattering of the projectile by close collisions with the target electrons, which results in the electrons being ejected with large kinetic energy, appears to have an important effect on $d\sigma/d\Omega$ for both single and double ionization [21,22]. It may be nonetheless that at very large scattering angles, well beyond the angle to which a proton can be scattered classically by a free electron at rest, the scattering of the projectile is dominated by the Coulomb interaction with the helium nucleus. If this is indeed the case, then the He^{2+}/He^{+} charge-state fraction observed in coincidence with these large projectile scattering angles (greater than 3 mrad) should agree with our calculated ratio of double to single ionization probabilities at small impact parameters. There is indeed fair agreement between our calculations and experiments. At $E = 1.0$ MeV the experimental ratio [19] is $1.3 \pm 0.2\%$; our calculation gives 0.9%. At $E = 0.30$ MeV the experimental result is $2.4 \pm 0.2\%$; our calculation gives 1.8% The reason for our calculations being somewhat low is probably the truncation of our basis to s and p states only. Similar measurements using antiprotons would be of considerable interest. Our calculations predict much larger values for this ratio for antiproton projectiles: 3.3% at $E = 1.0$ MeV and 5.5% at $E = 0.3$ MeV. Also of interest would be similar measurements for projectiles of larger charge; we predict a double to single ratio at small b (large projectile scattering angle) of 74% for C^{6+} at 0.64 MeV/u. Another possible experiment to probe the same effect would be a collision between neutral helium and a partially stripped ion, such as nitrogen, oxygen or neon. Double ionization of the helium in coincidence with electron removal from the projectile ion would select out small impact parameters, of the order of the orbital radius for the ion. Our calculations suggest substantial enhancement of the double to single ratio compared with that measured without the simultaneous removal of a projectile electron, if the projectile ion is stripped down to orbitals with radius much smaller than the helium Bohr radius.

A very interesting feature of the double ionization $d\sigma/d\Omega$ measurements of Giese and Horsdal [19], for 0.3 to 1.0 MeV protons, is that the double to single ratio as a function of projectile scattering angle shows a pronounced peak at about 0.9 mrad. The structure is undoubtedly due to the fact that the projectile can be scattered both by the target helium nucleus and by the target electrons. As noted above, scattering from the target nucleus must dominate the large angle scattering. But at angles near and below the maximum scattering angle (0.54 mrad) for a proton off a free electron at rest, scattering by the target electrons dominates. Such hard hits of the electron, that appreciably scatter the projectile, produce very energetic ionized electrons. If one electron is ionized in such a hard collision the other electron can be removed by the electron-electron interaction in a shakeoff-type process. Two hard hits, one with each electron, are also possible. And in fact classical trajectory Monte Carlo (CTMC) calculations of Olson et al [22] find this two-hard-hit mechanism to be responsible for the peak. Both electrons are removed by interaction with the projectile, so electron-electron interaction (correlation) is not involked. This result is compelling in that the peak is observed at about twice the maximum angle that the projectile can be deflected by one electron. The CTMC calculations give a peak for 1 MeV at about the same position and same shape as found experimentally. But the absolute value of the double to single ratio as calculated is about a factor of three larger than the experiment. Also, the CTMC results have been reported only for 1 MeV; experimental data exists for both lower [19] and higher [20] collison energies.

We have published a preliminary model calculation [21] that also shows a peak, allbeit at a somewhat smaller scattering angle than in the experiment. The calculation considered two incoherent double ionization mechanisms. One is scattering of the projectile from the nucleus, that produces two slow ionized electrons. The second is the scattering of the projectile by a hard collision with an electron, that produces one energetic ionized electron and one slow one via the electron-electron interaction (shakeoff). Both mechanisms rely at least in part on electron correlation to ionize one of the electrons. These processes are incoherent since they produce ionized electrons of different energies. The scattering from the nucleus dominates for both large and small angles; scattering from a target electron is important only around 0.5 mrad but produces a higher fraction of double ionization, hence the peak. This preliminary calculation did not include the two-hard-hit mechanism found to be important by Olson et al. And of course in any given collision the projectile is simultaneously deflected both by the electrons and by the nucleus. Further calculations are in progress. It is not simple to incorporate projectile deflection by the electrons into our semiclassical FIM calculations.

Still another mechanism for producing the 0.9 mrad peak, a Thomas type multiple scattering, has been suggested by Végh [23]. Further insight into the complicated nature of the scattering is given in the single ionization study of Dörner et al [24]. Calculations (CTMC) and experiment show that coupling between the electronic and nuclear degrees of freedom is essential to understand the collision dynamics even in this simpler case. Single and double ionization $d\sigma/d\Omega$ measurements using antiprotons or bare ions of larger charge would be very helpful in further understanding the complex dynamics of these collisions, as also would measurement of the energy and angular distributions of the ionized electrons.

There are of course other two-electron transitions beside double ionization. One is excitation ionization, in which one electron is ionized while the other is excited but remains in a bound state. Very recent measurements of this process in helium for electrons and protons have been carried out by Pedersen and Folkmann [25]. For collision energies between 0.92 MeV/u and 3.76 MeV/u they find that the electron cross section is about a factor of two larger than the cross section for protons. This result is very similar to the double ionization case, as one would intuitively expect. We know of no quantal calculation of this process at these energies, but the origin of the factor of two difference is undoubtedly the same as for double ionization: interference between first and second Born amplitudes. And again the first Born amplitude requires an electron-electron correlation interaction to be nonzero, so the proton versus electron difference is a result of correlation.

On the other hand, recent measurements of the two-electron transition to doubly excited resonance states of helium show no conclusive $Z_p = +1$ versus $Z_p = -1$ (electrons) difference. This result is quite surprising, since one would expect that this two-electron transition would behave qualitatively like the other two-electron transitions we have considered, namely double ionization and excitation ionization, where there is about a factor of two difference for the two projectiles. Pedersen and Hvelplund [26] measured the autoionization electron yields from the unresolved $(2s2p\,^1P + (2p)^2\,^1D)$ pair of states excited by 1.84 MeV/u protons and electrons. Giese et al [27] made similar measurements at 1.5 MeV/u, and resolved the individual $2s^2\,^1S$, $2p^2\,^1D$ and $2s2p\,^1P$ resonances. Preliminary calculations we have performed in the FIM agree qualitatively with these experiments in that they show little difference between $Z_p = +1$ and $Z_p = -1$. The coupled-channels calculations of Fritsch and Lin [28] also find little difference for exciting the ^1S and ^1P resonances, but they do find protons to be nearly twice as effective as electrons at exciting the ^1D state. Our cross sections calculated for specific projectile charge Z_p can be fit to a power series in Z_p, so that the various terms in the Born series can be identified. When we do this for our double excitation calculations at the experimental energy we find both Z_p^2 and Z_p^4 terms, corresponding to first and second Born, but a very small Z_p^3 interference term, hence the small $Z_p = +1$ versus $Z_p = -1$ difference. Both first and second Born amplitudes are of appreciable magnitude, but their interference term is very small. Model calculations by Pedersen [29] suggest that the lack of interference is due to different impact parameter dependence of the first and second Born amplitudes. He finds that the first Born amplitude is dominated by large impact parameters, whereas the second Born is dominated by small impact parameters. These model calculations have also been done for the double ionization [12] and excitation ionization [25] cases as well, and give a good account of the first and second Born contributions in all three cases. But the model is unable to calculate the Z_p^3 interference term. The model divides the collision into close and distant encounters. The distant collison contribution is then evaluated by the Weizsäcker-Williams method of virtual quanta, and the close collision contribution is calculated using a classical picture for collisions between free particles. More theoretical work is needed before quantative comparison can be made between collision calculations and the experiments, that measure the yield of ionized and autoionized electrons. The resonance states in question are embedded in the single ionization continuum, so there is a similar interpretation problem as that due to the overlapping single and double ionization continua.

There is another class of two-electron transitions in helium, those that involve at least one electron being captured into a bound state on the projectile. We have ourselves not yet done any calculations that include both electron correlation and rearrangement channels, so will just comment briefly on some of the recent experiments and theoretical work in this area. Correlation can play an important role in these processes.

One such process is transfer ionization, where one electron is captured and the other is ionized [30,31]. The relevant experimental quantity is R^*, the ratio of transfer ionization to total capture. As for double ionization there are two obvious mechanisms [31]. One is independent capture of one electron and ionization of the other due to the electron-projectile interaction. The second is capture of one electron and then ionization of the other due to the electron-electron interaction. In the literature the ratios due to these two mechanisms are denoted R^*_D and R^*_R, so

$$R^* = R^*_D + R^*_R. \tag{4}$$

At high collision energies the captured electron is guaranteed to leave with large velocity relative to the target, since it has the velocity of the projectile. Thus one expects a shakeoff mechanism to operate due to the change in screening seen by the second electron. This mechanism is expected to produce a value of R^*_R equal to the value of R_γ, the ratio of double to single ionization produced by high energy photons. In photoabsorption all the photon energy is given predominantly to one electron, which is then energetically ionized when the photon has high energy. The other ratio R^*_D, since it comes from an IPM process, can be calculated in terms of a simple product of independent probabilities for capture and ionization. The theoretical picture is complicated by the fact that at asymptotically high collision velocities the second Born rather than first Born dominates for capture, and one second Born mechanism involves scattering of the electron to be captured off the second electron, which is therefore invariably ionized. The experimental situation is that this model (eq.(4)) gives R^* in good agreement with experiment for He^{2+}, Li^{3+}, and O^{8+} projectiles, but overestimates R^* for H^+. For the former set of larger Z_p projectiles the R^*_D term dominates at energies where data is available whereas for H^+ both terms are comparable. For H^+ the model predicts that $R^*_R = R_\gamma$ dominates over R^*_D above about 1 MeV/u, and the available data at these energies agree with this. But the data continues to match R^*_R at lower energies, while the model predicts that R^* rises above this due to R^*_D. As noted by McGuire et al [31] one possible reason for the discrepancy for H^+ is that the simple model calculations do not allow for the possibility of interference between the R^*_D and R^*_R mechanisms. As these authors further note, what is needed is a calculation that treats all mechanisms at the same time and in a uniform way, rather than adding independent calculations.

A related experiment is the measurement by Pálinkás et al [32] of the ionized electron energy and angular distribution in the transfer ionization of helium by 1 MeV protons. Forward peaks are found and attributed to the mechanism where the proton interacts independently with each electron, capturing one and ionizing the other. Peaks near 90° from the beam direction are ascribed to a two-step process of proton-electron scattering in the first step and a localized interaction of electron-electron scattering (correlation) in a second step, in which one of the electrons is deflected into a bound state on the projectile while the other electron is ionized. Theoretical calculations by Briggs and Taulbjerg [33] and by Ishihara and McGuire [34] give the near 90° peaks seen in the experimental angular distributions, but there is some disagreement in absolute magnitude of the differential cross sections.

Finally, there is a large literature on simultaneous transfer and excitation, in collisions where the projectile enters carrying one or more electrons. In the process one electron is captured from the target and the original projectile electron is excited. See for example the recent calculations by Fritsch and Lin [35] for $He^+ + H$, that use a coupled-states expansion that includes two-electron correlated states. There are also several experiments that give evidence for strong electron correlation effects in double capture from helium [36].

IV. ACKNOWLEDGEMENTS

This work was supported by NSF Grant PHY-9009717.

REFERENCES

1. Stolterfoht N.: Spectroscopy and Collisions of Few-Electron Ions, SCOFEI'88 (Ed. by M. Ivascu, V. Florescu, V. Zoran), p.342, World Scientific (1989); Phys. Scr. **42**, 192 (1990)
2. McGuire J.: Phys. Rev. A**36**, 1114 (1987); Adv. At. Mol. Phys., in press
3. Pedersen J.O.P.: Phys. Scr. **42**, 180 (1990)
4. Reading J.F., Ford A.L.: Comments At. Mol. Phys. **23**, 301 (1990)
5. Reading J.F., Ford A.L.: Phys. Rev. A**21**, 124 (1980)
6. Becker R.L., Ford A.L., Reading J.F.: Phys. Rev. A**29**, 3111 (1984)
7. Becker R.L., Ford A.L., Reading J.F.: J. Phys. B **13**, 4059 (1980); Ford A.L., Reading J.F., Becker R.L.: Phys. Rev. A**23**, 510 (1981)
8. Knudsen H., Andersen L.H., Hvelplund P., Astner G., Cederquist H., Danared H., Liljeby L., Rensfelt K.-G.: J. Phys. B **17**, 3545 (1984); Shah M.B., Gilbody H.B.: J. Phys. B **18**, 899 (1985)
9. Heber O., Bandong B.B., Sampoll G., Watson R.L.: Phys. Rev. Lett. **64**, 851 (1990)
10. McGuire J.H.: Phys. Rev. Lett. **49**, 1153 (1982)
11. Haugen H.K., Andersen L.H., Hvelplund P., Knudsen H.: Phys. Rev. A**26**, 1962 (1982)
12. Andersen L.H., Hvelplund P., Knudsen H., Møller S.P., Sørensen A.H., Elsener K., Rensfelt K.-G., Uggerhøj E.: Phys. Rev. A**36**, 3612 (1987); Andersen L.H., Hvelplund P., Knudsen H., Møller S.P., Pedersen J.O.P., Tang-Petersen S., Uggerhøj E., Elsener K., Morenzoni E.: Phys. Rev. A**40**, 7366 (1989)
13. Reading J.F., Ford A.L.: J. Phys. B **20**, 3747 (1987); Ford A.L., Reading J. F.: J. Phys. B **21** L685 (1988)
14. Becker R.L.: private communication (1983)
15. Olson R.E.: Phys. Rev. A**36**, 1519 (1987)
16. Végh L.: Phys. Rev. A**37**, 992 (1988)
17. Végh L., Burgdörfer J.: Phys. Rev. A**42**, 655 (1990)
18. Ford A.L., Reading J.F.: J. Phys. B **23**, 2567 (1990)
19. Giese J.P., Horsdal E.: Phys. Rev. Lett. **60**, 2018 (1988)

20. Kamber E.Y., Cocke C.L., Cheng S., Varghese S.L.: Phys. Rev. Lett. **60**, 2026 (1988)
21. Reading J.F., Ford A.L., Fang X.: Phys. Rev. Lett. **62**, 245 (1989)
22. Olson R.E., Ullrich J., Dörner R, Schmidt-Böcking H.: Phys. Rev. A**40**, 2843 (1989)
23. Végh L.: J. Phys. B. **22**, L35 (1989)
24. Dörner R, Ullrich J., Schmidt-Böcking H., Olson R.E.: Phys. Rev. Lett. **63**, 147 (1989)
25. Pedersen J.O.P., Folkmann F.: J. Phys. B **23**, 441 (1990)
26. Pedersen J.O.P., Hvelplund P.: Phys. Rev. Lett. **62**, 2373 (1989)
27. Giese J.P., Schulz M., Swenson J.K., Schöne H., Benhenni M., Varghese S.L., Vane C.R., Dittner P.F., Shafroth S.M., Datz S.: Phys. Rev. A**42**, 1231 (1990)
28. Fritsch W., Lin C.D.: Phys. Rev. **41**, 4776 (1990)
29. Pedersen J.O.P.: Phys. Scr. **42**, 180 (1990)
30. Knudsen H., Andersen L.H., Hvelplund P., Sørensen J., Ćirić D.: J. Phys. B **20**, L253 (1987)
31. McGuire J.H., Salzborn E., Müller A.: Phys. Rev. A**35**, 3265 (1987)
32. Pálinkás J., Schuch R., Cederquist H., Gustafsson O.: Phys. Scr. **42**, 175 (1990)
33. Briggs J., Taulbjerg K.: J. Phys. B **12**, 2565 (1979)
34. Ishihara T., McGuire J.M.: Phys. Rev. A**38**, 3311 (1988)
35. Fritsch W., Lin C.D.: Phys. Rev. Lett. **61**, 690 (1988)
36. Andersen L.H., Frost M., Hvelplund P., Knudsen H., Datz S.: Phys. Rev. Lett. **52**, 518 (1984); Stolterfoht N., Havener C.C., Phaneuf R.A., Swenson J.K., Shafroth S.M., Meyer F.W.: Phys. Rev. Lett. **57**, 74 (1986); Tanis J.A., Schiwietz G., Schneider D., Stolterfoht N., Graham W.G., Altevogt H., Kowallik R., Mattis A., Skogvall B., Schneider T., Szmola E.: Phys. Rev. A**39**, 1571 (1989)

ORIENTATION PROPENSITY IN ELECTRON CAPTURE BY MULTIPLY CHARGED IONS

M. Barat, P. Roncin, C. Adjouri, N. Andersen+, M.N. Gaboriaud and L. Guillemot
Laboratoire des Collisions Atomiques et Moléculaires (URAD0281), Bât. 351,
Université de Paris-Sud, 91905 Orsay Cedex, FRANCE
+Physics Laboratory, H.C. Ørsted Institute, DK 2100 Copenhagen, DENMARK

I. Introduction

Orientation in atomic physics relates to a preferential sense of rotation of an active electron around the atomic core in heavy particle collisions in keV range. Orientation effect, with respect to the planar scattering plane can manifest itself in two different processes : in *excitation* (or deexcitation from an oriented initial state) or in *electron capture*. Propensity rules for orientation were first discovered for *excitation*[1]. Predictions were then confirmed both experimentally and theoretically.

To date for electron capture, orientation propensities have not so well been established. The first measurements concern the Na^+-Na(3p) resonant charge transfer below 100 eV[2]. This data shows small orientation effect. More recently this problem was studied both theoretically and experimentally in H^+-Na(3p) quasi resonant charge exchange[3,4]. However in all these cases involving singly charged ions, *direct* excitation channels are imbricated with exchange channels both being significantly populated, resulting in an intricate theoretical analysis[4]. In contrast collisions involving multiply charged ions do not suffer from this drawback since electron capture largely dominates over the other inelastic channels. We select the $B^{3+}(1s^2)$ + He, Ne systems[5] for which (i) only *two capture* channels are effective (fig 1) resulting in the $B^{2+}(1s^22s)$ and $B^{3+}(1s^22p)$ population. (ii) the *207 nm photons* emitted in the BIII 2^2S-2^2P transition is very workable and polarization analysis is possible with high efficiency (iii) with He target, the $He^+(1s)$ ion is left in a single atomic state leading to full coherence. However additional experiments were carried out with a Neon target, for which population of the B^{2+} ^2P state is dominant (contrarily to He target) as easily predicted by the over barrier model[6]. Finally during the preparation of the experiment, a theoretical analysis[7] appeared, predicting a *strong* orientation propensity in B^{3+}-He collisions.

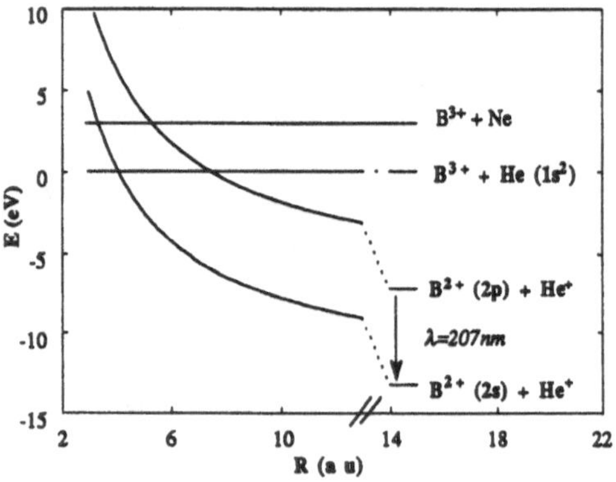

Fig 1 : Schematic potential energy curves for single electron capture in the B^{3+}+He system

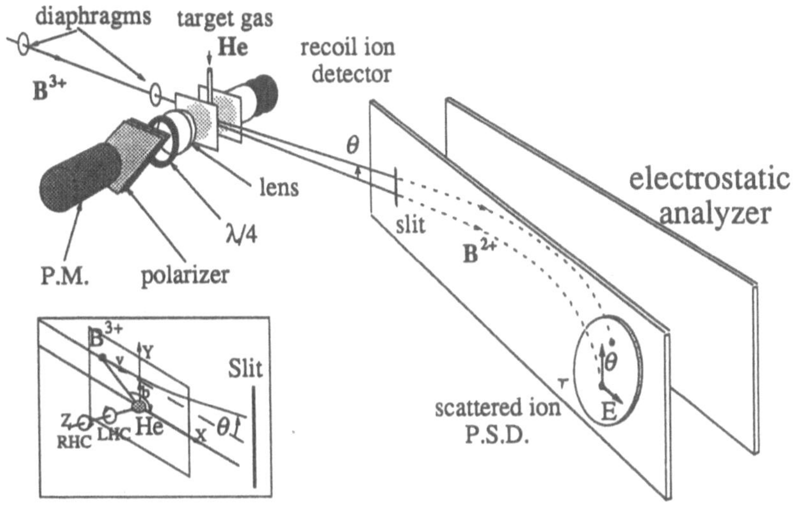

Fig 2 : Experimental set up. In insert, collision geometry used in the present experiment

II. Experimental data

The LCAM coincident energy gain-scattering angle spectrometer has been described several times (see e.g. 8). A Boron 3+ ion beam delivered by the ECR ion source of the LAGRIPPA facility in Grenoble, crosses an effusive target atomic beam before entering a parallel plate electrostatic analyzer (fig 2). The scattered ions are received on a position sensitive detector. The position on the X axis gives the energy and the charge state of the projectile and the vertical position the scattering angle θ. The recoil ions are also detected. The recoil ion charge states are determined by the time of flight difference between scattered and recoil ions. For the present experiment, a photon detection system has been placed in the opposite side of the recoil ion detector, facing the collision volume. Circular polarization is measured by a

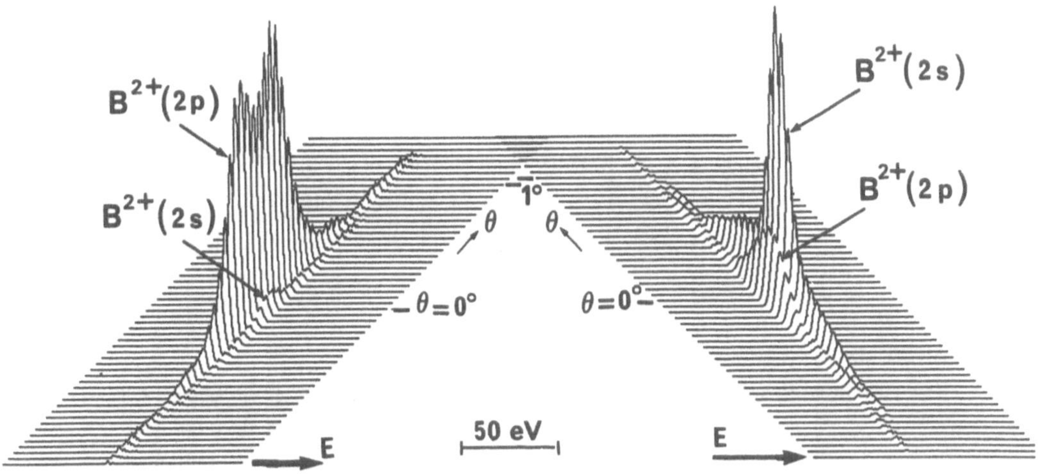

Fig 3 : Doubly differential Energy gain-angle coincident spectra of the B^{2+} ions scattered by (a) He target (b) Ne target, at 3 keV collision energy

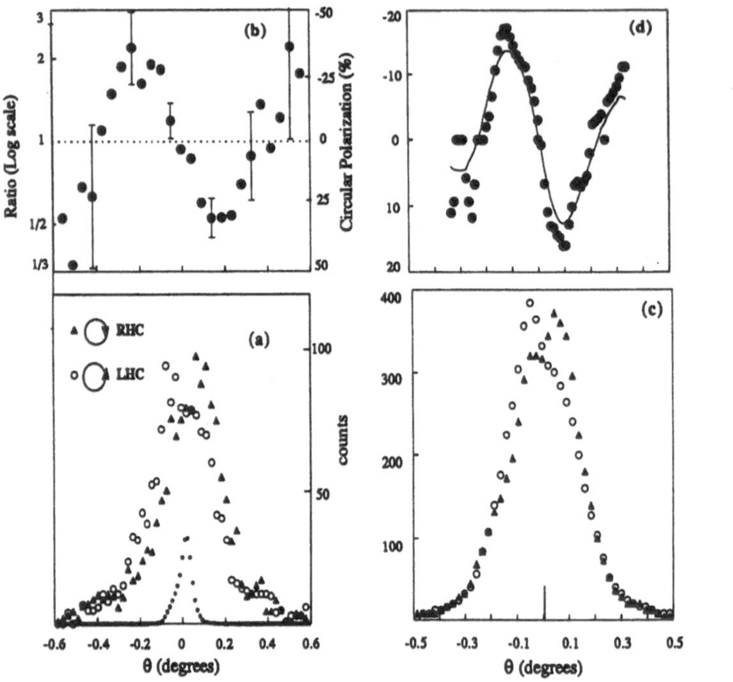

Fig 4 : Differential cross sections $\frac{d\sigma}{d\theta}$ of the $B^{2+}(1s^2\,2p)$ ions, at 1.5 keV, in coincidence with RHC or LHC polarized radiations (a) on He target (the small dots indicate the incident B^{2+} beam angular profile: FWHM=0.08°) (c) on Ne target, and the circular polarization (%) versus scattering angle (b) on He target (d) on Ne target.

λ/4 plate and a pile-of-plate Brewster angle linear polarizer. Typical *coincident energy gain spectra* (recoil-scattered ion coincidences) are shown on fig 3 for both He and Ne targets. These raw data (corrected for angular transmission) actually yield the differential cross sections (DCS) $\frac{d\sigma}{d\theta}$ for the various processes. Both spectra display two structures corresponding to capture into B^{2+} (2s) and B^{2+}(2p) states. The ^2P state is dominantly populated with Ne at variance with the He target case. Stueckelberg oscillations are also visible as expected in systems with a small number of channels.

Fig 4(a,c) shows the angular profiles for scattered ions in coincidence with polarized radiations. The number of right hand (RHC) and left hand (LHC) circularly polarized photons directly reflects the populations of the B^{2+} m=−1 and m=+1 magnetic substates respectively. Note that in this space fixed frame (insert fig 2) the m=0 substate (which incidentally is not the σ state of the rotating molecular frame) is not populated for symmetry reasons. These data clearly show for both targets, a *strong propensity for population of the m=−1* state. For He target this is in qualitative agreement with the theoretical predictions[7]. Theoretical DCS are not yet available for a quantitative comparison. It is noteworthy that this propensity extends at least up to 12 keV, the highest investigated energy.

III. Discussion

A- Orientation propensity in *direct excitation.*

The present work was initiated by the orientation propensity rules proposed by Andersen and Nielsen[1] for excitation processes. Considering a s-p excitation of a single ion-atom collision the system reduces to a 3 state problem (s, p+, p-) which can easily be solved in a first order perturbation approach in the impact parameter approximation (atomic expansion). At this level, the excitation amplitude is expressed as

$$a_{\pm 1}(+\infty) = \pm\frac{i}{v}\int_{-\infty}^{+\infty} V_{exc}(R).e^{i(\frac{\Delta E}{v}X \pm \varphi)}\, dX \qquad (1)$$

Where V_{exc}, the real part of the *coupling* term, is the same for (+) end (−) orientations. Therefore differences in amplitudes come *only* from the phase $\varphi(X)$. ΔE is equal to the energy difference $\Delta\varepsilon$ between the s and p states corrected for the perturbation by the ion at

$$\Delta E\, X = \Delta\varepsilon\, X + \int dX \langle p \mid V \mid p\rangle - \langle s \mid V \mid s\rangle \qquad (2)$$

for excitation occuring at large R.

The phase rapidly oscillates for "$+\varphi$" and $\Delta E > 0$ while these oscillations are much slower for "$-\varphi$" and even nearly cancel for $\frac{\Delta E}{v} a = \varphi$ (here a is the spatial extension of the coupling region). This effect originates from the strong enhancement for p_- excitation[1].

It is interesting to notice that Russek et al[9] could have reached similar conclusion in a *quasi-molecular* approach for a curve crossing situation. These authors show that expressing the conservation of the total angular momentum, the energy curves for p_+ and p_- states are shifted by $\pm\frac{d\varphi}{dt}$ depending, whether the electronic angular momentum $p_{\pm}$ is parallel or antiparallel to the nuclear orbital momentum. As seen in fig 5(a) this effect displaces the crossing towards larger R for the curve connected to the p_- state leading in principle to a larger cross-section. We therefore reach similar conclusion to the propensity rule discussed above. In the spirit, the quasi-atomic case discussed by Andersen and Nielsen[1] can be schematized in fig 5(b) and it can be easily understood that the energy defect is reduced for p_- state with a consecutive enhancement of the p_- cross section. A more definite connection between the two approaches remains to be done.

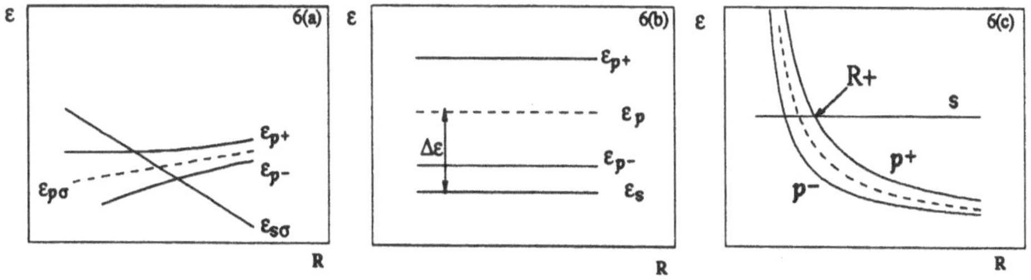

Fig 5: Molecular energy diagrams in body fixed system. $\varepsilon_{p\pm} = \varepsilon_p \pm \frac{d\varphi}{dt} \approx \varepsilon_p \pm \frac{vb}{R^2}$

B- Orientation propensity for Electron Capture

If we refer to the previous discussion, for an exothermic ($\Delta\varepsilon < 0$) process p_+ states should be preferentially populated in a direct transition at R_{+1} (Eq.1 and fig 5(c)) in contradiction with the theory[7] and the present experimental results. Actually Nielsen et al[10] have derived for electron capture an approximate formula similar to (1) and including electron translational factors (ETF) : the $p_{\pm}$ capture amplitudes are given by :

$$\hat{a}_{\pm}(+\infty) = -\frac{i}{v} \int_{-\infty}^{+\infty} G_{\pm}\ (b,v,R)\ e^{\left(i\,[\Delta\varepsilon+\frac{1}{2}v^2]\,\frac{X}{v} \pm \varphi\right)}\ dX \qquad (3)$$

where $\Delta\varepsilon$ X takes a form similar to (2)[10]. Therefore the v^2 energy term just adds a small correction to the phase (at low keV collision energy) and furthermore this correction does not depend on the $\pm$ 1 orientation. From these arguments, propensity could not be found in the phase term of (3). Orientation propensity of capture is therefore to be looked for in the $G_{\pm}$ term. If ETF are *neglected* Nielsen et al[10] have shown that, in the H^+-Na system as for excitation, $G_{\pm}$ are identical for $|p_+\rangle$ and $|p_-\rangle$ states. This strongly suggests that as in the H^+-Na system, the presently observed orientation propensity can only be explained considering inclusion of ETF. Orientation effects in electron capture is expected to be extremely sensitive to ETF even at low collision velocity. Actually the present data reveal a factor 2 in the " $\pm$" capture probabilities at v = 0.07 a.u.

Conclusion

This discussion clearly suggests that the orientation propensity rules have different physical origins for excitation and capture. For the latter case the strong connection with ETF reminds the naive idea of a capture enhancement when the two particles (projectile and electron) "travel together" in the same sense. Further work is required to better understand the very natures of these propensity rules.

References

(1) N. Andersen and S.E. Nielsen, Europhys.Lett. 1, (1986) 15
(2) R. Witte, E.E.B. Campbell, C. Richter, H. Schmidt and I.V. Hertel, Z.Phys.D 5, (1987) 101
(3) D. Dowek, J.C. Houver, J. Pommier, C. Richter, T. Royer, N. Andersen and B. Palsdottir, P.R.L.64, (1990) 1713
(4) see e.g. R.J. Allan, C. Courbin, P. Salas and P. Wahnon, J.Phys.B 23, (1990), L461
(5) P. Roncin, C. Adjouri, M.N. Gaboriaud, L. Guillemot, M. Barat and N. Andersen, submitted
(6) H. Ryufuku, K. Sasaki, T. Watanabe Phys. Rev A 21, 745 (1980)
(7) J.P. Hansen, L. Kocbach, A. Dubois and S.E. Nielsen, P.R.L. 64, (1990) 2491
(8) M. Barat, M.N. Gaboriaud, L. Guillemot, P. Roncin, H. Laurent and S. Andriamonje, J.Phys.B 20, (1987) 5771
(9) A. Russek, D.B. Kimball and M.J. Cavagnero, Phys.Rev.A 23, (1981) 139
(10) S.E. Nielsen, J.P. Hansen and A. Dubois, J.Phys.B, 23, (1990) 2595
see also A. Dubois, Thesis Copenhagen 1990

E Collision of Antiparticles with Atoms

Chairman: W.E. MEYERHOF

E.1 Invited Survey

E. MORENZONI
Collisions of Antiparticles with Atoms

E.2 Invited Contribution

V.V. BALASHOV, M.V. GORELENKOVA and A.I. MAGUNOV
Charge-Asymmetry Effects in Polarization Characteristics of Excitation of Atoms by Fast Protons and Antiprotons

E.3 Contributions (Posters)

L.L. BALASHOVA, N.M. KABACHNIK and CH. TRIKALINOS
Electronic Energy Loss and Channelling of Antiprotons in Crystals

K. TŐKÉSI and G. HOCK
Particle and Antiparticle – H Collisions

COLLISIONS OF ANTIPARTICLES WITH ATOMS

Elvezio Morenzoni

Paul Scherrer Institute (formerly SIN), CH-5232 Villigen PSI, Switzerland

ABSTRACT

Since a few years beams of antiparticles such as antiprotons and positrons have become available for atomic collision experiments. The use of these projectiles, together with protons and electrons, create a whole range of interesting possibilities. Results from investigations on single ionization, inner-shell ionization, double ionization and ion-solid interactions with antiprotons and positrons are summarised and, from the comparison with data obtained with protons and electrons, effects resulting from the sign of projectile charge and from the mass are discussed.

1. INTRODUCTION

Highly energetic beams of antiparticles are being widely used in the realm of nuclear and particle physics but for the atomic processes considered here these beams are generally too energetic (with the notable exception of channeling phenomena). In the field of atomic physics the use of antiparticle probes has concentrated up to now on the study of processes induced by stopped or almost stopped particles in matter. Examples of this class of experiment are the study of exotic atoms (muonic, pionic, antiprotonic, kaonic) and of the atomic and molecular processes leading to muon fusion [1].

For atomic collision experiments (where the projectile is inelastically scattered by the target) a new instrument has been offered at CERN by the availability of low energetic antiprotons ($<$ 5 MeV/amu) at the Low Energy Antiproton Ring (LEAR). This allows us for the first time to extend the study of ion-atom collision processes to a collision system with a heavy negatively charged projectile and an experimental program to investigate various aspects of $\bar{p}$- atom collisions has been undertaken by a Aarhus-CERN-PSI collaboration.

Parallel to this development beams of a light antiparticle, the positron, have also become available in various laboratories for atomic-collision studies. These, together with the traditionally available beams of protons and electrons, provide us with a complete set of heavy and light projectiles with positive and negative unit charge. The use of singly charged particles in comparison of cross sections or other relevant quantities provides fundamental tests of the theory of ion-atom collisions and the appropriate comparison (antiproton impact versus proton impact, e^+ versus e^-, heavy versus light) permits the isolation of the effects of varying sign of charge and mass. For instance, the study of the dependence of cross sections on the sign of the projectile is significant from the point of view of a perturbation treatment of the collision since even and odd terms of the Born series have opposite parity under charge conjugation. By comparing cross sections for antiproton impact with cross sections obtained for proton impact, deviations from Z_p^2 scaling (Z_p, projectile atomic number) expected on the basis of first order perturbation theory can be detected and the relative importance of higher order terms of the series can be determined. Furthermore, the effects peculiar to each particle may be discerned, for instance the exchange phenomena for electron impact and the capture channel for positron (positronium formation) and proton collisions. In the following we will give a brief survey of results obtained with antiparticles. In accordance with the subject of the workshop special emphasis will be placed on the results obtained with antiprotons and on a comparison with proton data.

The paper is organized as follows. In the next section the characteristics of the available antiparticle beams are described. In Section 3 the results of single ionization cross sections for two simple systems such as He and H_2 are discussed. Inner shell vacancy production cross section measurements are presented in Section 4 and multiple ionization results in Section 5. Results on antiparticle-solid interactions are then shown in Section 6. Concluding remarks and an outline of possible future experiments are given in the last part.

2. BEAMS FOR ANTIPARTICLES-ATOM COLLISION STUDIES

The class of feasible atomic collision experiments with antiparticles is determined by the quality and intensity of available beams at low energies (few MeV/amu and less). The creation of an antiparticle involves at some stage a large kinetic energy which is reflected in the large kinetic energy and energy spread of the antiparticle. To be used as a projectile for atomic collision experiments, the antiparticles have to be decelerated and the momentum spread has to be reduced (cooling). At CERN antiprotons are produced by 26 GeV/c protons from the proton synchrotron (PS) incident on an external target. They are collected at their production optimum of 3.5 GeV/c in the Antiproton Collector, cooled and stored in the Antiproton Accumulator and decelerated in the PS to 0.6 GeV/c before being transferred to LEAR. The antiprotons for LEAR typically occur in bunches of up to $3 \cdot 10^9$ particles. In LEAR, the beam will be either accelerated or decelerated and to improve the beam quality it will also be stocastically cooled. At present ejection momentum ranges from 105 MeV/c to 1700 MeV/c. The beam can be slowly extracted with spill times ranging from minutes to 5 hours. Under normal operating conditions the minimum emittances obtained are around $10 \cdot \pi \cdot mm \cdot mrad$ transversally and $\Delta p/p = \pm 1 \cdot 10^{-3}$ longitudinally. Especially interesting for atomic collision experiments is the lowest extraction momentum corresponding to an antiproton energy of 5.91 MeV. (A beam of ~2.6 MeV kinetic energy is presently under development). At this energy, with a spill duration time of 40 min., a DC beam with an average intensity of about 10^5 antiprotons per second is available. The filling time between spills is typically 20-30 minutes. Lower energies can be produced by the use of degrader at the expense of emittance and momentum resolution. For beam energies down to 0.5 MeV the momentum resolution is still acceptable for most experiments. Particles with energies down to 40 keV can be obtained in the tail of the energy distribution of the beam after passing through thick degrader, but in this case the energy of each incoming particle has to be measured on an event by event basis.

Positrons from radioactive sources (for instance ^{22}Na, produced by $\sim$ 1 GeV proton bombardment of Al) can be slowed down and cooled to a few eV with an efficiency of few per thousand in appropriate moderators such as annealed tungsten [2]. The positrons can then be accelerated to the desired energy up to tens of keV. In this way beams of typically 10^4 -10^5 e^+/sec have been produced and used for atomic collision experiments at various laboratories. A higher positron intensity (up to 10^8 e^+/sec) can be created by the process of bremsstrahlung and pair production using (pulsed) linear accelerators [2].

In principle, one could also consider as possible projectiles positive and negative muons (or even positive and negative pions), which are produced copiously at high momenta at meson factories. Both are heavy projectiles ($m_\mu/m_p \cong 1/9$, $m_\pi/m_p \cong 1/7$). Pions are created when energetic ($\sim$ 600 MeV) protons impinge on a thick target. Within 26 ns they decay emitting a muon. Although the primary pion beam is more intense than the muon beam, at low energies the short decay length strongly reduces the intensity of the pions. Therefore muons are a better choice. Present beams at 5.9 MeV/amu (same velocity as the antiprotons from LEAR) at a meson factory such as the Paul Scherrer Institute (PSI) have a maximum intensity of $\sim 9 \cdot 10^4$ μ^+/sec and $\sim$ 500 μ^-/sec over a beam spot of typically 4 cm diameter. Other meson factories have comparable or less intensity. Such beams are not competitive with the antiprotons for ionization cross section measurements (see Sect. 3), but they have been already used for investigations of energy loss differences between particles and antiparticles, which do not require such intense beams [3].

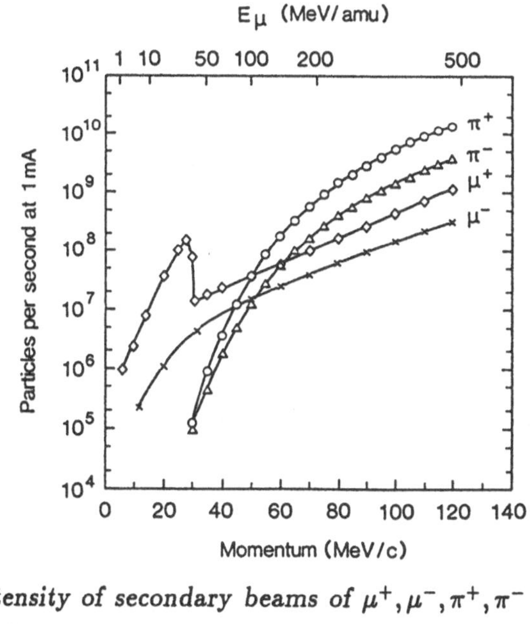

Figure 1: *Beam intensity of secondary beams of $\mu^+, \mu^-, \pi^+, \pi^-$ at PSI with a 1 mA primary proton beam .*

This situation will greatly improve in the near future. A machine upgrade currently underway at PSI for instance (higher proton current, new beam line) should deliver up to 6 $\cdot 10^6 \mu^+$/sec and 3 $\cdot 10^5 \mu^-$/sec at 5.9 MeV/amu (Fig. 1) [4]. Although the emittance will not attain the LEAR quality without cooling, this intensity will allow for a tighter beam definition at the expense of intensity. Muons can be more readily collimated than antiprotons since they produce less disturbing decay products. Another advantage is that one can switch from positive to negative muons by simply reversing the magnet polarity, leaving the experimental conditions and the beam properties unchanged. This greatly reduces systematic errors. Also, these muon beams have a 100% duty cycle since no storage is involved. In summary, in the near future muon beams will offer interesting possibilities to study sign of charge effects and mass effects in atomic collisions.

3. SINGLE IONIZATION OF He AND H_2

Single ionization can be described within the framework of an independent electron model where one electron is ionized while the others do not participate actively in the collision process. At high impact energy single ionization is well described by the first Born approximation, which predicts a Z_p^2 dependence for the cross section. From measurements with multiply charged ions it is known that the scaling breaks down at low energies indicating that higher order terms beginning with a Z_p^3 term must be taken into account. Particularly interesting for an understanding of ionization phenomena are simple targets like He, H_2,H since in this case a large variety of theoretical approaches beyond first Born approximations are available [5-10]. For He and H_2 target, single ionization data have recently become available for all four projectiles p, $\bar{p}$, e^- and e^+.

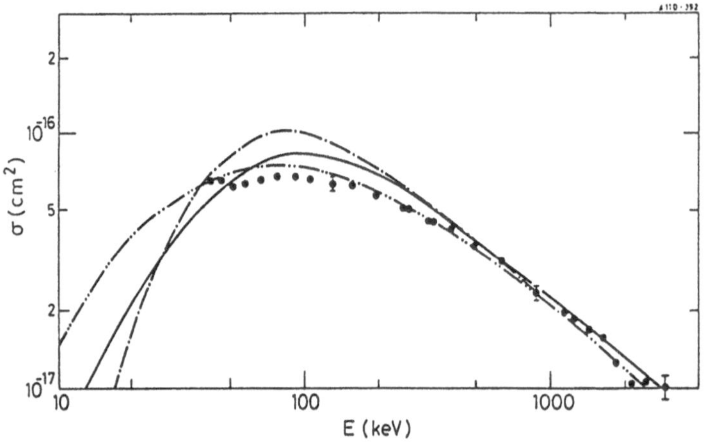

Figure 2: *Total cross sections for single ionization of He by p's and $\bar{p}$'s as a function of energy.* • : *$\bar{p}$ data [11]*; —, *p experimental data [12]*; —· · — *and* — · —, *CDW-EIS theoretical results for $\bar{p}$'s and p's [6].*

Fig. 2 shows the He cross section as a function of projectile energy for antiprotons [11] and protons [12]. From a previous measurement it is known that the single ionization cross section of He for fast equivelocity p's and $\bar{p}$'s is independent of the sign of

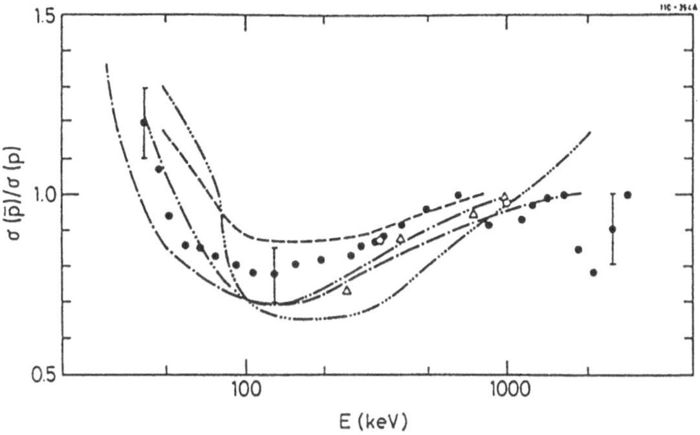

Figure 3: *Ratio between total single ionization cross sections of He for $\bar{p}$'s and p's as a function of energy. • : $\bar{p}$ results divided by p data. Theory : ∘ [5]; Δ [8]; - - - [6]; - ·· - [10]; - ···· - [7] (CTMC); - - -, [7] (CCIP).*

charge above 1.5 MeV [13]. Therefore in Fig.2 the data have been normalized at this energy. Both data sets are compared with the predictions from the Continuum Distorted Wave Eikonal Initial State (CDW-EIS) approximation [6]. Qualitatively there is a good agreement between theory and experiment. Below 200 keV the theory for antiprotons is in even better agreement than the theory for protons, perhaps reflecting the fact that for antiprotons the electron capture process is suppressed thus reducing the complexity of the problem.

The sign of charge effect can be best evidenced by forming the ratio between the cross sections for antiproton and proton impact (Fig. 3). From the ratio it is clear that above ~ 500 keV there is no difference between the two projectiles within the experimental uncertainty (± 7%). At energies below 500 keV the proton cross section is larger by up to 22%, whereas at energies below 50 keV the ratio becomes larger than unity.

In Fig.3 the experimental data for the ratio are compared with predictions from various theoretical calculations. The approaches range from quantum mechanical calculations involving a large set of coupled states such as the Forced Impulse Method (FIM) [5], to the CDW-EIS approximation [6], to the close coupled impact parameter method (CCIP) [7], to a classical description of the collision process where a sample of classical states approximate the quantum mechanical atomic distributions (CTMC: classical trajectory Monte Carlo) [8,10]. Despite the diversity of approaches, all the calculations show a similar behaviour of the ratio, and the experimental shape is well described.

An intuitive explanation of the behavior of the ratio can be given as follows. At high velocity, where the projectile velocity v_p is much larger than the target electron orbital velocity, a first order approximation theory is valid and there is no difference in cross sections for proton and antiproton impact. At intermediate velocities $v_p \sim v_e$ polarization of the target electron cloud plays a role. Due to the polarization, p's draw

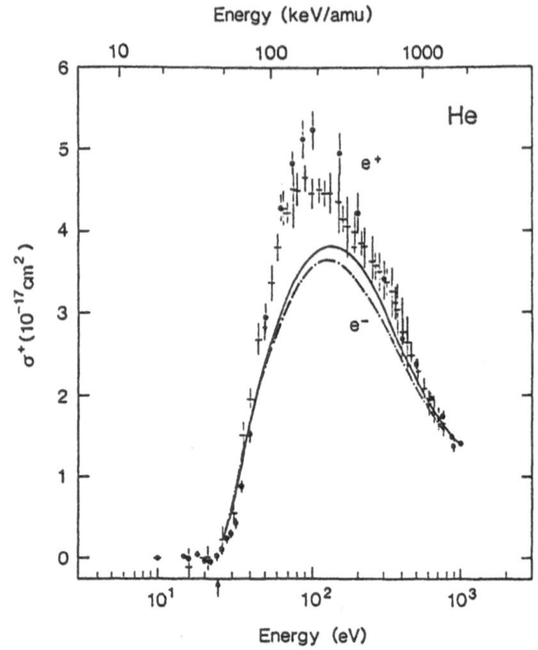

Figure 4: *Total cross sections for single ionization of He by $e^+$and e^-as a function of energy.* • *: e^+data from Ref. [14]; +: from [15]. The e^-data are from [16] (—) and [17] (–·—).*

the electrons closer whereas $\bar{p}$'s repel them (antipolarization). In the first case one has an increase in the ionizing interaction between protons and electrons, which cause an increase in the ionizing cross section; in the second case one has a decrease. Related to the polarization is another mechanism evidenced in the CTMC calculations, which is only present in the case of positive projectiles, namely the creation by the protons of the so called saddle point electrons. These electrons originate from the region of reduced binding centered around the midpoint of the atomic core and a charged positive projectile. Due to the force cancellation, electrons can be ionized with a velocity approximately one half of the projectile velocity. At lower energies where the average impact parameter for an ionizing collision is smaller than the radius of the electron, binding effects become important. A negatively charged particle can screen the Coulomb field of the target nucleus hence reducing the binding energy of the electrons and increasing the ionization cross section with respect to a positive one where the opposite effect is produced. Furthermore for positively charged particles the electron capture channel gets more and more important and the partitioning between ionization and charge transfer channels additionally accounts for the drop of the proton ionization cross section with respect to the antiproton.

Measurements of single ionization cross sections of He are also available for the light projectiles. Results are presented in Fig.4 [14,15]. Similarities related to charge effects and differences due to mass effects can be detected by comparing Fig.2 with Fig.4 at the same velocity. In the high energy region the positron data coincide with the electron data [16,17]. Furthermore at an energy above ~ 1.5 MeV/amu the single ionization cross section has the same value for heavy and light projectile impact in accordance with the expectation that the high energy behavior depends solely on the projectile charge squared. In the intermediate energy for decreasing projectile energy the positron cross section becomes increasingly larger than the electron cross section, reflecting as in the proton/antiproton case the enhancement for polarization and saddle point electron effects.

If the difference between $\sigma^+_{e^-}$ and $\sigma^+_{e^+}$ at intermediate energies is only charge dependent, then the ratio $\sigma^+_{e^+}/\sigma^+_{e^-}$ should be equal to $\sigma^+_p/\sigma^+_{\bar{p}}$. If we take the data from [14] which in the peak region are about 15% higher than previous data from Fromme et al. [15] the ratio $\sigma^+_{e^+}/\sigma^+_{e^-}$ is slightly higher than $\sigma^+_p/\sigma^+_{\bar{p}}$ around the maximum region. The difference could indicate the role of a process special to electrons namely the electron exchange process which lowers the electron cross section [15]. Also, because of the low mass, light particles are more likely to be affected by the Coulomb field of the target. Positrons are decelerated and electrons accelerated. At velocities above the maximum cross section, which is also not exactly at the same position for electrons and positrons, this would also lead to a larger $\sigma^+_{e^+}/\sigma^+_{e^-}$ ratio. At small velocities the cross section for light projectiles is much smaller than for heavy particles. This is due to a simple dynamic effect. Due to its smaller mass, low velocity e^+/e^- have much less energy above the ionization threshold. Consequently fewer light particles at low velocity succeed in ionizing the target than the equivelocity heavy particles.

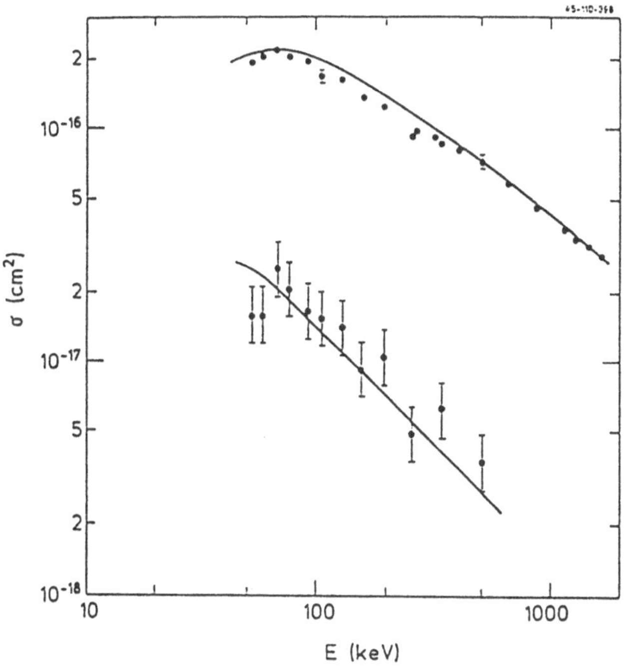

Figure 5: *Cross section σ_{ni} for non dissociative ionization, upper part, and σ_{di} for dissociative ionization, lower part.* •: *$\bar{p}$ results [18];* — : *results for p impact [19].*

The same experimental technique used for the $\bar{p}$+He measurement has been used by the Aarhus-CERN-PSI collaboration for cross section measurements in H_2 [18]. The processes $\bar{p}+ H_2 \rightarrow \bar{p} + H_2^+ + e^-$ (non dissociative ionization) and $\bar{p} + H_2 \rightarrow \bar{p} + H^+ + H + e^-$ (dissociative ionization) have been investigated. In the first process only one electron is active. The results are shown in Fig.5, where the antiproton points have been placed on an absolute scale by normalizing to high energy proton results of Shah and Gilbody ([19] also shown in the figure). The good agreement between proton and antiproton for non dissociative ionization in the energy interval 400-1500 keV is similar to the $\bar{p}/p$ agreement found for single ionization of He in the same energy interval. As for He the ratio $\sigma_{\bar{p}}^+/\sigma_p^+$ (Fig.6) shows polarization effects between 50 and 400 keV with an excess of proton ionization of up to ~ 14%. The theoretical predictions are for atomic hydrogen. The relative behavior for positron and electron impact is similar to the He case [14].

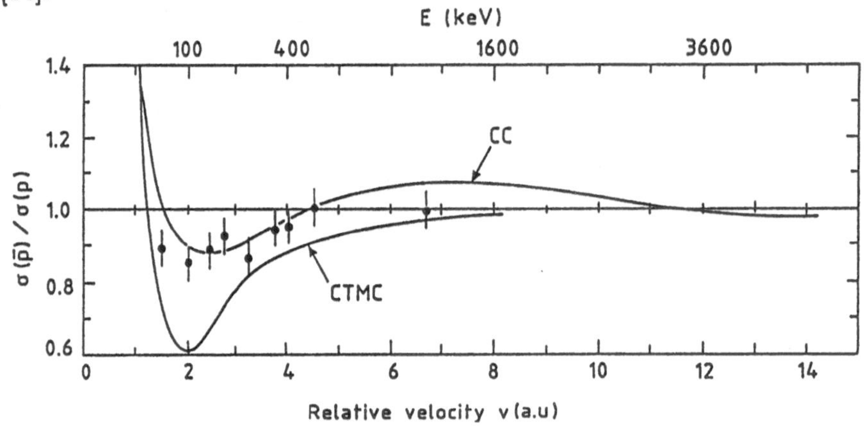

Figure 6: *Ratio between total single ionization cross sections of H_2^+ for $\bar{p}$'s and p's as a function of energy, compared with predictions from coupled channel (CC) [7] and Monte Carlo (CTMC) calculations [10] for atomic hydrogen.*

4. K-SHELL IONIZATION

Inner-shell ionization plays also a special role in the physics of atomic collisions because of its relative simplicity. The intrinsic complexity of a many electron atom is reduced by the fact that inner shell electrons can be considered in good approximation to move in a screened mean potential thus reducing the problem to a single electron process. For this reason it has been intensively studied theoretically and experimentally with a large variety of ions.

Deviations from the Z_p^2 dependence are known from heavy ion impact at low velocity and are interpreted on the basis of polarization, binding and Coulomb effects similar to the single ionization of He, and H_2. These corrections are put in evidence in an analytical way by the ECPSSR treatment of the inner-shell ionization [20]. PSS stands for Perturbed Stationary States method; it takes into account binding and polarization. Later work added terms to include the effects of Coulomb deflection (C), relativistic effects (R) and energy loss of the projectile (E) [21]. All these effects are incorporated into a PWBA framework through appropriate scaling of the variables. This gives an explicit dependence of the cross sections on the relevant quantities such as projectile velocity (v_p), inner-shell electron velocity (v_K), binding energy (E_K), target and projectile atomic numbers (Z_p, Z_t) and masses. For its computational facility the model is widely used as a basis for systematic studies of K-shell vacancy production cross sections in a broad energy range and for a large variety of projectile-target combinations [22]. The cross section for light ions can be predicted with reasonable accuracy by using the ECPSSR theory but residual deviations from the data especially at low energies have been found to be statistically significant. For heavier projectiles the increasing probability of the charge transfer channel renders the systematic studies more difficult. For this reason it is interesting to test our understanding of inner shell ionization with antiprotons independently.

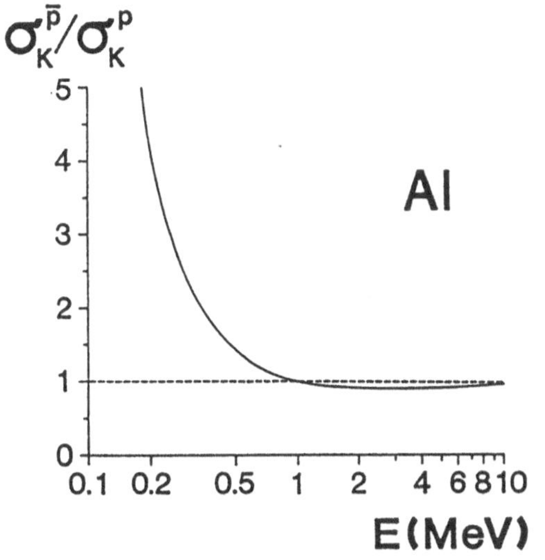

Figure 7: *ECPSSR predictions for the ratio of Al K-vacancy production cross section by antiproton to proton versus projectile energy.*

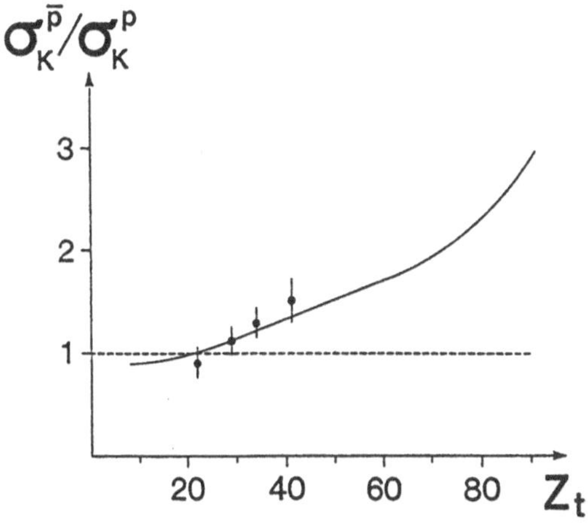

Figure 8: *Ratio of K vacancy production cross section by antiproton and proton impact at 3.5 MeV versus target atomic number. The solid line is the prediction from the ECPSSR model.*

Various theoretical calculations have been published for antiparticle impact [23-25]. Fig. 7 shows the prediction of the ECPSSR model for the ratio $\sigma_{\bar{p}}^K/\sigma_p^K$. Striking similarities with Fig. 3 are present, with a high velocity region insensitive to the sign of charge, an intermediate region where the polarization (smaller than in the He case) lowers the $\sigma_{\bar{p}}^K/\sigma_p^K$ ratio and a low velocity region where the binding effect sets in. However some qualitative differences are present. In the inner-shell vacancy case at very low velocities the Coulomb effect is more pronounced. The different Coulomb trajectories followed by particles and antiparticles in the field of the nucleus have an effect on the vacancy pro-

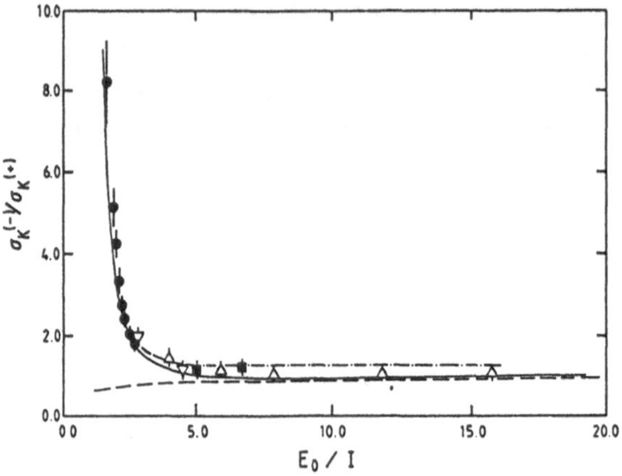

Figure 9: *Cross section ratio for K-shell ionization by $e^-$and e^+impact versus energy E_0 in units of the K-shell binding energy. Solid line : prediction from corrected PWBA [29]. Dashed line : PWBA. Experimental results for Ag (•) and Cu (■) from [27] and references therein ($\triangle, \nabla$).*

duction. The trajectory of the proton is deflected away from the target nucleus, whereas antiprotons are attracted by the nucleus thus increasing the ionization probability. Furthermore the attractive potential experienced by the antiprotons increases the velocity and hence the cross section. Also, differently from a light target, electron capture from the K-shell does not contribute significantly to the K-shell vacancy production. Of special interest in a measurement with antiprotons is the role of the Coulomb correction, which is the one affected by the largest uncertainty. Questions must be raised especially in the ECPSSR model when this correction is applied to the antiprotons, because the formal replacement $Z_p \rightarrow -Z_p$ gives a divergent integral, which must be cut off below the upper integration limit [26].

Measurement of the K-shell ionization by antiprotons are more difficult than single ionization measurements in light ions. The K x-ray yields are very low at low energies. Also special care must be taken to reduce the background arising from the annihilation products of the antiprotons [27]. In a first experiment dedicated to the inner-shell ionization we measured K X-rays production from Ti, Cu, Se and Nb targets by antiprotons and protons of 3.5 MeV. The ratio between K-vacancy cross section for antiproton impact and the K-vacancy cross section for proton impact is shown in Fig. 8 as a function of the target atomic number and is in agreement with the predictions from the ECPSSR model. At this energy binding and polarization effects dominate. Very recent measurements extending down to 1.5 MeV for Ti and Cu should be more sensitive to the Coulomb effect.

A few experimental studies have compared the inner-shell ionization by low energy positrons and electrons. Results for $\sigma^K_{e^-}/\sigma^K_{e^+}$ are shown in Fig. 9. The data for two different targets (Cu,Ag) seem to scale if the kinetic energy is expressed in unit of the K-shell ionization energy. Near the threshold region the ratio deviates strongly from unity increasing rapidly with decreasing impact energy. As for the heavy particles negatively charged particles produce more K-shell vacancy than the positive ones. Calculations have to take into account the wave nature of the light projectiles. A simple modification of the PWBA to include the acceleration of an incoming $e^-$and the deceleration of an e^+, gives a very good agreement with the available experimental data [29] (Fig. 9). Although the increase in the ratio is reminiscent of the increase at very low velocities in Fig. 8 it sets on for the light particles at much higher relative velocity due to the small mass of the projectile.

5. DOUBLE IONIZATION

Contrary to single and inner-shell ionization, double ionization is a process where the electron correlation due to the mutual Coulomb interaction plays an active role beyond the static screening of the nuclear charge. The field of electron correlation was strongly revived by the first experiment on antiproton-atom collision, the measurement of the ratio of double to single ionization in He (R= σ^{2+}/σ^{+}) [30]. The ratio for antiproton impact was found to be about a factor of two higher than the corresponding value for protons and equal to the value for e^- impact, demonstrating that the already known difference in σ^{2+} of He by e^- [31] as compared to p is a charge effect rather than a mass effect. The difference in R (Fig. 10) above 500 keV is due solely to σ^{2+}, since in this energy region σ^{+} is independent of the sign of charge (see Fig. 3). Similar effects have been observed in Ne, and Ar targets.

In the mean time the measurements of the ratio R have been extended down to impact energies of 65 keV and up to 20 MeV [32], and a number of theoretical interpretations have appeared based on some of the approaches discussed in Section 3 [5,8,10,33]. Although qualitatively they predict the $\bar{p}/p$ difference, they tend to underestimate the extra amount of double ionization by antiprotons. For instance the most complete calculation with the Force Impulse Method accounts for only 50% of the measured difference at higher energies, probably due to the neglect of higher order partial waves in the calculations. In the CTMC calculations the importance of the e^- -e^- correlation has been pointed out classically [8]. In the FIM method the difference in double ionization produced by $\bar{p}$ compared with p results from the combined effects of dynamical electronic correlations and higher order Born terms. In the absence of either the difference disappears. The interference appears between first and second order amplitudes and depends critically on non-dipole transitions since the second Born term requires a non-dipole transition to interfere with the first [5]. McGuire [34] was the first to suggest that the observed effect is caused by interference between two double ionization mechanisms. In this case $\sigma^{2+} = |a_I Z_p + a_{II} Z_p^2| = R_I Z_p^2 - 2R_{int} Z_p^3 + R_{II} Z_p^4$ and a Z_p^3 term responsible for the sign of charge effect appears.

The interference idea was further developed by Sørensen [13], who theoretically estimated R_I from a one-step projectile electron interaction and R_{II} from a two-step process. R_I, R_{II} and R_{int} were also determined from measurements of R for antiprotons, protons and He^{++}. A good agreement was found in the energy range from 1 to 10 MeV/amu. This analysis can also explain the recent results of a larger high energy value for R by N^{7+} projectiles [35]. Using the velocity dependence of R_I, R_{II} and R_{int}

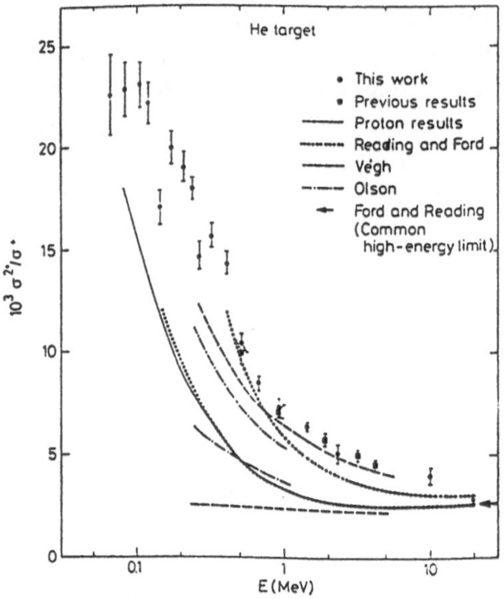

Figure 10: *Ratio* $R=\sigma^{2+}/\sigma^{+}$ *in He for* $\bar{p}$*'s (■ and ●) and for p's (solid line). Theoretical estimates are drawn for comparison. Upper curves relate to* $\bar{p}$*'s, lower to p's.*

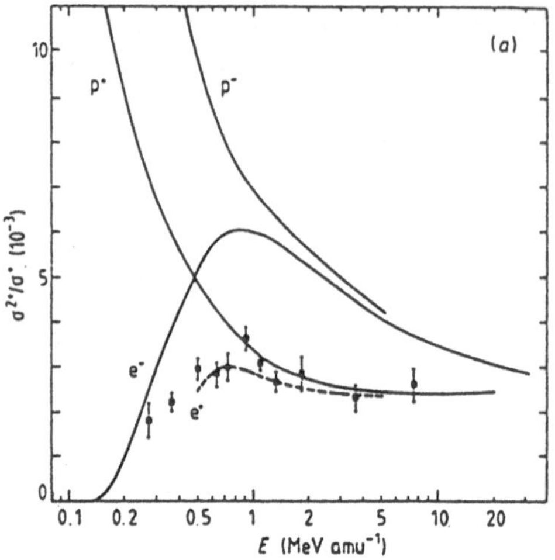

Figure 11: *Ratio* $R=\sigma^{2+}/\sigma^{+}$ *in He for* $\bar{p}$*, p,* e^+*,* e^- *impact at various energies (from [37]). Solid curves represents experimental trends. Dashed curve: prediction for* R_{e^+} *from* $R_{e^+} = R_p(R_{e^-}/R_{\bar{p}})$.

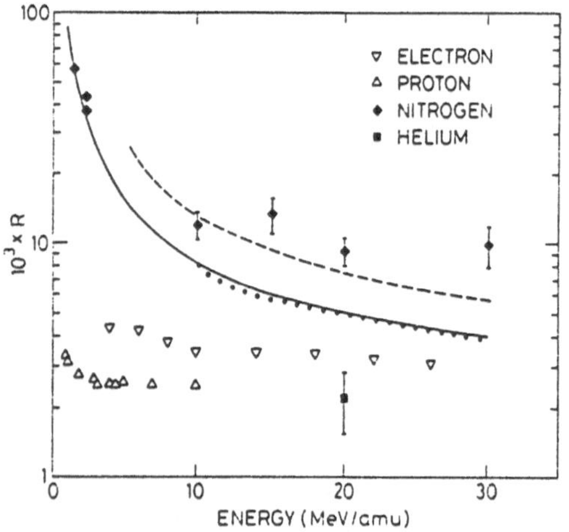

Figure 12: *Ratio* $R=\sigma^{2+}/\sigma^{+}$ *in He for various projectiles. The dashed line represents the R ratio estimated from an interference analysis (see text). The solid curve is the prediction from a semiempirical calculation and the dotted curve from a FIM calculation (see [35,36] for more details).*

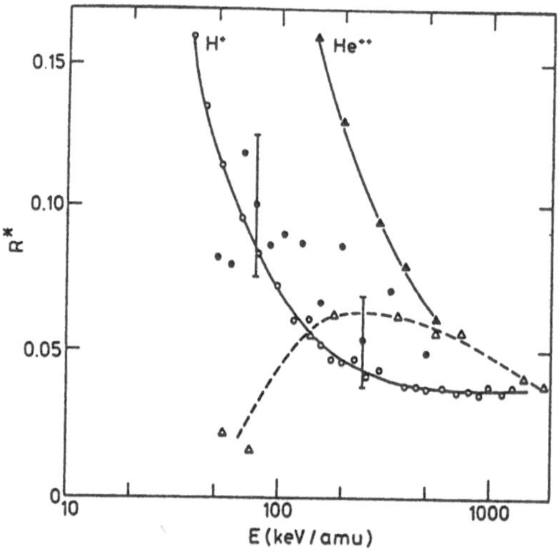

Figure 13: *Ratio* R^* *between dissociative and non dissociative ionization of* H_2 *(from [18]). Results for ●:* $\bar{p}$*, ○: p, ▲:* He^{2+}*, △ :* e^-.

obtained with the low charge a good agreement is found with the N^{7+} value [36] (Fig. 11).

Recently the double ionization of He for e^+ impact has been measured [37], so that we have now a complete picture of σ^{2+}/σ^+ with the four singly charged projectiles (Fig. 11). The charge effect is evident as well as the common high energy limit insensitive to the sign of charge. As for single ionization the small mass of e^+ and e^- produces the pronounced decrease at kinetic energies approaching the threshold.

Other two-electron processes have been investigated in He with e^+, e^- and p to separate the interference effect by using particle antiparticle projectiles. Besides double ionization (DI), double excitation (DE) [38] and ionization excitation (IE) [39] were studied with the conclusion, that when close collisions play an important role as they are believed to do in DI and IE, interference effects are important and the cross section for negative particles are up to a factor of two larger than for positive particles. This has been observed in DI and IE but not in DE.

DE and IE have not yet been investigated with antiprotons, but a process similar to IE is the dissociative ionization in H_2, previously mentioned (see Fig. 5). It is known that dissociative ionization results from a repulsive excited H_2^* intermediate state where two electrons are active and it can be interpreted as an IE process. Since non dissociative ionization is a one electron process and dissociative ionization a two electron process one can form the ratio $R^*=\sigma_{di}/\sigma_{ni}$ in analogy with the ratio R for He. Fig. 13 shows R^* for electrons, protons , He^+ and antiproton projectiles. R^* behaves in a qualitative way like the ratio between double and single ionization of He, i.e. at high velocities, the ratio is larger for negative than for equivelocity positive particles. This indicates that the two-electron process is also influenced by interference between a one-step process and a two-step process. An analysis, analogous to the He case, in terms of R_I^*, R_{II}^* and R_{int}^* supports this interpretation [18].

6. ANTIPROTON-SOLID INTERACTIONS

Sign of charge effects appear also in the interaction of charged particles with solids. It is well known that the energy spectra of electrons ejected from thin foils by protons in the forward direction show a cusp-shaped peak of convoy electrons at $v_e \approx v_p$ due to electron capture to the continuum (ECC) [40]. The attractive final-state interaction between the electron and the positive ions leads to the cusp like enhancement, while for $Z_p < 0$ a pronounced dip (anticusp) is expected in a gas target [41]. For a foil target the dip is more or less smeared by cascading electrons [42]. This behaviour has been recently confirmed experimentally [43]. Fig. 14 shows electron spectra for a 2 $\mu g/cm^2$ C foil bombarded by 600 keV protons and 610 keV antiprotons. For protons the well known cusp is clearly visible whereas for for antiprotons no cusp is recognizable in agreement with a Monte Carlo calculation taking into account solid-state effects [42].

An additional structure is predicted in the $v_e \approx v_p$ part due to electrons bound in the oscillatory wake potential created by the fast projectile in the solid [44]. Due to the absence of the cusp and the fact that the electron capture probability into wake riding states is dramatically enhanced for antiprotons, the observation of wake riding electrons is more likely for antiprotons than for positive ions. In the measurement of Ref. [43] a bump is recognized at an energy position consistent with the energy of an electron released from a wake riding state, but a confirmation must await data with more statistics.

Differences in the stopping power of particles and antiparticles at high velocities have been known as Barkas effect. Barkas et al. [45] found the range of negatively charged particles (Σ^-,π^-) greater than for positive particles under otherwise equal conditions. This behavior is related to a Z_p^3 term in the stopping power formula, whose first term is given by the Bethe formula:

$$-\frac{dE}{dx} = \frac{4\pi e^2}{m v_p^2} N Z_t Z_p^2 (L_0 + L_1 Z_p + L_2 Z_p^2)$$

(v_p: projectile velocity, N: target density, Z_t: target atomic number). The Z_p^3 and Z_p^4 terms have been studied by comparing stopping powers for p, α particles and Li [46], but the advantage of a particle-antiparticle measurement is that, apart from the suppression of electron capture for $\overline{p}$, the pure Z_p^3 contribution can be singled out and it is not obscured by even terms. The interest in stopping power is not only practical (a detailled understanding of the stopping power and range is required in many domains from accelerator physics to ion-implantation etc), but also because the physics of the Barkas effect is related to basic atomic processes.

For distant collisions the Barkas effect is interpreted as a result of the polarization of the electron medium, much like the difference in ionization by $\overline{p}$ and p mentioned in Sect. 3. This effect increases the ionization cross section and therefore the stopping power for positive particles. The contribution of close collisions has been subject of a long standing discussion, where most authors assert that close collisions are essentially those of free particles in a Coulomb field, giving an exact Z_p^2 dependence. On the other hand it has been argued that there is an equal contribution to the Z_p^3 term from close collisions, since they are not exactly Coulomb like due to dynamical screening [47]. Recently the antiproton stopping power in Si has been determined with an accuracy of about 1% [48]. It has been found to be 3%-19% lower than that for p over the energy range 3-0.5 MeV. The Barkas term extracted from these data is shown in Fig. 15 plotted as a function of the velocity in units of the Bohr velocity. The measured Barkas term is about a factor of two larger than that calculated by Jackson and McCarthy [49] for the distant collisions only and is in close agreement with an estimate with equal contributions from close and distant collisions [50] The measurements alone do not provide a direct evidence for a significant close-collision contribution to the Barkas effect since by changing the impact parameter cutoff in distant collision theories equally good agreement with the data can be obtained. A continuation of the $\overline{p}$ stopping power measurements to energies down and below the stopping power maximum (at 100 keV/amu) outside the limit of validity of the Bethe-Bloch-Barkas formula is being pursued [51]. No direct measurements for antiparticle are available in this region, where a recent calculation [52] predicts the energy loss for antiprotons to be about one half of that for protons independent of velocity.

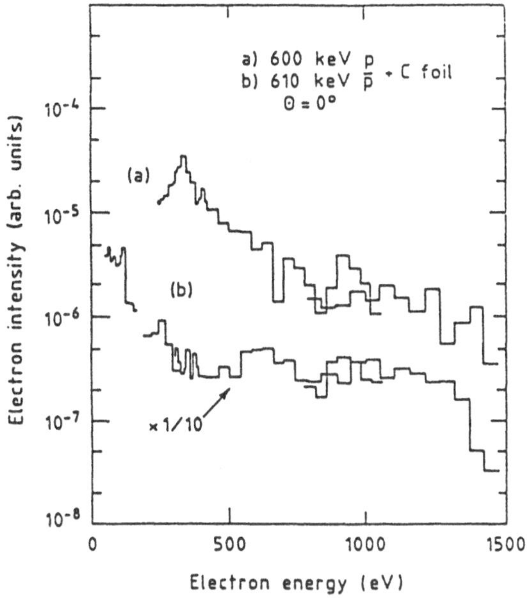

Figure 14: *Electron spectrum versus electron energy in the forward direction for a) 600 keV p's and b) 610 keV $\overline{p}$'s incident on a C foil [43].*

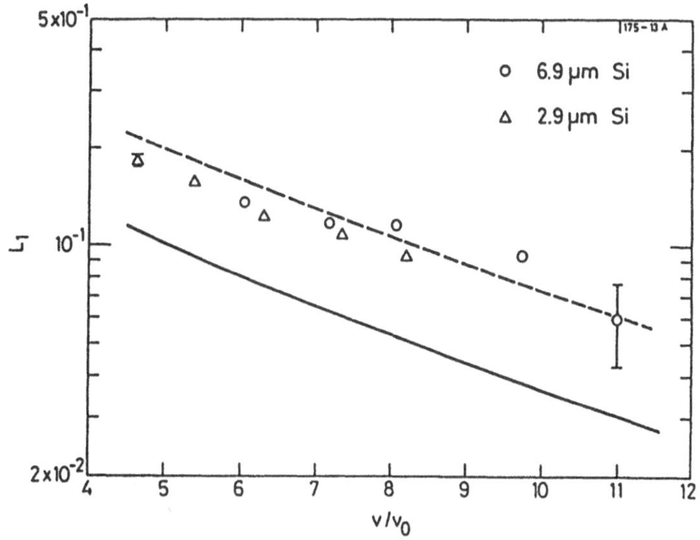

Figure 15: *The Z_p^3 contribution L_1 to the stopping power, obtained from $p/\overline{p}$ measurements [48]. The solid line and the dashed curve represents the Jackson McCarthy results [49] and twice this value respectively [50].*

7. CONCLUSIONS

The availability of low energetic antiparticle beams for atomic collisions experiments has provided a powerful instrument to obtain new insight into the dynamics of collision processes. Comparisons between particles and antiparticles give evidence for the importance of polarization, binding and Coulomb effects in single and inner-shell ionization and that of electron correlations and interference in two-electron processes. The sign of charge dependence manifests itself also in ion-solid interactions such as energy loss and electron emission from thin foils.

Up to now , due to the limited intensity of antiparticle beams, the investigations have concentrated on total cross section measurements, but differential measurements have been proven to be accessible with the present beams (for instance the measurement of the energy spectrum of forward emitted electrons) and they will represent the next generation of experiments in antiparticle-atom collisions [53]. Here various theoretical predictions are already available, ranging from impact parameter dependence of ionization processes [25,54], to angular and energy distribution of the emitted electrons [55,56]. Experimental data in this respect would greatly improve our understanding of the details of collision mechanisms. The basic experiment of ionization in $\bar{p}$+H is envisaged (for e^++H see for instance [57]) It would also be interesting to extend the investigations to energies below 1 a.u.. For instance for He at antiproton velocities much smaller than the orbital velocity, in the quasimolecular regime , adiabatic ionization via electron promotion into the continuum should take place [9,58]. This effect should cause the $\sigma_{\bar{p}}/\sigma_p$ ratio to attain very large values at low projectile velocities. Measurements at antiproton energies below 20 keV are not feasible at present, but the possibility of obtaining very low energy antiproton beams down to eV is being pursued [59]. This in combination with positron beams would offer the possibility to create antihydrogen atoms, and to open a new world of fundamental investigations and precision experiments in antimatter [2]. Atomic collision experiments will also profit from the improvement of low energetic μ^+ and μ^- beams, which will become available in the near future.

ACKNOWLEDGEMENTS

The experiments of the Aarhus-CERN-PSI collaboration described in this paper have been carried out over the last years with L.H. Andersen, K. Elsener, P. Hvelplund, H. Knudsen, S.P. Møller, R. Medenwaldt, J.O.P. Pedersen, S. Tang-Petersen, E. Uggerhøj and T. Worm. The author is grateful for the collaboration and the stimulating discussions with his colleagues.

REFERENCES

[1]. Electromagnetic cascades and chemistry of exotic atoms, Ettore Majorana Institute (1989), Erice, Science Series, Phys. Sciences, Vol 52.

[2]. Antimatter Symposium 1987, Hyperfine Interactions, 44, (1988).

[3]. W. Wilhelm, H. Daniel, F.J. Hartmann, Phys. Lett. B23, 33 (1981).

[4]. R. Abela, D. Renker, PSI Internal Report (1990).
[5]. J.F. Reading and A. L. Ford, J. Phys. B 20, 3747 (1987).
[6]. P.D. Fainstein, V.H. Ponce and R.D. Rivarola, Phys. Rev. A36, 3639 (1987).
[7]. A. M. Ermolaev, Phys. Lett. A (1990) in print.
[8]. R.D. Olson, Phys. Rev. A36, 1519 (1987).
[9]. A. Müller-Grehling, G. Soff, Z. Phys. D9, 223 (1988).
[10]. D.R. Schultz, Phys. Rev. A40, 2330 (1989).
[11]. L.H. Andersen, P. Hvelplund, H. Knudsen, S.P. Møller, J.O.P. Pedersen, S. Tang-Petersen, E. Uggerhøj, K. Elsener, E. Morenzoni, Phys. Rev. A41, 6536 (1990).
[12]. M.B. Shah and H. B. Gilbody, J. Phys. B18, 899 (1985).
[13]. L.H. Andersen, P. Hvelplund, H. Knudsen, S.P. Møller, A. H. Sørensen, K. Elsener, K. G. Rensfelt, E. Uggerhøj, Phys. Rev A36, 3612 (1987).
[14]. H. Knudsen, L. Brun-Nielsen, M. Charlton, M.R. Poulsen, J. Phys. B (1990) in print.
[15]. D. Fromme, G. Kruse, W. Raith, G. Sinapius, Phys. Rev. Lett. 57, 3031 (1986).
[16]. E. Krishakumar, S.K. Srivastava, J. Phys. B21, 1055 (1988).
[17]. R.G. Montague, M.F.A. Harrison, A.C.H. Smith, J. Phys. B17, 3295 (1984).
[18]. L.H. Andersen, P. Hvelplund, H. Knudsen, S.P. Møller, J.O.P. Pedersen, S. Tang-Petersen, E. Uggerhøj, K. Elsener, E. Morenzoni, J. Phys. B23 , L395 (1990).
[19]. M.B. Shah and H. B. Gilbody, J. Phys. B15, 3441 (1982).
[20]. G. Basbas, W. Brandt and R. Laubert, Phys. Rev. A7, 983 (1973) and Phys. Rev. A17, 1665 (1978).
[21]. W. Brandt and G. Lapicki, Phys. Rev. A23, 1717 (1981).
[22]. H. Paul, J. Sacher, At. Data and Nucl. Data Tables, 42, 105 (1989).
[23]. M. H. Martir, A. L. Ford, J. F. Reading, R. L. Becker, J. Phys. B15, 1729 (1982).
[24]. W. Brandt and G. Basbas, Phys. Rev. A27, 578 (1983) and A28, 3142 (1983) (E).
[25]. G. Mehler, B. Müller, G. Greiner and G. Soff, Phys. Rev. A36, 1454 (1987).
[26]. L. Sarkadi, T. Mukoyama, to be published (1990).
[27]. E. Morenzoni, X90: X-ray and Inner Shell Processes, AIP Proceedings 1990.
[28]. F. Ebel, W. Faust, C. Hahn, M. Rückert, H. Schneider, A. Singe, T. Tobehn, Phys. Lett. A40, 114 (1989) and References therein.
[29]. R. Hippler, Phys. Lett. A144, 81 (1990).
[30]. L.H. Andersen, P. Hvelplund, H. Knudsen, S.P. Møller, K. Elsener, K. G. Rensfelt, E. Uggerhøj, Phys. Rev. Lett. 57, 2147 (1986).
[31]. L. J. Puchett, D. W. Martin, Phys. Rev. A1, 1432 (1970).
[32]. L.H. Andersen, P. Hvelplund, H. Knudsen, S.P. Møller, J.O.P. Pedersen, S. Tang-Petersen, E. Uggerhøj, K. Elsener, E. Morenzoni, Phys. Rev. A40 , 7366 (1989).
[33]. L. Vegh, Phys. Rev. A37, 992 (1988).
[34]. J.H. McGuire, Phys. Rev. Lett. 49, 1153 (1982).

[35]. O. Heber, B. B. Bandong, G. Sampoll, R.L. Watson, Phys. Rev. Lett. 64, 851 (1990)

[36]. L.H. Andersen, P. Hvelplund, H. Knudsen, S.P. Møller, J.O.P. Pedersen, S. Tang-Petersen, E. Uggerhøj, K. Elsener, E. Morenzoni, Phys. Rev. Lett. 65, 1687 (1990) (C).

[37]. M. Charlton, L. H. Andersen, L. Brun-Nielsen, B. I. Deutch, P. Hvelplund, F. M. Jacobsen, K. Knudsen, G. Laricchia, M. R. Poulsen, J.O.P. Pedersen, J. Phys. B21, L545 (1988).

[38]. J.O.P. Pedersen, P. Hvelplund, Phys. Rev. Lett. 62, 2373 (1989).

[39]. J.O.P. Pedersen, F. Folkmann, J. Phys. B23, 441 (1990).

[40]. M. Breinig, S.B. Elston, S. Huldt, L. Liljeby,C. R. Vane, S.D. Berry, G.A. Glass, M. Schauer, I. A. Sellin, G.D. Alton, S. Datz, S. Overbury, R. Laubert, M. Suter, Phys. Rev. A25, 3015 (1982).

[41]. P. D. Fainstein, V. H. Ponce, R. D. Rivarola, Phys. Rev. A40, 2828 (1989).

[42]. J. Burgdorfer, J. Wang, J. Müller, Phys. Rev. Lett. 62, 1599 (1989)

[43]. K.I. Komaki, K. Kuroki, Y. Yamazaki, S. Fujimoto, L.H. Andersen, E. Horsdal-Pedersen, P. Hvelplund, H. Knudsen, S.P. Møller, E. Uggerhøj, K. Elsener, J. Phys. Soc. of Japan, in print (1990).

[44]. V.N. Neelavathi, R. H. Ritchie, W. Brandt, Phys. Rev. Lett. 33, 302 (1974).

[45]. W.H. Barkas, N.J. Dyer and H.H. Heckman, Phys. Rev. Lett. 11, 261 (1963).

[46]. H.H. Andersen, J. F. Bak, H. Knudsen, B.R. Nielsen, A16, 1929 (1977).

[47]. For a recent discussion see S. P. Møller, Nucl. Instr. Methods B48, 1 (1990) and H. Bichsel, Phys. Rev. A41, 3642 (1990) and ref. therein.

[48]. L.H. Andersen, P. Hvelplund, H. Knudsen, S.P. Møller, J.O.P. Pedersen, S. Tang-Petersen, E. Uggerhøj, K. Elsener, E. Morenzoni, Phys. Rev. Lett. 62, 1732 (1989).

[49]. J. D. Jackson, R. L. McCarthy, Phys. Rev. B6, 4131 (1972).

[50]. J. Lindhard, Nucl. Instr. Methods 132, 1 (1976).

[51]. R. Medenwaldt, S. P. Møller, E. Uggerhøj, T. Worm, P. Hvelplund, H. Knudsen, K. Elsener, E. Morenzoni in preparation (1990).

[52]. A. H. Sørensen, Nucl. Instr. Methods B48, 10 (1990).

[53]. R. Dörner,F. Foberich, H. Schmidt-Böcking, J. Ullrich, R. E. Olson, Letter of Intent, CERN-PSCC/90-15, (1990).

[54]. A. L. Ford, J. F. Reading, J. Phys. B23, 2567 (1990)

[55]. R.E. Olson and T.J. Gay, Phys. Rev. Lett. 61, 302 (1988).

[56]. P.D. Fainstein, V.H. Ponce and R.D. Rivarola, J. Phys. B22, L559 (1989).

[57]. G. Spicher, B. Olsson, W. Raith, G. Sinapius, W. Sperber, Phys. Rev. Lett. 64, 1019 (1990).

[58]. M. Kimura and M. Inokuti, Phys. Rev. A38, 3801 (1988).

[59]. J. Eades and L.M. Simons, Nucl. Instr. and Methods A278, 368 (1989).

CHARGE ASYMMETRY EFFECTS IN POLARIZATION CHARACTERISTICS OF EXCITATION OF ATOMS BY FAST PROTONS AND ANTIPROTONS

V.V.Balashov, M.V.Gorelenkova, A.I.Magunov
Institute of Nuclear Physics, Moscow State University,
Moscow 119899, USSR

In the LEAR experiments with low-energy antiprotons [1] a charge asymmetry effect was observed in the total ionization cross section of atoms by protons (p) and antiprotons ($\bar{p}$). A similar effect should be expected [2] for the excitation of discrete atomic levels. The first calculations on this problem have been performed in [2] for the $1^1S \to 2^1P$ and $1^1S \to 3^1D$ transitions in helium excited by fast particles with opposite charges (μ^+ and μ^-, π^+ and π^-, p and $\bar{p}$). It has been shown that the effect predicted could be expected especially pronounced at excitation of two-step transitions. The relation between the excitation cross sections by positively charged particles $\sigma^{(+)}$ and by negatively charged ones $\sigma^{(-)}$ has been found to be in agreement with the sign of the Z^3- correction to the ionization energy loss formula for fast particles (Barkas effect [3]) as well as with the first direct measurements [4] of the antiproton stopping power.

It's well known from modern atomic physics that polarization and correlation characteristics of inelastic scattering of electrons and ions from atoms are especially sensitive to details of the excitation mechanisms as well as of the atomic structure. It is naturally to expect that the charge sign effect for excitation of atoms by protons and antiprotons can be even more pronounced in such characteristics,in particular, in the angular momentum alignment of excited atomic states.

We demonstrate here our calculations of the total cross sections as well as the polarization characteristics of excitation of helium and neon atoms by protons and antiprotons, based on the multichannel diffraction approximation by Feshbach and Hufner [5] which has been successfully used in atomic physics for the description of alignment effects in electron - heavy atom inelastic collisions [6]. The amplitude of inelastic scattering $\vec{k}_i \to \vec{k}_f$ with excitation of the $|i\rangle \to |f\rangle$

transition in atom is

$$F_{fi}(k_i,\theta) = -i^{\Delta M+1}k_f\int_0^\infty J_{\Delta M}(k_f b\sin\theta)\int_{-\infty}^{\infty} e^{ik_f(1-\cos\theta)}\frac{\partial u_f(b,z)}{\partial z}dzbdb, \qquad (1)$$

where $\Delta M = M_f - M_i$ -is the change of the angular momentum projection on the beam direction; $J_n(x)$- cylindrical Bessel function; θ - scattering angle of the projectile. Functions $u_f(b,z)$ are obtained by solving the system of N coupled equations:

$$u_n(b,z) = \delta_{ni} - i\frac{\mu}{h^2 k_n}\sum_{m=1}^{N}\int_{-\infty}^{z} V_{nm}(b,z')e^{i(k_m-k_n)z'}u_m(b,z')dz', \qquad (2)$$

where μ is the reduced mass of the colliding particles, $V_{nm}(b,z)$ - matrix element of the atom-projectile interaction in scattering plane ($\varphi = 0$).

The excitation amplitudes $F_{fi}(k_i,\theta)$ give the partial cross sections as well as the angular momentum density matrix for the excited state $|f\rangle$ or, equivalently, the set of the statistical tensors of this state:

$$\langle J_f M|\hat{\rho}|J_f M'\rangle = \sum_{kq}(-1)^{J_f-M'}(J_f M J_f -M'|kq)\ \rho_{kq}(J_f,J_f). \qquad (3)$$

The alignment parameter of the excited state is defined as $A_{20}=\rho_{20}/\rho_{00}$. In the case of the $^1S\to{}^1P$ collisionally induced transition followed by the $^1P\to{}^1S$ spontaneous deexcitation process the anisotropy parameter of the angular distribution of radiation $W(\theta)\sim 1+a_2P_2(\cos\theta)$ (relative to the incident beam direction) and its linear polarization P_L (when measured perpendicular to the incident beam direction) are given by A_{20} as

$$a_2 = \frac{A_{20}}{\sqrt{2}}, \qquad P_L = -\frac{3A_{20}}{2\sqrt{2} - A_{20}}. \qquad (4)$$

To test an applicability of this method for the specific problem of physics of ion-atom collisions discussed here we calculate a number of characteristics of excitation of atoms by protons. One can see from Figures 1 to 3 a rather close agreement between these calculations and those performed within other approaches. Comparison of these results with experimental data shows that the method used is quite adequate at least for proton energy higher than 200 keV (the charge exchange prosess should be taken into account at lower energies). For details of calculations see References [2,7].

Figures 4 to 6 show that the charge sign effects discussed are expected to be more stronger in alignment characteristics of excitation of atoms by fast protons and antiprotons than in the total cross sections. Note that in some energy region (approximately from 100 to 400 keV) the linear polarization fraction P_L of the radiation induced by protons and antiprotons is predicted of opposite sign.

The consideration shows that the experiments suggested to investigate angular momentum alignment effects under atomic excitation by antiprotons can give an important contribution to future development of the low energy antiproton physics.

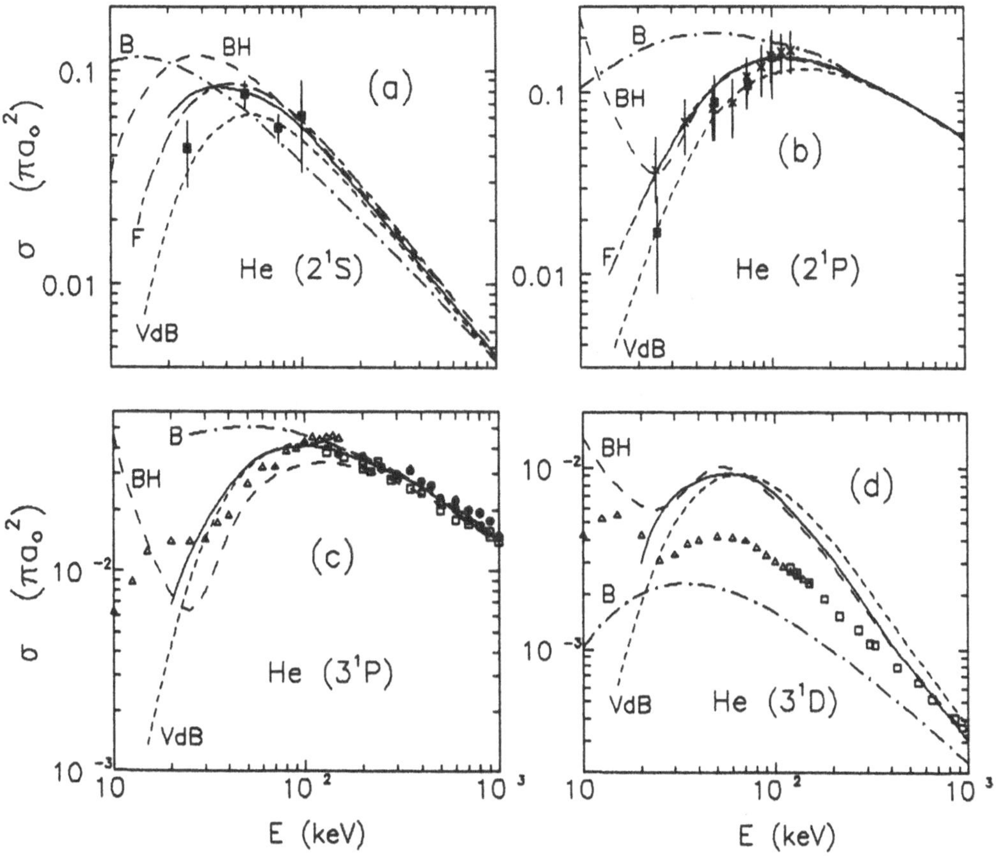

Figure 1:

Total cross sections for the excitation of helium by proton impact. Calculations: solid lines - our results for 16 channels (1^1S to 4^1D states); impact parameter treatment: F - 4 states (1^1S to 2^1P) approximation by Flannery [8], VdB- 9 states (1^1S to 3^1D, excluding 3^1D) approximation by Van den Bos [9], BH - 22 states (1^1S to 5^1D, excluding 4^1F) by Baye and Heenen [10]; B - Born approximation. Experimental data: (a) 2^1S - ■ [11]; (b) 2^1P - ■ [11], × [14]; (c) 3^1P - Δ[12], □[13], o[15], ●[16]; (d) 3^1D - Δ [12], □ [13].

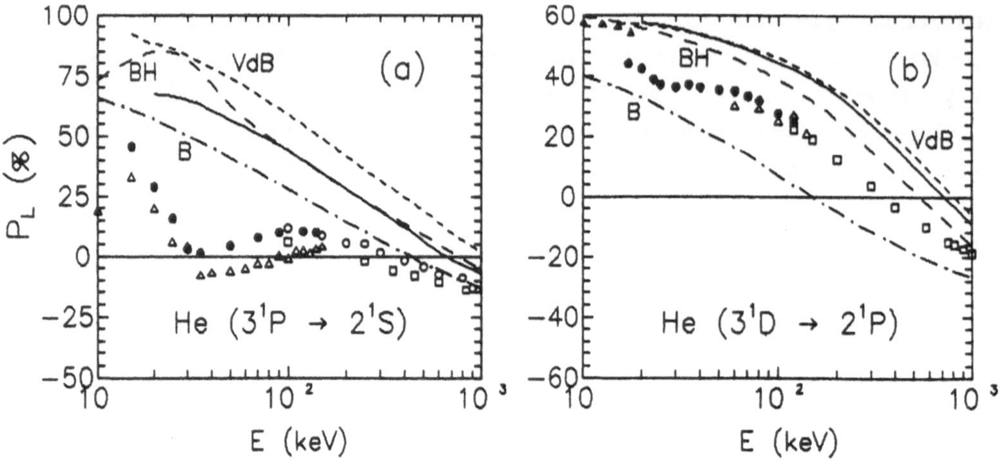

Figure 2:

Linear polarization fraction of the radiation induced by the excitation of helium by proton impact. Designation of curves as in Figure 1. Experimental data: (a) $3^1P \to 2^1S$ transition - Δ [12], □ [17], o [18], ●[19]; (b) $3^1D \to 2^1P$ transition - Δ[12], □[17], ●[19], ▲[20].

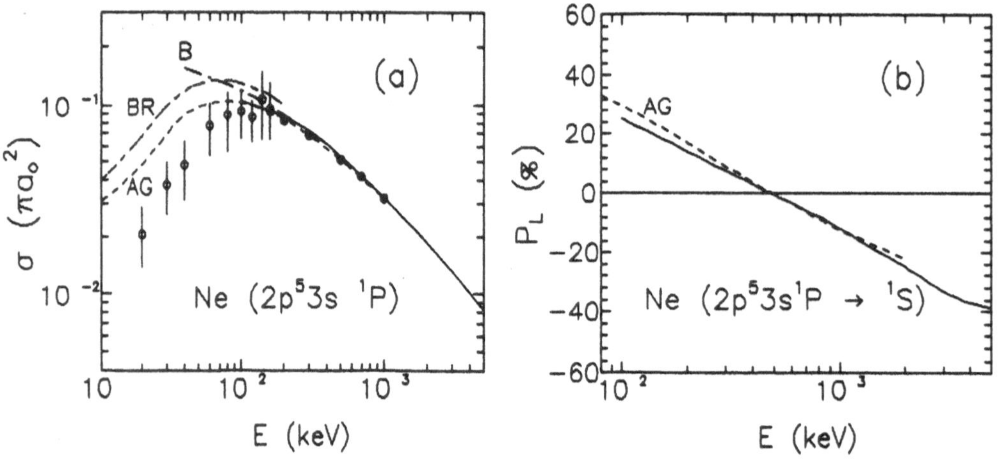

Figure 3:

(a) Total cross sections for the excitation of $2p^53s^1P$ state in neon by proton impact. (b) Linear polarization fraction of the $2p^53s^1P \to {}^1S$ radiation. Calculations: solid line - our results for 4 channels (^1S, $2p^53s^1P$, $2p^53p$ ^1S); AG - impact parameter method for 4 states ($2p^6$ ^1S, $2p^53s^1P$, $2p^53p$ ^1S) by Albat and Gruen [21]; BR - eikonal distorted wave approximation by Briggs and Roberts for $2p^53s$ configuration [22]; B - Born approximation. Experimental data: o [23], ● [24].

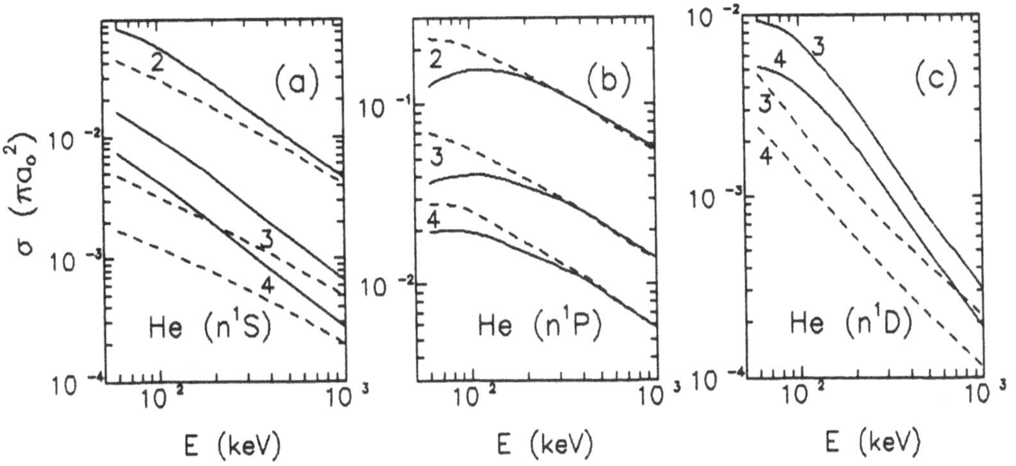

Figure 4:

Total cross sections for the excitation of helium by proton and antiproton impact. Solid lines for protons and dashed lines for antiprotons. (a) n^1S states (n=2,3,4); (b) n^1P states (n=2,3,4); (c) n^1D states (n=3,4).

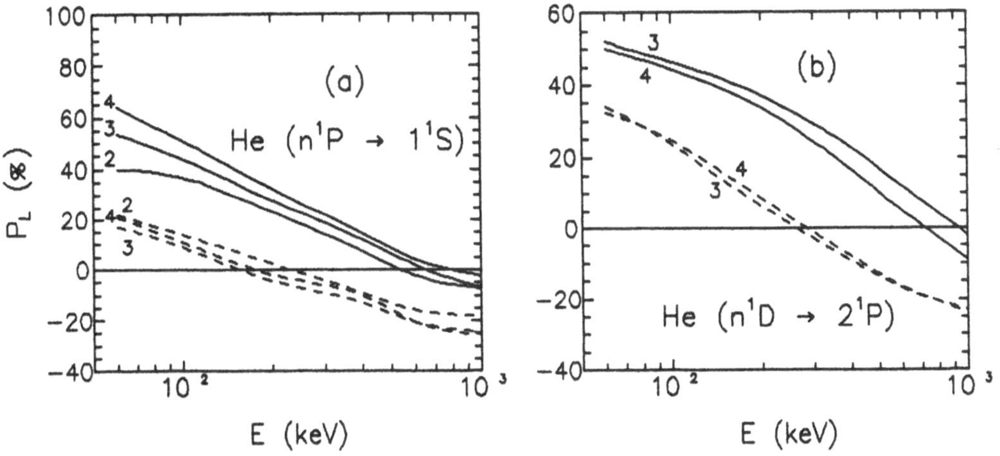

Figure 5:

Linear polarization fraction of the radiation induced by the excitation of helium by proton and antiproton impact. Solid lines for protons and dashed lines for antiprotons. (a) $n^1P \rightarrow 1^1S$ transitions (n = 2,3,4); (b) $n^1D \rightarrow 2^1P$ transitions (n = 3,4).

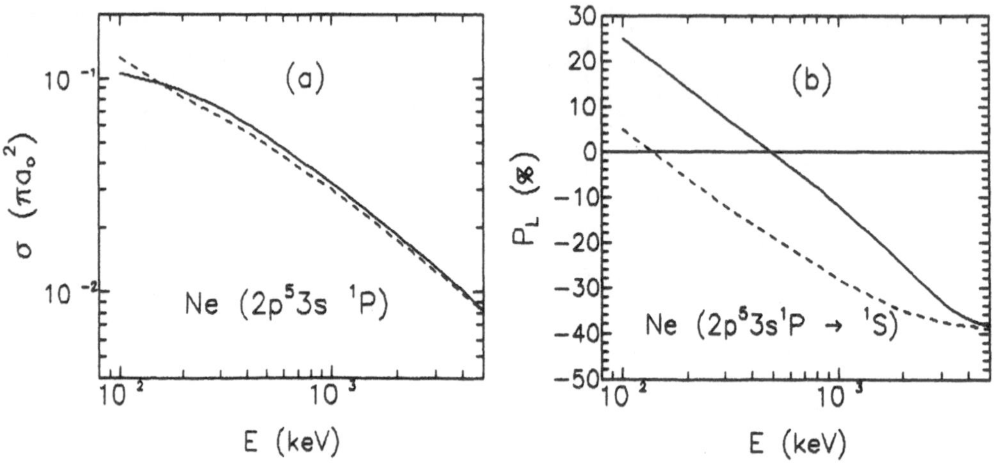

Figure 6:

(a) Total cross sections for the excitation of $2p^5 3s\ ^1P$ state in neon by proton and antiproton impact. (b) Linear polarization fraction of the $2p^5 3s\ ^1P \rightarrow {}^1S$ radiation induced by proton and antiproton impact. Solid lines for protons and dashed ones for antiprotons.

References

[1] Andersen L.H., et al.// Phys. Rev. Lett., 1986, V. 57, P. 2147.
[2] Balashov V.V., Gorelenkova M.V., Magunov A.I.// Preprint INP MSU 88-01637, 1988 (in Russian); Abstracts XVI ICPEAC, New York,1989, P. 494.
[3] Smith F.M.,Birnbaum W.,Barkas W.H.// Phys. Rev.,1953, V.91, P.765.
[4] Andersen L.H., et al.// Phys. Rev. Lett., 1989, V. 62, P. 1731.
[5] Feshbach H., Hufner J.// Ann. Phys., N.Y., 1970, V. 56, P. 268.
[6] Balashov V.V., Kozhevnikov I.V., Magunov A.I.// J. Phys. B, 1981, V. 14, P. 2059.
[7] Balashov V.V., Gorelenkova M.V., Magunov A.I.// Phys. Lett. A, 1990, V. 147, P. 223.
[8] Flannery M.R.// J. Phys. B, 1970, V. 3, P. 306.
[9] Van den Bos J.// Phys. Rev. A, 1969, V. 181, P. 191.
[10] Baye D., Heenen P.-H.// J. Phys. B, 1973, V. 6, P. 1255.
[11] Kvale T.J., et.al.// Phys. Rev. A, 1985, V. 32, P. 1369.
[12] Van den Bos J., Winter G.J., de Heer F.J.// Physica, 1968, V. 40, P. 357.
[13] Scharmann A., Schartner K.-H.// Z. Phys.,1969, V. 228, P. 254.
[14] Park J.T., Schowengerdt F.D.// Phys. Rev., 1969, V. 185, P. 152.
[15] Hasselkamp D.// Z. Phyz., 1971, V. 248, P. 254.
[16] Thomas E.W., Bent G.D.// Phys. Rev., 1967, V. 164, P. 143.
[17] Scharmann A., Schartner K.-H.// Phys. Lett. A, 1967, V.26, P.51.
[18] Hasselkamp D., Scharmann A., Schartner K.-H.// J. Phys. B, 1978, V. 11, P. 1975.
[19] Schartner K.-H., et.al.// Phys. Rev. A, 1983, V. 27, P. 816.
[20] Belanger M.G., et.al.// Abstracts V ICPEAC, Leningrad,1967, P.318.
[21] Albat R., Gruen N.// J. Phys. B, 1975, V. 8, P. 959.
[22] Briggs J.S., Roberts A.G.// J. Phys. B, 1974, V. 7, P. 1370.
[23] York G.W., et.al.// Phys. Rev. A, 1972, V. 6, P. 1497.
[24] Sharpton F.A., et.al.// Phys. Rev. A, 1970, V. 2, P. 1305.

F Ion–Solid Collisions

Chairman: K.-O. GROENEVELD

F.1 Invited Surveys

J. BURGDÖRFER
Electronic Excitations in Ion–Solid Collisions

S. ANDRIAMONJE, M. CHEVALLIER, C. COHEN, J. DURAL, R. GENRE, Y. GIRARD, K.O. GROENEVELD, J. KEMMLER, R. KIRSCH, A. L'HOIR, R. MAIER, J.C. POIZAT, Y. QUÉRÉ, J. REMILLEUX, D. SCHMAUS and M. TOULEMONDE
Charge Exchange Processes of High Energy Heavy Ions Channeled in Crystals

F.2 Invited Contributions

S.B. ELSTON
The Development of Projectile Excitation during Ion–Solid Collisions Observed through Convoy Electron Emission

H.D. BETZ, R. SCHRAMM and W. OSWALD
Convoy and Rydberg Electrons in Ion–Solid Collisions

F.3 Contributions (Posters)

J. WAGNER, R. HARDER, M. LOTTER, M. TEMPEL, H.-G. TOEWS, B. HERTEL, J. MAERITZ and H. KUIPER
Fast-Ion Rydberg State Production in Carbon Foils

L.L. BALASHOVA
Molecular Ion Reflection from Single Crystals

O. BENKA
Application of Secondary Electron Emission for Particle Identification in ERDA

Electronic excitations in fast ion-solid collisions

Joachim Burgdörfer
Department of Physics and Astronomy,
University of Tennessee, Knoxville, TN 37996
and
Oak Ridge National Laboratory, Oak Ridge, TN 37831, USA

Abstract

We review recent developments in the study of electronic excitation of projectiles in fast ion-solid collisions. Our focus will be primarily on theory but experimental advances will also be discussed. Topics include the evidence for velocity-dependent thresholds for the existence of bound states, wake-field effects on excited states, the electronic excitation of channeled projectiles, transport phenomena, and the interaction of highly charged ions with surfaces.

I. Introduction

Since the advent of the so-called "beam foil" spectroscopy almost three decades ago, an impressive amount of data[1] has been accumulated on the production of charge states and excitation states, both bound and continuum states, produced by the transmission of atoms and ions through solids. A complex array of interaction processes produces a variety of excited, sometimes "exotic", configurations not easily accessible by other means. This is to be contrasted with excitation by photons which is limited by stringent selection rules. Despite the extensive application of the ion-solid interaction as a spectroscopic tool and as an efficient ionizing agent ("stripper"), a microscopic understanding of the dynamics of the excitation process, of the evolution and transport of electrons accompanying fast ions, has been rather limited. Many fundamental questions concerning the existence, the modes of formation, and the lifetime of excited states first noted by Bohr and Lindhard[2] in their classic paper from 1954 are still unanswered. These difficulties result from the complexity of the interaction, from the importance of multiple scattering and from the fact that relevant perturbations are often sufficiently strong as to preclude a perturbational treatment. In recent years, however, considerable progress has been made in the theory of ion-solid collisions, in particular, since high-speed large scale computations allow an increasingly realistic simulation of the transient electronic dynamics. While these efforts are still in their infancy, they have already provided considerable insight into "what is going on" inside the solid.

To set the stage, Fig. 1 shows the evolution of the electron density as a proton approaches, penetrates, and finally exits a thin "foil." This calculation is performed in a time-dependent Hartree Fock (TDHF) approximation[3]. The foil is modelled by a slab of "jellium", i.e., an electron gas confined by a uniform positive background potential with sharp edges located at $z = \pm 10$a.u. with 20 a.u. thickness and a lateral extension of 40 a.u. The speed of the protons equals the Fermi velocity

$$v_F = (3\pi^2 N_e)^{1/3}, \tag{1}$$

the characteristic speed of the electron gas having density N_e. The unperturbed electron gas displays density oscillations (Friedel oscillations) inside the slab and an exponentially decaying tail of tunneling electron into the vacuum (Fig. 1a). Both are consequences of the quantum mechanical response of the electronic density to the presence of the surface potential step. As the proton approaches the surface (Fig. 1b) charge exchange by resonant tunneling into the Coulomb well of the proton sets in at rather large distances. At the same time density fluctuations in the target ("plasma oscillations") with plasmon

$$\omega_p = (4\pi N_e)^{1/2} \tag{2}$$

frequency are being excited. Inside the foil the proton drags along a charge cloud of screening electrons with a bow wave in front and a wake of density oscillations trailing

a)

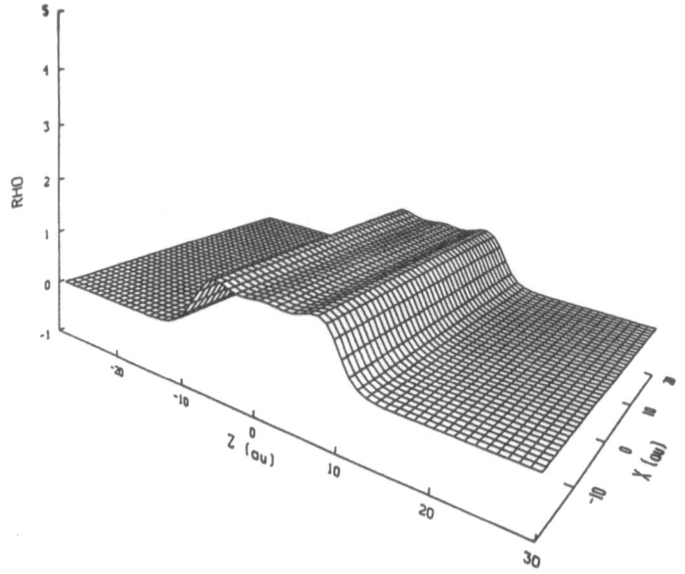

b)

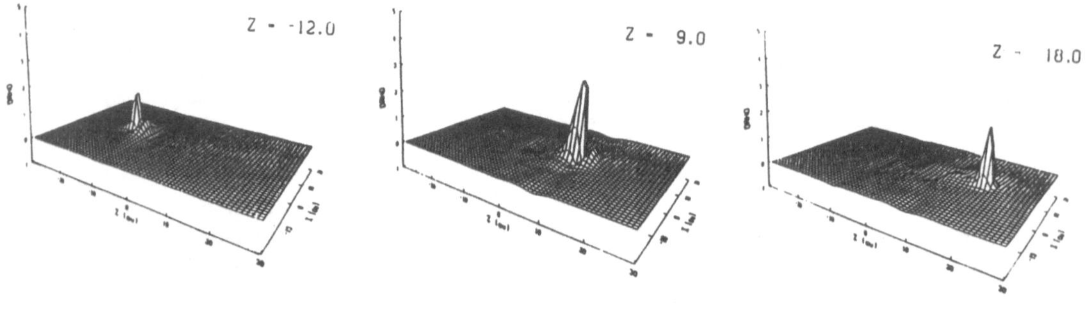

Fig. 1 a) Electron density of a jellium slab, $(-10 \leq z \leq 10)$, in ground state; **b)** electronic density fluctuations induced by a proton with $v_p = v_F$: at the positions $Z = -12$; $Z = 9$ and $Z = 18$a.u.; from Ref 3.

the ion. Finally, after the ion exits, part of the screening charge remains well localized around the protons, presumingly forming projectile centered bound states. Plasma oscillations in the medium still persist which will be eventually damped away. Thus, ion-solid collisions lead to excitation of the target and of the projectile.

Our focus will be primarily on the description of projectile centered states since here both experiment and theory are far more advanced. However, the recent observation[4] of electronic shock waves in the target is noteworthy. The density fluctuations trailing the ion are expected[5] to be confined within a Mach cone with cone angle

$$\theta_{cone} \simeq \cos^{-1}(v_F/v_P). \tag{3}$$

As the cone approaches the surface preferential low energy electron emission with $v_e \lesssim v_F$ is expected in a direction perpendicular to the shock wave front. The peak in the angular distribution of slow secondary electrons (Fig. 2) emitted by N^+ penetrating a thin foil of thickness $d = 1890$ a.u. ($v_P = 2.4$ a.u.) have been attributed to the emission of shock electrons. Their identification relies on the dependence of the peak angle on the collision velocity (Eq. 3). These data appear to be the first direct and detailed evidence for "collective" target excitations by fast ions. The well known plasmon loss peaks have been observed, up to now, only for electron transmission[6].

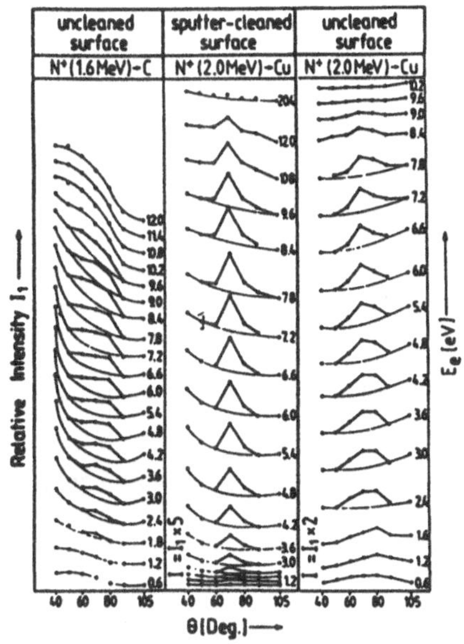

Fig. 2. Angular distributions ($40° \leq \theta \leq 105°$) of low-energy secondary electrons at different electron energies ($0.072 \leq E \leq 0.75$) from a C foil ($d =$ 1890 a.u., left), and a Cu foil ($d =$ 1890 a.u.) sputter-cleaned (center) and uncleaned (right), from Ref. 4.

In Sect. II we introduce basic concepts of the theory of projectile excitation in dense media and present simple-minded order-of-magnitude estimates in order to extract characteristic parameters of the problem. In later sections we will fill in on this physical picture with help of a few examples such as velocity dependent thresholds, dynamical screening, channeling, transport, and surface effects. In accordance with the guidance given to the authors by the organizers of this conference I will attempt to deemphasize the presentation of work from our group. Nonetheless, my selection of examples is far from being representative and inadvertently biased towards topics of my own interest. Atomic units will be used throughout unless stated otherwise.

II. Basic Concepts

The electronic evolution associated with the penetration of fast ions through matter is an extremely complex many-body scattering problem for which even the construction of the appropriate Schrödinger (or Lippmann-Schwinger) equation is a highly non-trivial task, let alone its solution at any satisfactory level. Drastic simplification based on intuition and hand-waving arguments have to be invoked in order to set up a tractable problem which, in turn, may be accessible to an approximate solution. The first step along the route of simplification concerns the translational degrees of freedom of the projectile. Since the mass ratio of electronic to nuclear mass $1/M \leq 10^{-3}$ is exceedingly small, the projectile ion is assumed to follow a classical constant-velocity trajectory

$$Z = v_p t, \tag{4}$$

neglecting thereby rare hard collisions with the ionic cores of the target. The projectile is assumed to be fast, i.e.

$$v_p >> v_F \simeq 1. \tag{5}$$

In cases where we want to employ perturbation theory for electronic transitions, the criterion (5) is sufficient only when electronic inner shell processes can be neglected. More rigorously, we should require

$$v_p \gtrsim v_K \tag{6}$$

where $v_K(\simeq Z_{p,T})$ is the orbital velocity of the K shell of the projectile or target, respectively.

Turning now to important parameters characterizing the bound state of the projectile, the binding energy (in hydrogenic approximation) is given by

$$\epsilon = -q^2/2n^2 \tag{7}$$

where q is the charge state of the ion. Heavy ions possess a charge state distribution characterized by a mean charge state $\bar{q}$ which may be different inside and outside the solid. Using a Thomas-Fermi model a phenomenological relation between $\bar{q}$, the nucleus charge Z_p and v_p can be found[7]

$$\bar{q} = Z_p(1 - \exp(-v_p/Z_p^{2/3})). \tag{8}$$

Other important parameters are the classical orbital frequency (or, quantum mechanically, the level spacing),

$$\omega_n = \frac{d\epsilon}{dn} = \frac{q^2}{n^3}, \tag{9}$$

and the orbital radius

$$< r >_n \simeq 3n^2/2q. \tag{10}$$

Projectile states are, in fact, not bound states but quasi-bound states, i.e. resonances in the continuum since they are subject to collisional destruction or radiative decay. Decay rates (inverse lifetimes) for radiative decay are of the order of

$$\Gamma_r = \frac{1}{\tau_r} \simeq \alpha^3 q^4/n^3, \tag{11}$$

($\alpha = 1/137$ the Sommerfeld fine structure constant) while an order-of-magnitude estimate (upper limit) for the collisional broadening of the state is given by the collision frequency of a free electron,

$$\Gamma_c = \frac{1}{\tau_c} = \frac{v_p}{\lambda_f} \tag{12}$$

where λ_f is the mean free path (mfp) of the free electron in the medium. For deeply bound states (Eq. 12) overestimates the collisional loss rate. The mfp is the characteristic parameter for the collisional perturbation by the solid. It contains contributions from elastic scattering at the ionic target cores (λ_{el}) and the inelastic electron-electron scattering (λ_{in}):

$$\lambda_f^{-1} = \lambda_{el}^{-1} + \lambda_{in}^{-1}. \tag{13}$$

As an estimate for λ_{el} we can use[8]

$$\lambda_{el} \simeq \frac{v_p^2/a_{TF}^2 + \epsilon_n^2}{4\pi Z_T^2 N_T} \tag{14}$$

where $a_{TF} \approx 0.885\, Z_T^{-1/3}$ is the Thomas-Fermi screening length (Z_T= nuclear charge of the target atom) and N_T is the number density of the target. (For monovalent targets, $N_T = N_e$). Eq.(14) is derived under the assumption of a quasi-free electron and is therefore not valid for small n. In the Rydberg limit, $n \to \infty$, Eq.(14) reduces to

$$\lambda_{el} \simeq v_p^2/(4\pi Z_T^2 a_{TF}^2 N_T) \tag{15}$$

The contribution to the mfp originating from inelastic collisions between the projectile electron and the valence electrons of the target can be estimated to be of the order of

$$\lambda_{in} \simeq \frac{v_p^2}{\omega_p ln\left(\sqrt{\frac{2}{\omega_p}} v_p\right)}. \tag{16}$$

In addition to the mfp (or, equivalently, the collision frequency), individual collisions are characterized by the collision time, the duration of the collisional perturbation which is of the order of

$$\Delta t_c \simeq a_{TF}/v_p. \tag{17}$$

with $\Delta t_c << \tau_c$. The nearest neighbor spacing of target atoms is of the order of

$$d \simeq N_T^{-1/3} \simeq 5\, a.u. \tag{18}$$

For propagation in amorphous media or in "random" directions of a crystal, the sequence of collisions is a stochastic random process with no well-defined time structure. For channeling in single crystals the (soft) collisional interaction with the lattice sites becomes periodic with frequency

$$\omega_{res} = 2\pi\, v_p/d \tag{19}$$

while the collisions with the electrons in the channel remain a stochastic process. Under channeling conditions $\lambda_{el}^{-1} \simeq 0$.

The interaction with the target electrons (treated as an "electron gas") has another profound effect: The collective response of the electron gas to the Coulomb field of the projectile leads to dynamical screening characterized by the dynamical screening length

$$\lambda_D = v_p/\omega_p. \tag{20}$$

For low v_p, the projectile velocity should be replaced by the Fermi velocity v_F in (20) defining the static screening length of the electron gas. The latter is somewhat larger than a_{TF} reflecting the fact that the valence electron density is smaller than the total electron density near the core. The effective potential due to dynamical screening, V_{sc}, (Fig. 3) reduces the binding energy of the projectile states and supports only a finite number of bound states due to its asymptotically short-ranged character. Furthermore it possesses an oscillatory wake which allows for the existence of exotic, non-atomic quasi-bound states of "wake-riding electrons."[9–11] The qualitative similarity between the dynamical screening potential (Fig. 3) calculated in plasmon-pole approximation for the dielectric function for an infinite medium and the density fluctuations calculated in the TDHF approximation for a jellium slab (Fig. 1), is quite remarkable in view of the vast differences in the underlying approaches.

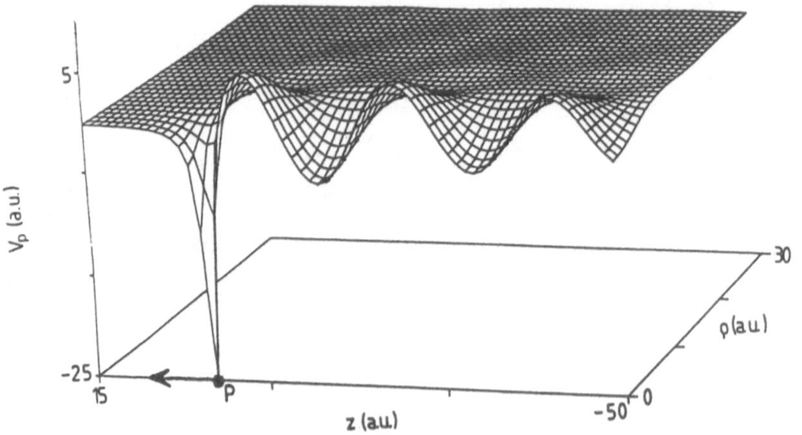

Fig. 3. Dynamical screening potential of S^{16+} in Al calculated in plasmon-pole approximation to the dielectric function ($v_p = 1$ a.u.).

We can now give a few order-of-magnitude estimates characterizing the production and evolution of projectile states inside the solid. Answering the most fundamental question to what extent excited states exist, if at all, several criteria have to be considered. The criterion most frequently discussed in literature is geometric in nature. The size of the orbital (Eq. 10) should not exceed the lattice spacing (Eq. 18),

$$3\frac{n^2}{q} \lesssim d, \tag{21}$$

thereby defining a cut-off quantum number

$$n_g \simeq \sqrt{dq/3} \tag{22}$$

above which bound states no longer "fit" inside the solid. We emphasize that (22) is neither necessary or sufficient since the stability of the electronic motion is determined by the dynamical processes, in particular, by the strength of the perturbation and the availability of open decay channels. The well known exciton states in semiconductors (in the static limit $v_p \to 0$) with diameters encompassing many lattice sites attests to the dynamic nature of stability limits.

Taking dynamical screening into account quasi-bound states cease to exist when the dynamical screening length and the radius of the orbital become comparable. Using (20) as characteristic length within which the Coulomb potential is altered we find

$$\frac{v_p}{\omega_p} \simeq \frac{3}{2}\frac{n^2}{q^2}, \tag{23}$$

or a critical quantum number

$$n_b \simeq (2v_p\, q/3\omega_p)^{1/2} \tag{24}$$

above which bound states no longer exist because of screening. In addition we can find a more stringent criterion for the existence of well-defined discrete bound states. We require that collisional broadening should not exceed the level spacing. Using for the latter the distance between energy shells (Eq. (9)), we find with help of Eqs. (12, 13),

$$\omega_n = \Gamma_c \tag{25}$$

or

$$n_c \simeq \left[\frac{q^2\, v_p}{\omega_p} ln \left(\sqrt{\frac{2}{\omega_p}} v_p \right)^{-1} \frac{1}{1 + \frac{3.1\pi Z_T^{4/3}\, N_T}{\omega_p} ln \left(\sqrt{\frac{2}{\omega_p}} v_p \right)^{-1}} \right]^{1/3} . \tag{26}$$

Eqs. (24) and (26) allow therefore to divide the projectile state space into different regions (Fig. 4): discrete core states ($n < n_c$) which closely resemble moderately perturbed atomic states, a continuum of negative energy states ($n_c < n < n_b$) which, in general, will be strongly perturbed having little resemblance to asymptotic stationary states, and finally a positive-energy continuum ($n > n_b$). On grounds of the Bohr correspondence principle, classical dynamics is expected to be valid for large quantum numbers, $n >> 1$ (more rigorously, all quantum numbers involved should be large). In the present case of a bound state spectrum embedded into a continuum via strong collisional coupling ($n > n_c$) , more complex considerations concerning the noise-driven quantum dynamics apply[12]. As a crude estimate we can use n_c (Eq. 26) as the classical-quantum border.

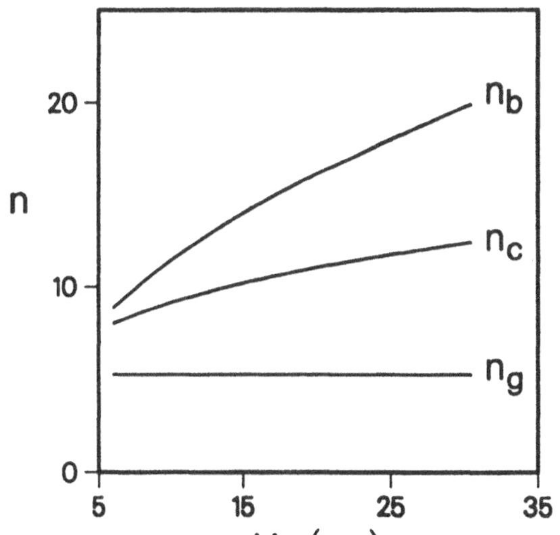

Fig. 4. Decomposition of excited state manifolds around a fast sulphur ion S^{16+} in a carbon foil as a function of v_p; n_c : classical-quantum border (Eq. 26); n_b: border of bound (negative energy) states (Eq.24); n_g: geometric border (Eq. 22). Only an order-of-magnitude estimate is given.

For core states we can, furthermore, distinguish between states quenched primarily by collisions and states decaying radiatively. The border is given by

$$\Gamma_c \simeq \Gamma_r. \tag{27}$$

or

$$n_r \simeq \left[\frac{\alpha^3 q^4 v_p}{\omega_p} ln \left(\sqrt{\frac{2}{\omega_p}} v_p \right)^{-1} \frac{1}{1 + \frac{3.1\pi Z_T^{4/3} N_T}{\omega_p} ln \left(\sqrt{\frac{2}{\omega_p}} v_p \right)^{-1}} \right]^{1/3} . \tag{28}$$

For $n < n_r$ radiative decay dominates over collisional destruction.

The evolution of core states can be described by rate equation models[13,14] (i.e., discrete master equations) with only a modest number of states included. The transition rates (W_{ij}) in the rate equation for the probability distribution P_i,

$$v_p \frac{dP_i}{dZ} = \sum_j (W_{ij} P_j - W_{ji} P_i) \tag{29}$$

are given in terms of atomic cross sections $\sigma_{j,i}$

$$W_{ij} = v_p N_T \sigma_{j,i}. \tag{30}$$

The channel indices (ij) in (29) refer to either different excited states of a given charge state or different charge states. Accordingly, $\sigma_{j,i}$ refer to the corresponding atomic cross sections for excitation, ionization, electron capture etc. The loss of phase coherence underlying the master equation (29) is due to the stochastic character of the collision process. This description has been applied to a large number of collision systems an example of which is shown in Fig. 5. The evolution of several charge states in 125 MeV

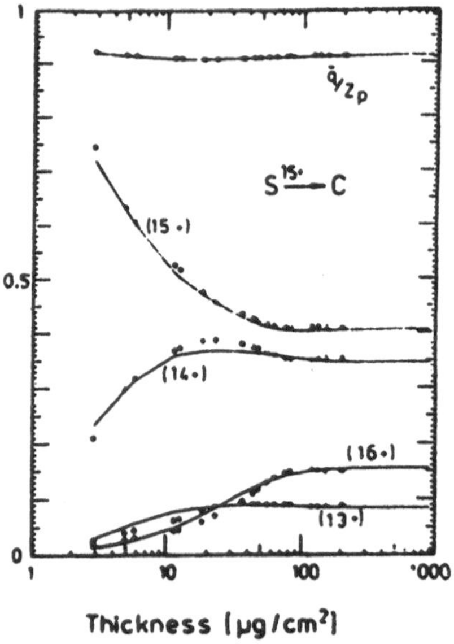

Fig. 5. Charge state of 125 MeV sulphur ions with initial charge state 15+ as a function of the thickness of carbon foil (o experiment; —— calculation), from Ref. 13.

sulphur as a function of foil thickness, i.e., as a function of distance of propagation can be traced quite accurately[13]. Limitations, however, have become apparent. Chetioui et al.[14] have reported on a perturbation of the ℓ distribution predicted by (29) due to the effective electric field produced by the anisotropic dynamical screening charge density. The atomic states in the solid are therefore Stark states. Furthermore, since this "wake" field is, to a good degree of approximation, time independent in the frame of the projectile, the characteristic interaction time is much longer (of the order of the dwell time in the solid) compared to the collision time for binary atomic collisions (Eq.17). Phase coherences due to field mixing may therefore persist and have been, indeed, observed[14].

For the formation of highly excited states beyond the border of existence of discrete quantum states ($n > n_c$) a different description of the continuous phase space distribution is needed. Within the framework of classical dynamics the electronic evolution is described by a phase space master equation for the distribution function ρ[15],

$$v_p \frac{\partial}{\partial Z} \rho(\vec{r}, \vec{v}, Z) = (\hat{L} + \hat{R}) \rho(\vec{r}, \vec{v}, Z), \tag{31}$$

where the classical Liouville operator

$$\hat{L} = -\vec{v} \cdot \vec{\nabla}_r + \vec{\nabla} V_p \cdot \vec{\nabla}_v \tag{32}$$

describes the phase space flow ("drift") due to the effective electron-projectile interaction V_p and $\hat{R}$ is the collision (integral) operator describing stochastic collisions of the electrons in the solid. The phase-space coordinates $(\vec{r}, \vec{v})$ refer to the projectile frame. One conceptual drawback of the classical dynamics approach is that the initial state of the electron, usually a tightly bound electron of the target or projectile, is a true quantum state. The somewhat surprising success of classical trajectory Monte Carlo (CTMC) calculations[16] in describing electronic transitions between low-lying states in ion-atom collisions indicates, however, a broader applicability of classical dynamics. An example for the application of (31) will be given in Section V.

Since in most cases the experimental observation refers to asymptotic states in the vacuum, the modification of the projectile state population upon exit from the solid plays an important role. The characteristic parameter for the transition is the transit time through the surface. For a particle exiting with angle δ relative to the surface normal the transit time t_s is of the order

$$t_s \simeq \frac{2}{v_p \cos \delta}. \tag{33}$$

For transmission problems with an outgoing trajectory close to the surface normal ($\delta \approx 0$) and high speeds the transit time is short ($t_s << 1$) compared to the orbital period ($\sim \omega_n^{-1}$, Eq. 9), except for deeply bound states. A second characteristic parameter is the strength of the surface potential. For a jellium model we have

$$V_o = W + \frac{1}{2} v_F^2 \tag{34}$$

in terms of the work function W and the Fermi energy. V_o has to be compared with the change of potential energy due to the breakdown of dynamical screening ($\sim q/\lambda_D$). The maximum strength for the time-dependent perturbation due to transit through the surface is therefore $\sim (V_o - q/\lambda_D)$. According to (33), for highly excited states the influence of the potential step can therefore be treated in sudden approximation by which the electron jumps from the potential surface of the dynamical screening potential to the bare atomic Coulomb potential in the vacuum[15]. As the transit time increases, e.g. for large angles of exit ($\delta \to \pi/2$), the influence of the surface potential becomes important. The latter has profound effects on electronic excitation in ion-surface scattering at grazing incidence (see Sec. VI). For low-lying states, the sudden approximation may not be valid. However, in this case the time-dependent perturbation is weak. Since

$$|V_o - q/\lambda_D| << \frac{q^2}{2n^2} \tag{35}$$

for $q >> 1$ and small n, core states are little affected during the penetration of the exit surface[17].

III. Velocity dependent thresholds for the existence of excited states.

After the above qualitative overview of fundamental interaction phenomena, we turn to specific examples illustrating recent progress.

Eq. (24) implies the existence of a characteristic threshold above which a bound state no longer exists because of dynamical screening. This threshold will depend on the projectile velocity, and the charge state involved as well as on the properties of the medium. The existence of a dynamical threshold has been, in fact, predicted some time ago[7,18]. Recently, Chevallier et al.[19] have found evidence for a velocity dependent threshold for the $n = 3$ manifold in He^+ (Fig. 6). The relatively abrupt change of the velocity dependent relative population fraction, $P(He^+(3p)/P(He^+)$, in collisions with a carbon foil near $v_p = 3.7$ a.u. was attributed to a threshold for the existence of a $He^+(3p)$ state inside the solid. These authors also find that in the same velocity region, $(2 \leq v_p \leq 5)$, the $2p$ state exists everywhere while the $4p$ state is unstable over the whole interval.[20] These results can be compared with a recent comprehensive theoretical study of atomic eigenstates in the dynamical screening potential by Müller et al[21]. Using both

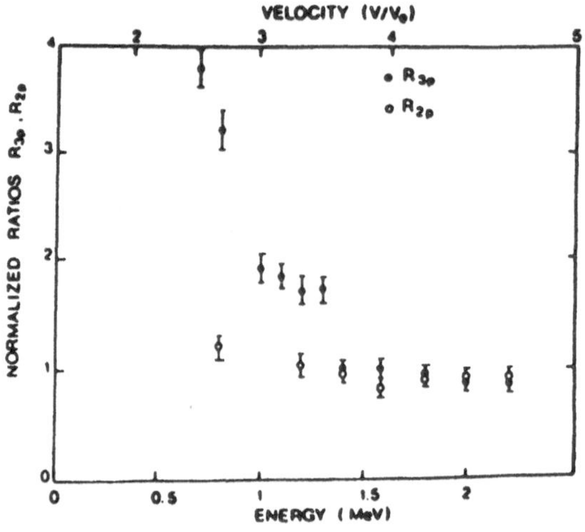

Fig. 6. Velocity dependence of ratios R, $P(He^+(3p))/(P(He^+)$ and $P(He^+(2p))$ $/P(He^+)$, normalized to unity for $v_p = 3.7$, from Ref. 19.

semiclassical Einstein-Brillouin-Keller (EBK) quantization, and diagonalization of the nonrelativistic hydrogenic Hamiltonian matrix in the truncated Hilbert space of ≈ 600 hydrogenic eigenstates of the same exact symmetry, a large number of eigenvalues have been calculated. Fig. 7 displays energy levels of the $n = 3$ manifold of the He^+

spectrum as a function of the inverse projectile velocity (i.e. proportional to the perturbation parameter q/λ_D). An excited state ceases to exist when the level is promoted to threshold

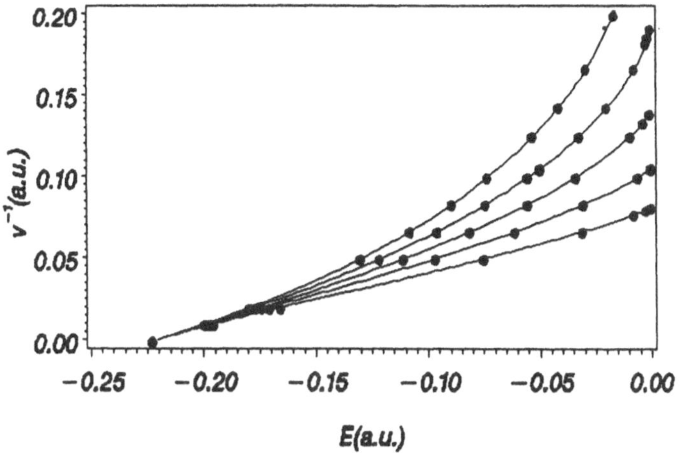

Fig. 7. Level splitting of the $n = 3$ manifolds of He^+ in carbon as a function of v_p^{-1} (proportional to perturbation parameter q/λ_D), approximate electric quantum number $k = -2, -1, 0, 1, 2$, from top to bottom.

($\epsilon = 0$). Because of a large number of avoided crossings in the vicinity of the threshold and the sudden transition upon exit (Eq. (33)) the diabatic extrapolation of the levels (avoided crossings are treated as real crossings) was used. The threshold for a given manifold is defined by the critical velocity v_c at which the last member of the manifold becomes unstable, i.e., enters the positive-energy continuum. Fig. 8 displays the critical

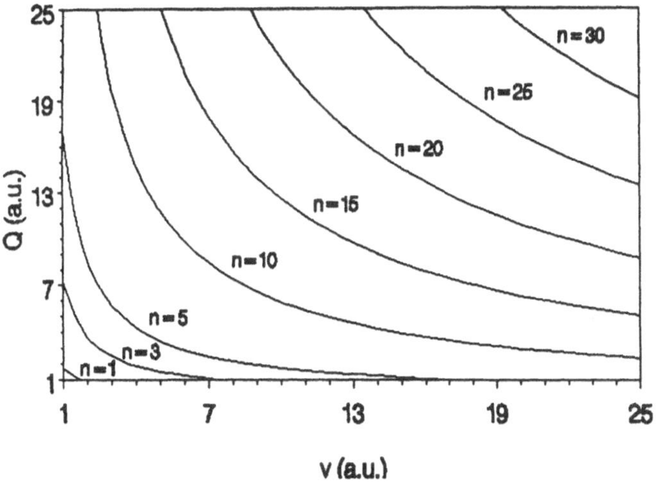

Fig. 8. Thresholds for existence of bound hydrogenic n manifolds for different charge states q and projectile speeds v_p.

threshold lines for the existence of a given manifold for different combinations of charge states and projectiles. For $He^{+}(q = 2)$ and $n = 3$ the threshold in carbons lies at $v_p = 3.6$a.u. in amazing agreement with the experiment. We expect this agreement to be, in part, fortuitous for several reasons: $n = 3$ manifold lies above n_c (Eq. 26). Sinc the collisional broadening (Eq. (12)). is expected to smear out any sharp threshold and poses a complication in determining an accurate threshold value. Furthermore, it appears that the experimental determination of sharp changes in population is subject to considerable uncertainty due to cascade contributions and the obstruction of the photon signal near the foil.

For small perturbation parameters (high speeds) the splitting of energy levels closely resembles the linear Stark effect (Fig. 7). The anisotropic screening charge cloud concentrated behind the ion creates an electric field with the magnitude[21]

$$F = \frac{\omega_p^2}{2v_p^2} q \ln(v_p/v_F) \tag{36}$$

near the nucleus thereby inducing Stark splitting of hydrogenic core states. This splitting has been first observed in experiments of resonant coherent excitation of channeled ions[22]. For core states of very highly charged ions, the Stark splitting

$$\Delta E_{Stark} = 3nkF/q \tag{37}$$

with the electric quantum number ($k = 0, \pm 1, \ldots, \pm(n-1),$) exceeds the level broadening $\Gamma = \Gamma_r + \Gamma_c$. The phase coherence leads in this case to quantum beats ("Stark beats") observable as intensity fluctuations of the $Ly\alpha$ radiation[14]. It should be noted that the absolute value of Stark splitting (Eq. 37) is, to first approximation, independent of q. The observation of Stark beats in highly charged ions becomes possible because of the reduction of the collisional broadening for deeply bound states (Eq. (14)).

IV. Electronic excitation of channeled ion

The most profound change in the excitation dynamics of channeled ions as compared with those propagated along "random" directions is the strong suppression of collisions between projectile centered electrons and target cores. This implies several important consequences:

(a) The elastic mean free path (Eq. 14) becomes very large (ideally, infinite). Only collisions between valence target electrons and the projectile electrons are possible.
(b) Collisional excitation and ionization displays threshold and resonance phenomena. Since the ionic cores of the target no longer act as a "reservoir" of a broad distribution in momentum and energy, only the energy ($\frac{1}{2}v_p^2$) and momentum (v_p) of valence electrons as seen in the projectile frame of reference can mediate excitation, thereby limiting the number of open channels. This property has been utilized by Datz et al.[22] to measure dielectric excitation and ionization of highly charged hydrogenic ions. The valence electron density in the channel acts like an extremely dense electron beam, as seen in the rest frame of the projectile. The space charge problem limiting the formation of a laboratory electron beam of high density is elegantly taken care of by the positive background potential provided by the target cores. If beam energy lies below the first excitation threshold,

$$\frac{1}{2}v_p^2 < \epsilon_{n=2} - \epsilon_{n=1} \tag{38}$$

the direct excitation channel is closed. However, the dielectronic excitation channel may be open near a resonance by which the energy defect to the $n = 2$ level is compensated by the capture of the impacting electron into a negative energy state of the projectile.

Experimental signatures of this two-electron process are either $2p \rightarrow 1s$ X rays or a change of the charge state ($q \rightarrow q-1$). Fig. 9 presents recent experimental data[23] for hydrogenic Titanium (Ti^{21+}) together with a solution of the rate equation (Eq. (29)). This data allows a test of the calculated dielectric capture cross section (proportional to the inverse Auger rate) for gas-phase collisions entering the rate equation. The width of the dielectronic excitation resonance is determined by the width of the momentum distribution of the valence electrons ($\approx 2v_F$). It should be noted that the second step in the dielectronic recombination, the radiative stabilization by X ray emission is suppressed relative to gas-phase collisions due to the collisional loss rate for excited states determined by λ_{in}. Consequently, dielectronic excitation in channels may result in dielectronic ionization (i.e. loss of the remaining 1s electron along the sequence $1s \rightarrow n\ell \rightarrow$ continuum) or in relaxation back to the initial charge state $Ti^{21+}(1s)$ via emission of one photon followed by collisional loss rather than in capture (Ti^{20+}, via emission of two photons). The availability of a high density electron beam in channels has been recently used by Claytor et al.[24] for a measurement of the electron impact excitation cross sections for H- and He-like Uranium at GeV energies.

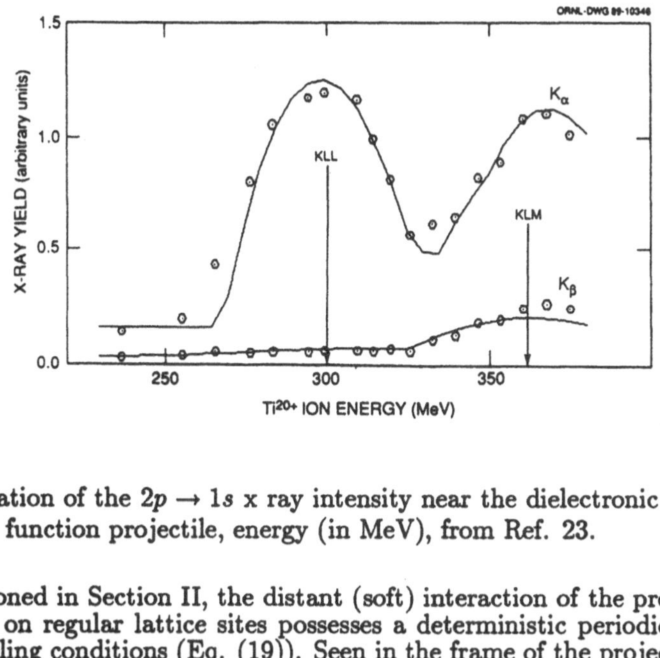

Fig. 9. Variation of the $2p \rightarrow 1s$ x ray intensity near the dielectronic resonance ($2p^2$) in Ti^{20+} as a function projectile, energy (in MeV), from Ref. 23.

As mentioned in Section II, the distant (soft) interaction of the projectile with the target atoms on regular lattice sites possesses a deterministic periodic time structure under channeling conditions (Eq. (19)). Seen in the frame of the projectile the interaction with the lattice structure acts as a coherent electric field capable of driving resonant transitions. This effect, predicted by Okorokov[25], was first unambiguously observed[22] in the resonantly modulated charge state fractions of Ni^{7+} ions as a function of the projectile when ω_{res} (or one of the higher harmonics) coincides with the excitation frequency $\omega = \epsilon_{2p} - \epsilon_{1s}$. The enhanced fraction of bare ions results from the collisional ionization prior to radiative relaxation (i.e, $\Gamma_c >> \Gamma_r$).

Recently, Kimura et al[26] have found evidence for resonant coherent excitation (RCE) in the yield of "convoy electrons". Convoy electrons correspond to near-threshold states in the continuum ($\epsilon \simeq 0$) and manifest themselves as a sharp peak ("cusp") in the forward spectrum for electron emission. The yield of convoy electrons in coincidence with carbon ions normalized to the number of exiting hydrogenic ions (C^{5+}) displays a pronounced enhancement when the ion beam velocity is tuned to the 2nd harmonic resonance of a gold < 100 > channel (Fig. 10b). The observation of enhanced convoy electron emission for RCE allows detailed insight into the production mechanism. We note first that the thickness of the gold crystal ($\approx$ 1600Å) is large compared to the mean free path λ_{in}. Any transient enhancement has therefore decayed as the projectile reaches the exit surface. The change in convoy electron production is due to changes in preequilibrium composition of the beam near the RCE resonance.

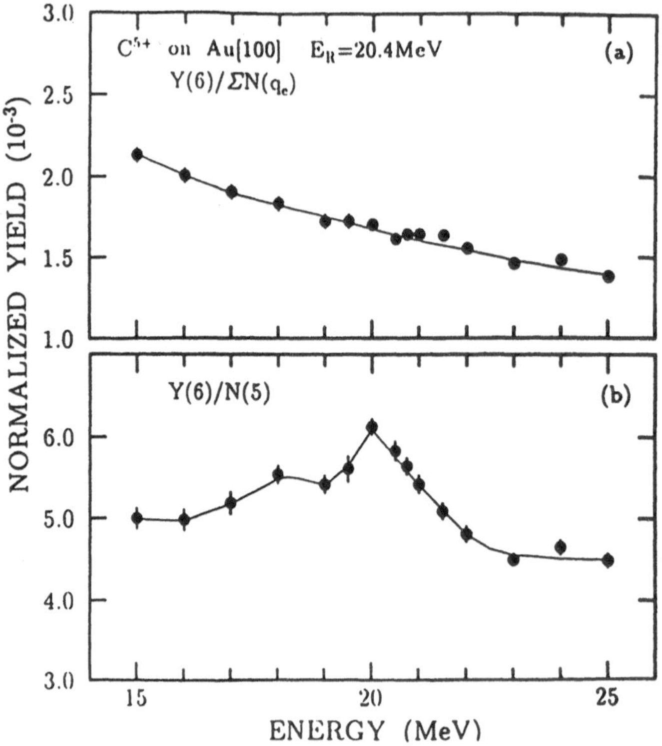

Fig. 10. Yield of convoy electrons in coincidence with outgoing C^{6+} emitted by incident C^{5+} on Au[100] near the RCE resonance. **a)** normalized to the incident beam. **b)** normalized to exiting C^{5+}, from Ref. 26.

Since the charge state distribution is far from equilibrium at the exit surface, we use the non-equilibrium solution to the master equation for the initial condition, $P_{1s}^{5+}(Z=0)=1$,

$$P_{1s}^{5+} = e^{-N_e Z(\sigma_{1s}^{5+}+\sigma_{1s-n=2}^{5+})}$$

$$P_{n=2}^{5+} = \frac{\sigma_{1s-2}^{5+}}{\sigma_{1s}^{5+}+\sigma_{1s-2}^{5+}-\sigma_{2}^{5+}} \left[e^{-N_e Z\sigma_2^{5+}} - e^{-N_e Z(\sigma_{1s}^{5+}+\sigma_{1s-2}^{5+})}\right], \qquad (39a\&b)$$

where N_e refers to the electronic density in the channel, σ_n^{5+} is the electron impact ionization cross section for the n shell of C^{5+} and σ_{1s-2}^{5+} is the excitation cross section $1s \to 2s, 2p$. The later will be modified by the additional RCE cross section σ_{RCE},

$$\sigma_{1s-2} \to \sigma_{1s-2} + \sigma_{RCE} \tag{40}$$

near the resonance. Eq.(39) follows from (29) if one takes only $n = 1$ and $n = 2$ core states into account and neglects all capture processes, which are small under channeling conditions, as well as deexcitation processes. The neglect of higher shells is only justified to the extent that they are not directly influenced by the resonance but only contribute to the incoherent yield.

The longstanding hypothesis[27] that convoy electron emission proceeds preferentially through electron loss from excited states can now be directly verified: Tuning through the RCE resonance, the changes in the yield of convoy electrons per incident ion (Fig. 10a) should, to a good degree of approximation, mirror the behavior of excited state fraction, F_2^{5+} (Fig. 11), but should display no correlation to either the bare ion fraction F^{6+}, or the ground state, F_{1s}^{5+}. The-at first glance-surprising observation that the "pumped" final state, $F_{n=2}^{5+}$, as well as the convoy yield (Fig 10a) show no resonant enhancement is consequence of the fact that both the total loss from excited states as well as ELC has a much larger cross section than from the ground state. Consequently, the resonant "pumping" of the $C^{5+}(n = 2)$ effectively pumps the C^{6+} charge state because of the rapid loss. Furthermore, since ELC cross sections for excited states are much larger, the depletion of the ground state is inconsequential for the convoy yield.

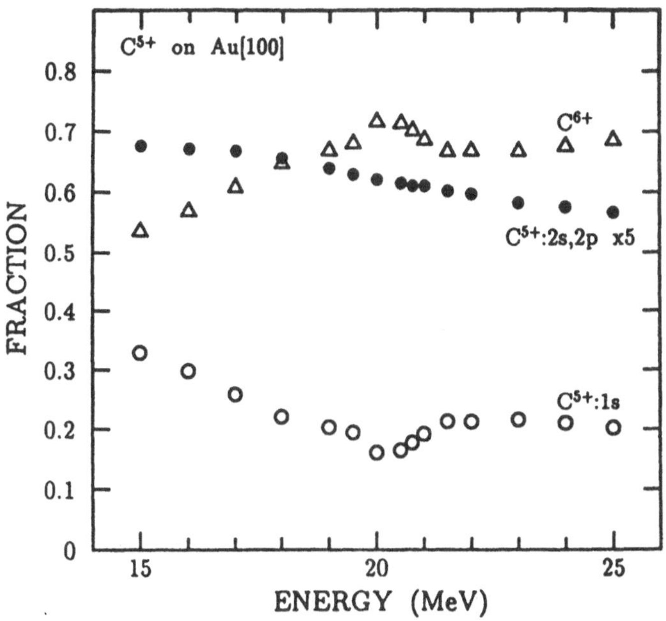

Fig. 11. Charge state (C_{1s}^{5+}, C^{6+}) and excited state $C^{5+}(n = 2)$ fraction near the RCE resonance, from Ref. 26.

We also note that the population fractions as well as the convoy yield displays a broadened resonance which is due to both collisional broadening and the Stark splitting of the $n = 2$ manifold. A comparison between convoy electron spectra for random and non-resonant channeling conditions has been recently performed at GANIL by Andriamonje et al.[28] for bare Xenon ions($q = 54$) and projectile velocities $v_p = 40\,a.u.$ One remarkable result was the expected strong suppression of electron capture to continuum under channeling conditions since mechanical capture of (quasi) free electrons is forbidden.

V. Transport processes and the production of high-ℓ states

For highly excited states beyond the border of discrete quantum states, the changes in population are described by the transport equation for the probability flow in phase space (Eq. (31)). Direct evidence for multiple scattering effects in the final projectile state population have been found in the abundance of high ℓ states absent in ion-atom collisions under otherwise similar conditions[29].

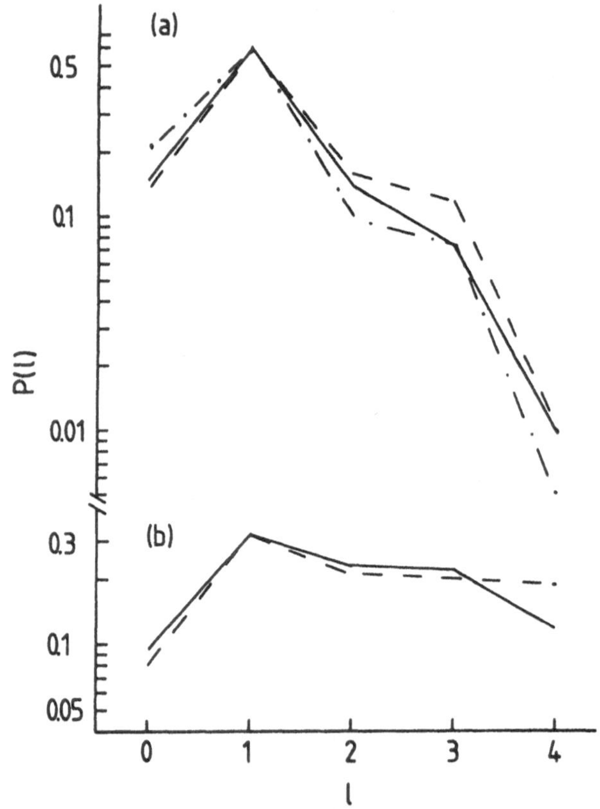

Fig. 12. ℓ-distribution $P(\ell)$ in $n = 5$ of $C^{2+}(2p5\ell)$ ($v_p = 2.25$ a.u.). **a)** - - - , experimental data (Ref. 30) for $C^{3+} + He \rightarrow C^{2+}(2p5\ell) + He^{+}$, —·—·—·—, experimental data (Ref. 30) for $C^{2+} + He \rightarrow C^{2+}(2p5\ell) + He^{+}$; ——, simulated initial distribution for random walk. **b)** - - -, experimental data (Ref. 30) for $C^{2+}(2p5\ell)$ on carbon foil; ——, escape-depth-averaged (steady state) solution of the Langevin equation, from Ref. 31.

A detailed comparison between experiment and theory has recently become available for doubly excited states in carbon $C^{2+}(2p5l)$ produced in collisions with carbon foils[30]. According to Eq. (26) the $5l$ orbital lies at the experimental collision velocity $v_p = 2.2\,a.u.$, well above n_c (in fact, in the positive-energy continuum according to (24)), and the classical phase-space master equation should be applicable. Comparison with the simultaneously measured ℓ distribution for both excitation $C^{2+} \rightarrow He$ and electron capture $C^{3+} \rightarrow He$ in gas-phase collisions clearly display the shift to high-ℓ states (Fig. 12). In order to treat the ℓ evolution within the framework of transport theory[31] we have chosen an initial distribution in ℓ, also shown in (Fig. 12), which is modeled after the experimental ion-atom collision data (Fig 12). The mapping of the continuous classical ℓ distribution onto discrete quantum numbers was made with use of equally spaced bins centered around the semiclassical value $\ell + 1/2$. An equivalent method has been used for n. The resulting asymptotic stationary ℓ distribution in $n = 5$ (averaged over different escape depths) after foil exit shows astonishingly good agreement with the data. The only statistically significant deviation, by a factor $\simeq 1.5$, from the experimental data appears to occur in the g-state population.

The physical picture underlying the dramatic shift of the ℓ distribution to high ℓ states as compared to ion-atom collisions under single collision conditions is that multiple scattering of electrons during their correlated motion with the projectile embodied, in the transport equation, leads to a "diffusive"-like redistribution among ℓ states. Several experimental groups have reported on an enhanced transport depth for convoy electrons[32]. A similar diffusion in n quantum number, i.e., energy, in H^o has been recently observed by Mohagheghi et al.[33] in transmission through thin foils of a relativistic H^-beam with an energy of 800 MeV. A more detailed discussion of the solution to the transport equation for projectile electrons can be found in Ref. 15.

VI. Ion-surface collisions

Electronic excitation processes of ions approaching surfaces (ideally without even penetrating the surface) have become a field of great interest, both from the viewpoint of fundamental interactions between atomic systems and a semi-infinite many-body system and from the viewpoint of a large number of applications, notably, the properties of a confined plasma near the wall of the container and the production of level inversions with potential applications to x-ray lasers.

Recent experimental progress in this field is closely linked to the advent of sophisticated techniques for surface preparation and surface diagnostics which allow controlled scattering experiments at microscopically clear surfaces. In addition, the development of ECR and EBIS ion sources has expanded the range of available charge states of low to intermediate energy projectiles.

The dynamics of neutralization of a highly charged ion has been studied in several laboratories[34–37]. The rapid neutralization of a highly charged ion by transfer of 10 or more electrons is a true multi-electron process for which not even a first-order approximation is currently available. This flow of electrons is assumed to occur via resonant charge transfer from the conduction band of the solid to high Rydberg states, followed by stepwise relaxation by sequential Auger transitions. However, the characteristic time constant for relaxation as estimated from conventional two-electron Auger matrix elements is too slow to fill inner shells prior to impact. The disparity of time scales for resonant transfer and relaxation leads to a "bottleneck" and thereby to the formation of "hollow" atoms. Fig. 13 displays the complex satellite structure of a He-like X ray transition for Au^{17+} impinging on a Ag surface with an energy of $\approx 20\,keV$ per charge. States with up to 7 holes in the K and L shells but 5 electrons in the M shell have been observed [38]. A theoretical description of this process is, as of now, missing. Apart from the obvious difficulty in describing such a violent n electron process, a more subtle problem consists in reconciling vastly different time scales. If one accepts the notion[36] that the multiple-step neutralization process occurs on the incoming part of the trajectory before the ion reaches the surface, the limited available time suggests that the relaxation process in a hollow atom in proximity of a surface differs dramatically from Auger decay of an isolated atom with only one or a few vacancies in inner shells. If at all realistic, electron-electron interactions ("correlations") and electron-surface interactions must

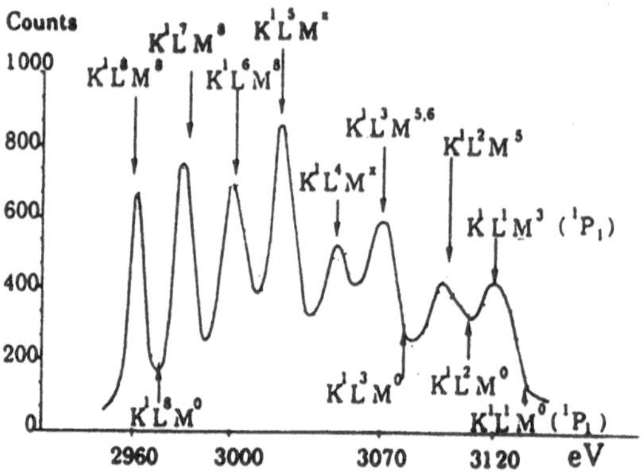

Fig. 13. $(1s2p \rightarrow (1s)^2$ x ray spectrum observed for Ar^{17+} ions impinging on Ag target, the identification of satellite lines is indicated, from Ref. 38.

play a decisive role in such a fast relaxation. Very likely, an adequate theoretical description should include quasi-resonant charge transfer between "intermediate" shells of the target and lower-lying excited states of the projectile at small distances from the surface ($\approx$0.5 to 3 a.u.). One may speculate that a classical n-body dynamics calculation which allows non-perturbative treatment of the interactions may be a promising approach to a theoretical description. The high density of states and the short time scale involved suggest that the quantal spread of the wavepacket should be of minor importance. Work along those lines is in progress[39].

Electron emission in grazing incidence collisions has recently been used in exploring the long-range interactions near the surface. Electrons emitted in ion-surface collisions have been predicted[17] to be subject to acceleration by the image potential of the accompanying ion, provided the characteristic interaction time t_s (Eq.(33)) is sufficiently long. For convoy electron emission at grazing incidence with $\delta_{ion} \approx \delta_{electron} \approx 89^o$, strong distortions due to the image potential are expected: the convoy electron peak is shifted[40,41] and broadened[42,43].

The electron in close proximity to the projectile receding from the surface is subject to the projectile potential and to the dynamical surface screening potential. The latter is given in a strongly simplified approximation by[40,44]

$$V_{image}(\vec{r}) = (2q\omega_s/v)\sin((z-vt)\omega_s/v_p)K_o\left[(\omega_s/v)[(x+x_o)^2+y^2]^{1/2}\right]$$
$$\exp((z-v_pt)\gamma/4v)\theta(v_pt-z), \qquad (41)$$

where K_o is the modified Bessel function, $\omega_s = \omega_p/\sqrt{2}$ is the frequency of the surface plasma, γ is the damping constant of the plasma, and θ is the step function. The projectile is assumed to travel parallel to the surface at distance x_o with velocity v_p. The effect of the electronic self-image can be, to lowest order, taken into account by replacing $q \to (q-1)$ in (41), assuming that the average position of the electron coincides with that of the ion. Accordingly, the line shift of the convoy electron is given by

$$\Delta E \simeq \int_{x_c}^{\infty} \frac{dx_o}{\sin\theta_{out}} F_Z(x_o), \tag{42}$$

where $F_Z = -\frac{\partial}{\partial Z} V_{image}$. Fig. 14 displays recent experimental data[40] together with the estimate(Eq.(42)) for the fractional change in energy of the convoy peak $(E_o + \Delta E)/E_o$, where $E_o = v_p^2/2$ is the expected position of convoy peak. Projectile ions (H^+, He^{++}) impinge on a SnTe (001) surface at an angle of incidence of $(\theta_i = \pi/2 - \delta_i)$=6 mrad. While the quantitative agreement is poor, this model reproduces the qualitative features of the data quite well. For H^+ no peak shift is expected since $q - 1 = 0$. For He^{++} the projectile energy dependence of the peak shift is reproduced, while the magnitude is overestimated. The agreement can be improved by an ad hoc increase in the angle of the outgoing trajectory(θ_{out}) from 6 to 25 mrad. This is in accord with the observation that the electronic and projectile scattering angles do not coincide. Clearly, better quantitative agreement with the data can be expected for a theoretical treatment employing the phase space master equation (Eq. 31). The study of convoy peak shifts has been recently extended by Koyama et al.[41] to highly charged projectiles with q up to 27. Here the peak shift reaches 250 eV.

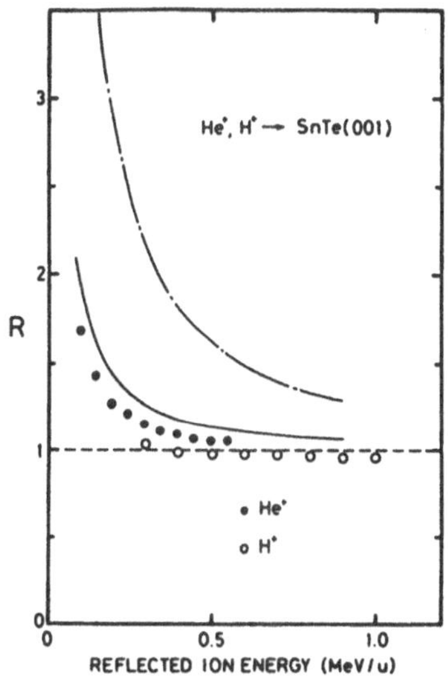

Fig. 14. Ratio of convoy electron energy position to expected convoy energy $(E_o + \Delta E)/E_o$ as a function of projectile energy (in MeV/u), angle of incidence θ_{in} = 6m mrad; o H^+ projectiles; • He^+calculations for exit angle relative to the surface $- \cdot - \cdot -$ $\theta_{out} = 6$ mrad; and —— $\theta_{out} = 25$ mrad, from Ref. 40.

Along similar lines, the broadening of the convoy peak can be explained in terms of an image-charge induced dipole final state interaction[42]. For large distances between the electron and the projectile, the electron moves in an effective dipole field generated by the projectile ion and its image. Employing the threshold law for a dipole $(\sim 1/r^2)$

potential, the cross section near threshold (as seen in the rest frame of the projectile) behaves as[43]

$$\frac{d\sigma}{d\epsilon} = const\, v^{-1+2\beta}, \tag{43}$$

where β depends on the angular momentum of the potential and the dipole moment. Transforming (43) to the lab frame gives a broadened cusp relative to the Coulomb cusp for all $\beta > 0$. Using β as free parameter, good agreement with the experimentally observed shape of the convoy peak has been found for $\beta \approx 0.4$, almost independent of the collision velocity.[43] This result underscores the importance of long-range potentials for the electronic excitation process in ion-surface collision.

VII. Summary

This short overview was intended to illustrate progress in the theoretical understanding of ion-solid collisions. At the same time, recent experimental advances point to the need for and to the directions of future investigations. They are motivated by both the many unsolved fundamental questions of properties of atomic systems in dense media and by their potential applications in the field of surface diagnostics, radiation damage, and coherent x ray light sources.

The author acknowledges important contributions made by many colleagues and collaborators. They include C. Bottcher, S. Elston, J. Gibbons, J. Kemmler, K. Kimura, J. Müller, I. Sellin, J. Wang and H. Winter. This work was supported in part by the National Science Foundation and by the U.S. Department of Energy, Office of Basic Energy Sciences, Division of Chemical Sciences, under Contract No. DE-AC05-840R21400 with Martin Marietta Energy Systems, Inc.

References

1. For an overview see recent proceedings of the conference series "Atomic Collisions in Solids," Nucl. Instr. Meth. B2 (1984); B13 (1986), B33 (1988), B48 (1990).
2. N. Bohr and J. Lindhard, K. Dan. Vidensk. Selsk. Mat. Fys. Medd. 28 (1954) No.7.
3. K. Schafer, N. Kwong, and J.D. Garcia, Computational Physics Communications (1990, in press).
4. H. Rothard, M. Burkhard, J. Kemmler, C. Biedermann, K. Kroneberger, P. Koschar, O. Heil, and K.O. Groeneveld, J. de Physique, 48-C9, 211 (1987).
5. W. Schäfer, H. Stöcker, B. Müller, and W. Greiner, Z. Physik A288, 349 (1978).
6. J. Ashley, J. Cowan, R. Ritchie, V. Anderson, and J. Hoelzl, Thin Solid Films 60, 361 (1979), and ref. therein.
7. See e.g., W. Brandt in Atomic Collisions in Solids, eds. S. Datz et al., (Plenum, N.Y., 1975) p. 261.
8. D. Röschenthaler et al., J. Phys. B16, L233 (1983).
9. V. Neelavathi, R. Ritchie, and W. Brandt, Phys. Rev. Lett. 33, 302 (1974).
10. J. Burgdörfer, J. Wang, and J. Müller, Phys. Rev. Lett. 62, 1599 (1989).

11. Y. Yamazaki, in Proc. of the NATO Advanced Study Institute on the Interaction of Charged Particles with Solids and Surfaces, Alicante, 1990 (Plenum Press, N.Y., 1991) in press.
12. R. Blümel et al., Phys. Rev. Lett. 62, 341 (1989).
13. H.D. Betz, R. Höppler, R. Schramm, and W. Oswald, Nucl. Instr. Meth. B33, 185 (1988).
14. J. Rozet, A. Chetioui, C. Stephan, A. Touati, D. Vernhet, and K. Wohrer, in *Proceedings of the Third Workshop on High Energy Ion-Atom Collisions*, edited by D. Berenyi and G. Hock, Lecture Notes in Physics Vol. 294 (Springer-Verlag, Berlin, 1988) p. 307.
15. J. Burgdörfer, in Ref. 14, p. 344; J. Burgdörfer and J. Gibbons, Phys. Rev. A42, 1206 (1990).
16. R. Abrines and I.C. Percival, Proc. Phys. Soc. (London) 88, 861 and 873 (1966).
17. J. Burgdörfer, Nucl. Instr. and Method. B24/25, 139 (1987).
18. J. Burgdörfer, Diploma Thesis, Freie Universität Berlin, 1978 (unpublished).
19. M. Chevallier et al., Phys. Rev. A41, 1738 (1990).
20. B. Mazuy et al., Nucl. Instr. and Meth. B31, 387 (1988).
21. J. Müller, J. Burgdörfer and D. Noid,, submitted for publication to Phys. Rev. Lett. (1990); J. Müller and J. Burgdörfer, to be published.
22. S. Datz et al., Phys. Rev. Lett. 40, 843 (1978); S. Datz, J. de Physique (Paris) C1-335 (1979).
23. S. Datz et al., Phys. Rev. Lett. 63, 742 (1989).
24. N. Claytor et al., Phys. Rev. Lett. 61, 2081 (1988).
25. V. Okorokov, JETP Lett. 2, 111 (1965).
26. K. Kimura et al., submitted for publication in Phys. Rev. Lett. (1990).
27. H. Hülskötter, J. Burgdörfer, and I. Sellin, Nucl. Instr. and Meth. B24/25, 147 (1987) and references therein.
28. S. Andriamonje et al., these proceedings.
29. H.-D. Betz, D. Röschenthaler, and J. Rothermel, Phys. Rev. Lett. 50, 34 (1983); H.-D. Betz, in Forward Electron Ejection in Ion Collisions, edited by K.O. Groeneveld et al., *Lecture Notes in Physics* Vol. 213 (Springer-Verlag, Berlin, 1984), p. 115.
30. Y. Yamazaki et al., Phys. Rev. Lett. 61, 2913 (1988).
31. J. Burgdörfer and C. Bottcher, Phys. Rev. Lett. 61, 2917 (1988).
32. I.A. Sellin, et al., in Forward Electron in Ion Collisions, *Lecture Notes in Physics*, Vol. 213 (Springer, Berlin, 1984), p. 109; R. Schramm, P. Koschar, H.-D. Betz, M. Burkhard, J. Kemmler, O. Heil, and K.O. Groeneveld, J. Phys. B18, L507 (1985); G. Schiwietz, D. Schneider, and J. Tanis, Phys. Rev. Lett. 59, 1561 (1987).
33. A.H. Mohagheghi et al., in *XVI International Conference on The Physics of Electronic and Atomic Collisions*, New York, 1989, Book of Invited Papers, eds. A. Dalgarno et al., (American Institute of Physics, N.Y., 1990), p. 487.

34. E. Donets, Nucl. Instr. and Meth. 9, 522 (1985).

35. M. Delaunay, M. Fehringer, R. Geller, P. Varga and H. Winter, Europhys. Lett.4, 377 (1987).

36. L. Folkerts and R. Morgenstern, J. de Physique 50-C1, 541 (1989).

37. D. Zehner et al. Surf. Science 178, 359 (1986).

38. J.P. Briand et al., Phys. Rev. Lett. 65, 159 (1990).

39. J. Burgdörfer to be published.

40. M. Hasegawa, K. Kimura and M. Mannami, J. Phys. Soc., Japan 57, 1834 (1988); M. Hasegawa, T. Fukuchi, T. Mizuno, K. Kimura, and M. Mannami, to be published.

41. A. Koyama et al., preprint (1990).

42. L.F. de Ferrariis and R. Baragiola, Phys. Rev. A33, 4449 (1986).

43. H. Winter, P. Strohmeier, and J. Burgdörfer, Phys. Rev. A39, 3895 (1989).

44. N. Takimoto, Phys. Rev. 146, 366 (1966).

CHARGE EXCHANGE PROCESSES OF HIGH ENERGY HEAVY IONS CHANNELED IN CRYSTALS

S. Andriamonje[1], M. Chevallier[2], C. Cohen[3], J. Dural[4], R. Genre[2], Y. Girard[3], K.O. Groeneveld[5], J. Kemmler[2], R. Kirsch[2], A. L'Hoir[3], R. Maier[5], J.C. Poizat[2], Y. Quéré[6], J. Remillieux[2], D. Schmaus[3], M. Toulemonde[4]

[1] *Centre d'Etudes Nucléaires de Bordeaux-Gradignan and IN2P3, Le Haut-Vigneau 33175 Gradignan Cedex - France*
[2] *Institut de Physique Nucléaire de Lyon and IN2P3, 43, Bd du 11 Novembre 1918, 69622 Villeurbanne Cedex - France*
[3] *Groupe de Physique des Solides Université Paris VII 75251 Paris Cedex 05 - France*
[4] *Centre Interdisciplinaire de Recherche avec les Ions Lourds 14040 Caen Cedex - France*
[5] *Institut für Kernphysik der Universität Frankfurt August-Euler-Str. 6, D6000 Frankfurt/Main 90, W Germany*
[6] *Laboratoire des Solides Irradiés, Ecole Polytechnique 91128 Palaiseau - France*

ABSTRACT

The interaction of moving ions with single crystals is very sensitive to the orientation of the incident beam with respect to the crystalline directions of the target. The experiments show that high energy heavy ion channeling deeply modifies the slowing down and charge exchange processes. In this review, we describe the opportunity offered by channeling conditions to study the charge exchange processes. Some aspects of the charge exchange processes with high energy channeled heavy ions are selected from the extensive literature published over the past few years on this subject. Special attention is given to the work performed at the GANIL facility on the study of Radiative Electron Capture (REC), Electron Impact Ionisation (EII), and convoy electron emission. Finally we emphasize the interest of studying resonant charge exchange processes such as Resonant Coherent Excitation (RCE), Resonant Transfer and Excitation (RTE) or Dielectronic Recombination (DR) and the recently proposed Nuclear Excitation by Electron Capture (NEEC).

I - INTRODUCTION

Channeling phenomena have been extensively studied in the last twenty five years and have shown to be very helpful for the more general study of ion-matter interaction. However, most of experimental studies have been performed with light projectiles. Channeling investigations with heavy ions have been mainly performed with moderately heavy ions at Tandem energies, particularly by the Oak Ridge group in successful experiments [1]. It can be expected that channeling studies of very fast heavy ions, then in high charge states, will reveal specific aspects of their interaction with matter.

The recent availability of high energy heavy ion beams at GANIL (Caen), BEVALAC (Berkeley), GSI-Darmstadt (UNILAC and SIS) makes it possible to undertake such studies. The feasibility of such experiments has been first demonstrated at GANIL, where $Ar^{(16+)}$ ions of 60 MeV/u have been channeled in a 100 μm thick Ge crystal [2]. This experiment has shown that it is possible to study new aspects of charge exchange processes when high energy heavy ions interact with solids. As a matter of fact, when a swift ion beam passes through a solid target its charge state is continuously changed due to successive electron capture, loss or excitation events. Channeling conditions constrain the incident ions to interact mainly with valence and/or conduction electrons. In this case the interaction of fast heavy ions with solid is strongly modified. Particularly the charge exchange process is found to be completely different from what it is in random conditions or in an amorphous medium.

After a short description of heavy ion channeling we describe channeling effects on:

- the emerging ion charge state distribution for low charge state and high charge state of the incoming ion beam
- electron capture and particularly Radiative Electron Capture (REC)
- electron Impact Ionisation (EII)
- convoy electron production.

At the end we present two resonant processes that allow to capture quasi-free electrons, the Resonant Dielectronic Capture [3], [4] and the recently proposed Nuclear Excitation by Electron Capture (NEEC) [5]. We describe also a resonant excitation effect, the coherent excitation of channeled ions (the so called Okorokov effect), specific of the channeling geometry.

II - ENERGETIC HEAVY-ION CHANNELING IN BRIEF

Channeling of positive ions in a crystal occurs when the direction of the incident beam is close to a planar or axial direction of the crystal [6], [7]. In perfect axial alignement for example, the ions entering the crystal with an impact parameter to an atomic row, p, larger than the thermal vibration amplitude, are repelled by the atomic row as a whole and can be considered as submitted to a continuum potential V(p) which is the periodic ion-atom potential averaged along the axial direction. A complete discription of axial channeling must take more than one atomic string into account. In figure 1 we show the maps in a Silicon crystal of the electron density averaged along the $< 110 >$ direction (a), and of the continuum potential for a single charged projectile (b). The electron density has been calculated by means of the electron wave functions in solid silicon, and the potential deduced from the electron density. The entrance conditions, incidence angle to the axis direction and impact location, define what is called the transverse energy of the incident projectile, that can be considered to be constant, at first approximation, during the passage through a thin crystal. The value of the transverse energy determines the closest distance of approach to the atomic rows. The best channeled particles have the smallest transverse energy and then are maintained quite far from the atomic rows. In particular some particles have such a small transverse energy that they are trapped between the same set of atomic strings all along their pathlength. They are said to be hyperchanneled.

An example of very good channeling conditions is given in figure 2. This figure shows the orientation dependences of the Xe Ly$_\alpha$ yield (resulting from Xe K-shell excitation in close collisions with Si atoms) and of the Xe^{37+} fraction in the transmitted beam when 27 MeV/u Xe^{35+} ions are aligned along the < 110 > direction of thin crystal. Whereas the Xe Ly$_\alpha$ yield presents an ordinary channeling dip, the Xe^{37+} component presents a sharp peak, which spectacularly shows that only the very best channeled ions, i.e. those with a very small transverse energy, are able to lose only two electrons during the traversal of the crystal (instead of about ten for most of the channeled ions).

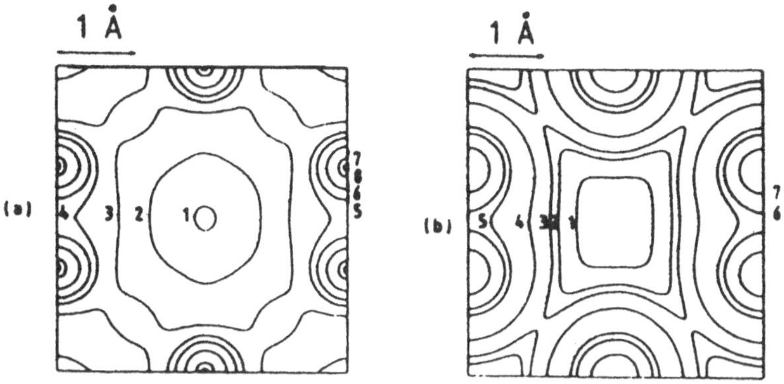

Figure 1 : (a) Map of the electron density in a silicon crystal, averaged along the < 110 > direction. Contour line 1 to 8 : 0.032, 0.1, 0.24, 0.5, 1, 3, 10 and 30 (electrons per Å^3) (b) Map of the continuum potential for a unit charge along the < 110 > direction of Si. Contour line 1 to 7 : 0.5, 1.5, 2.25, 5, 15, 21, 50(eV) (from Ref. [20])

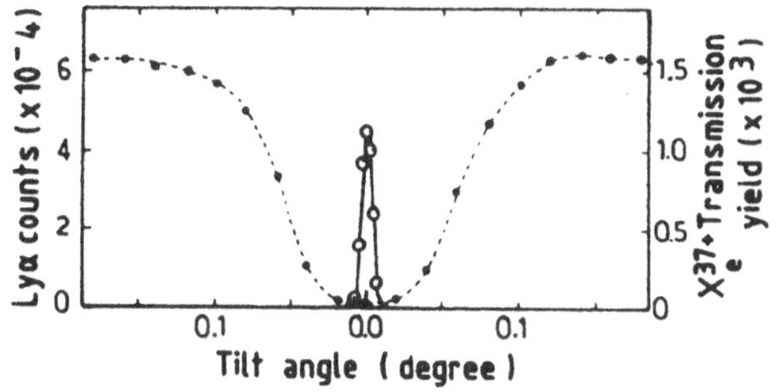

Figure 2 : Angular scans across the [110]-axis. Full circles: Xe Lyman α photon yield. Open circles: fraction of emerging Xe^{37+}. The solid line through the open circles is the result of a Monte Carlo simulation (from Ref. [20]).

III - EXPERIMENTAL SET-UP

A typical experimental set up, for the study of charge exchange processes in channeling conditions, is essentially constituted by :

- a beam monitor for dose determination
- a very good single crystal placed in a high resolution goniometer
- one (or several) X ray detector viewing the crystal
- an electron spectrometer

- a charge state analyser placed after the target chamber for measurements of the charge state distribution and energy loss of emergent ions, associated with a wire chamber or position sensitive detector

- a microchannel plate (or channeltron) or a position sensitive detector for single (or multiple [8]) coincidence measurements between X ray/or electron emitted at 0° and transmitted ions of a given (or various) charge state.

And finally a high quality incident beam of the appropriate charge state with very low energy and angular dispersion.

IV - CHARGE STATE DISTRIBUTIONS

It is well known, for random orientation of the crystal, that electron loss and capture leads into an equilibrium charge state distribution of the emerging beam if the target is thick enough. This distribution depends on the nature and velocity of the ion but is independent of the initial charge state.

In channeling conditions the situation is very different. That effect was observed, for the first time, by the Oak Ridge group [1] in the study of the emerging charge state distribution of oxygen ions, at Tandem energies (4.45 to 40 MeV) in a gold crystal. An example of their results [1], [16] is reported in figure 3.

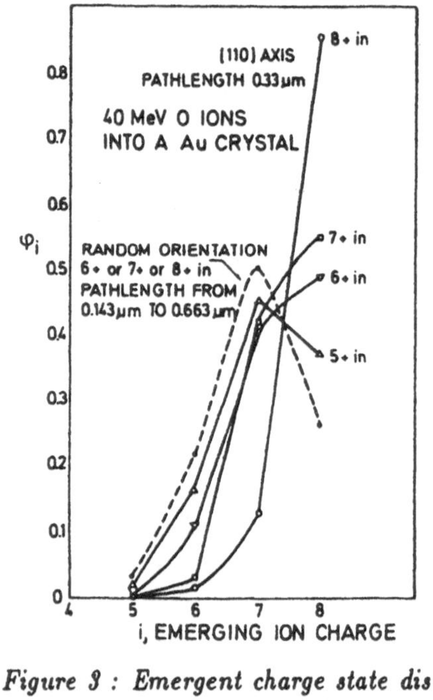

Figure 3 : Emergent charge state distributions obtained for random and [110] channeled oxygen ions with an input energy of 40 MeV. Input charges and pathlengths indicated. (from ref. [1])

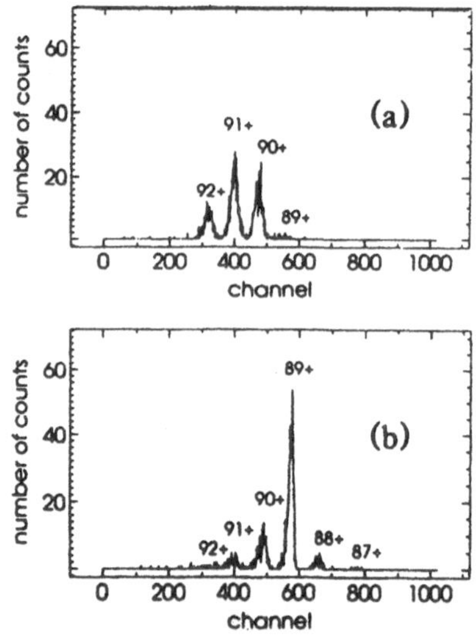

Figure 4 : Observed charge state distributions from 405 MeV/u U^{89+} exiting the 0.37 mm thick Si single crystal. (a) The ions pass through a random direction of the crystal. (b) The ions are aligned with the < 110 > axis of the crystal (from ref. [17])

This work has been extended by the GANIL group for 25 MeV/u and 44 MeV/u Xenon ions with different incoming charge states channeled in a Si crystal, and also by Claytor et al [17] at BEVALAC for 405 MeV/u Uranium ions channeled in a Si crystal (cf. figure 4).

The emerging charge-state distribution extracted from works performed recently at GANIL [9, 10, 18] are reported in figure 5 : 27 MeV/u Xe^{35+} incident ions (a), 25 MeV/u Xe^{50+} to Xe^{54+} incident ions (b) and 44 MeV/u Xe^{54+} incident ions (c) in random conditions and for < 110 > alignment after a pathlength of 21 μm through a Si crystal.

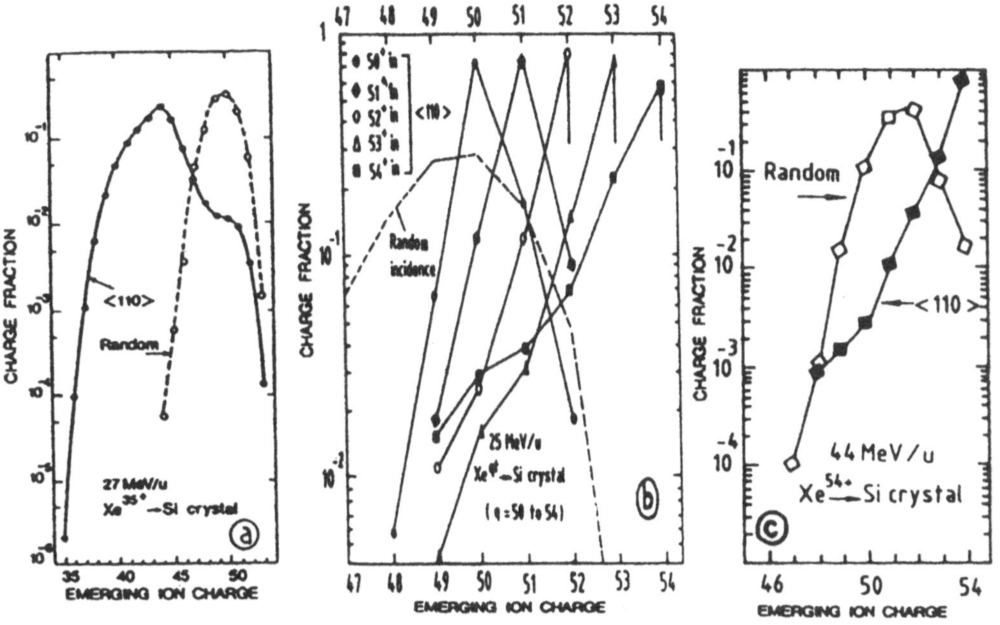

Figure 5 : Charge state distributions after the passage of incident (a) 27 MeV/u Xe^{35+} (b), 25 MeV/u Xe^{q+} (q = 50 to 54) (c), 44 MeV/u Xe^{54+} ions through a 21 μm Si crystal in random and in < 110 > *alignment conditions, respectively*

For incoming Xe^{35+} ions (figure 5a) the distribution obtained in axial alignment extends from q = 35 to q = 53 which reveals a large dispersion in the charge changing process of aligned projectiles. As it will be confirmed by the analysis of energy losses, this broad distribution results from the transverse energy distribution of the aligned particles. It is clear in particular that the small fraction of ions which have kept their initial charge state are the best channeled ones, with a very small transverse energy, that have travelled through regions of very low electron density. The only part of the distribution that is charge equilibrated is the highest charge side, which contains the unchanneled component ($\sim$ 2 %) of the beam.

For the incoming highly stripped Xe ions (q = 50 to 54) represented in figure 5b, the charge distributions obtained for random incidence of the various charge states are nearly identical and this distribution is equilibrated around the mean value 49.5 (the same value is also found for 27 MeV/u incident Xe^{35+}). On the contrary the distributions obtained for axial alignment are quite different from each other and then quite far from being equilibrated.
They exhibit the same essential feature, which is the "freezing" of the incident charge state : in the five cases, 60 to 85 % of the incident ions have kept their initial charge state. Moreover the detailed observation of the energy distribution of the transmitted ions shows that "frozen" ions were channeled in the crystal, as revealed by their reduced energy loss.
The distributions obtained with 52^+ and 53^+ incident ions (figure 5b), i.e He-like and H-like species, show that these two species cannot lose their K-shell electron(s) if they are channeled, which is easily understood since collisions with quasi-free target electrons, the only collisions allowed to channeled particles, cannot transfer enough energy to ionize K electrons. In both cases a fraction of the incident ions have been able to capture one electron, and again the energy distribution of these ions shows that most of them are well channeled. The simultaneous measurements of the charge state distribution and X ray emission show that the capture process which is involved here is Radiative Electron Capture (REC) [9] (cf. below).
The channeling charge state distributions obtained with 50^+ and 51^+ incident ions (figure 5b) are different from the previous ones because the loss of L-shell electrons becomes possible for these channeled ions. The electron loss probability is of course higher for 50^+ ions than for 51^+ ions. They also may capture electrons by means of REC. The capture probability is seen on figure 5b to be higher for 51^+ ions than for 50^+ ions. Even though these two charge distributions are not equilibrated, the balance between electron loss and electron capture for each of the two incident species indicates that the mean charge state of the equilibrated distribution for channeled ions would be about 50.5, since a mean charge state at equilibrium corresponds to the charge state for which electron loss and capture have the same probability to occur. These results are explained in details by Poizat et al [18]. We comment only here the still unpublished results for 44 MeV/u reported in figure 5c. For incident Xe^{54+}, 95 % of the ions are in a "frozen" state, this result can be compared with the result obtained for the same incident charge state at 25 MeV/u where only 60 % are "frozen" (figure 5b). For that channeled ions the capture process involved is also Radiative Electron Capture and REC cross sections are proportional to v^{-5}, where v is the projectile velocity.

V - RADIATIVE ELECTRON CAPTURE

Two mechanisms are usually responsible for charge tranfer in the interaction between swift ions and solid targets : Mechanical Electron Capture (MEC) and Radiative Electron Capture (REC).
The MEC process is radiationless : three particles (the projectile, the captured electron and the target-core) share out the initial energy [8], [11]. In the REC process, the electron-projectile dipole allows the emission of a photon which takes away the excess energy and momentum [12], [13]. MEC is dominant for impact parameters of the order of the radius of the initial electronic orbital in the target atom. On the other hand a

much larger range of impact parameters contributes to REC [14]. And for relativistic ions in low-atomic number targets, REC has been shown to be the dominant process for charge transfer [15].

When ions are channeled through single crystals the MEC process is strongly inhibited and a detailed study of REC is possible. The first experiment designed for the study of REC using channeling conditions was performed at Oak Ridge [16] using medium heavy ions at Tandem energies. A sample of these results for 27.78 MeV 0^{7+} ions channeled along the < 110 > axis of an Ag single crystal is reported in figure 6.

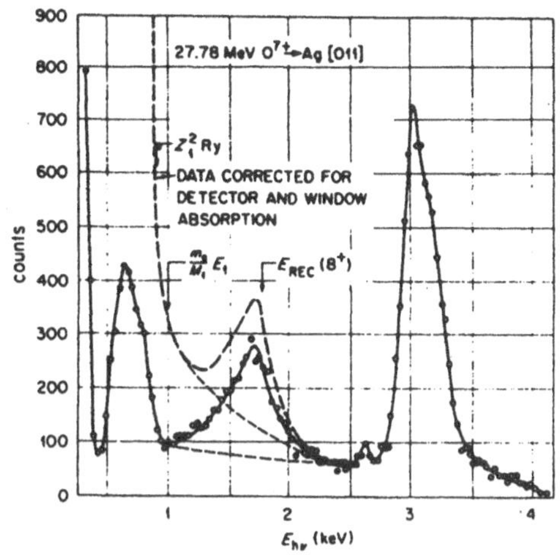

Figure 6 : Measured and absorption corrected photon spectrum resulting from 27.78 MeV O^{7+} ions transmitted parallel to the [011] axial channels of a thin Ag single crystal (from ref. [16])

This work has been extended at GANIL for higher energies and for heavier ions [9]. Using 25 MeV/u Xe ions (with incident charges 52^+, 53^+, 54^+) in a Silicon crystal we have shown that channeling measurements with fast stripped heavy ions incident on thin crystals allow not only an observation of REC into K, L and M shells but also a detailed study of the process.

The study of X-ray spectra obtained with the three incident charge states of the Xe ions have been published in references [9] and [10].

An example is presented in figure 7, where is shown the X-ray spectra obtained from the interaction between 25 MeV/u Xe^{53+} ions and a Silicon crystal. The photons are detected by a solid state Ge detector viewing the crystal. Two spectra are given in figure 7a, for random and < 110 > incidence, respectively and for the same number of incident ions. Three peaks corresponding to K, L and M-REC respectively have been observed. They are located at energy : $E^i_{REC} \simeq E^i_B + (m/M)E_o$ where E^i_B is the electron binding energy in shell i(= K, L, M) after capture, m and M are the electron and projectiles masses, respectively, and E_0 the Kinetic energy of the projectile.

The L and M-REC intensities are observed to be about the same in channeling and in random conditions although they appear more clearly in the channeling spectrum as a result of the decrease of bremsstrahlung in channeling conditions. On the other hand, in random conditions, the K-REC contribution is small because the first capture that fills the (single) K hole is most often a non radiative event, and also because, at equilibrium the K-shell is almost always filled. For < 110 > alignment, the K-REC intensity is markedly larger, due to the fact that most of the incident 53^+ ions are "frozen" and a target electron capture into their K-shell is possible.

In brief the use of channeled swift heavy ions permit a fine study of REC. The cross sections can be measured [9], the line shapes appear very clearly and can be analyzed in great detail. As it can be seen in figure 7a, the REC lines are wider than Lyman lines, and this is due to the fact that the target electrons captured by the REC process are not at rest, but have a momentum distribution, usually called the Compton profile. The study of orientation dependence around the target axial direction permit to measure this effect with good precision.

The first measurement has been published in reference [9] where is shown the increase of the K-REC line width as a function of the tilt angle around the Si < 110 > direction. The particles with very low transverse energy can only capture valence electrons, since these electrons have small momenta the line associated to this process is small. When increasing the tilt angle, the transverse energy of the particles increases and the capture of L shell target electrons (with much higher momenta) becomes more and more probable leading to broader line widths.

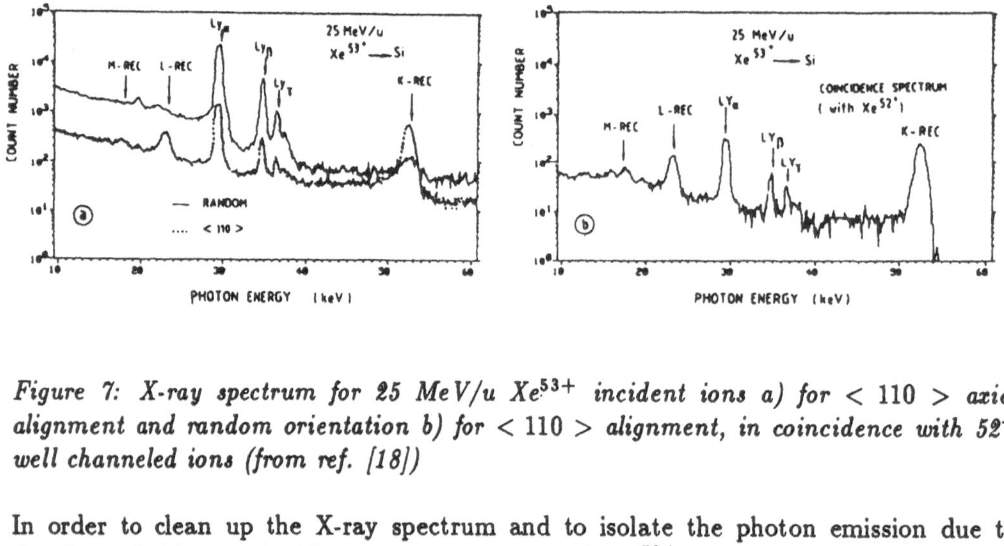

Figure 7: X-ray spectrum for 25 MeV/u Xe^{53+} incident ions a) for < 110 > axial alignment and random orientation b) for < 110 > alignment, in coincidence with 52^+ well channeled ions (from ref. [18])

In order to clean up the X-ray spectrum and to isolate the photon emission due to the channeled ions, coincidence measurements with Xe^{53+} incident ions have been performed. The X-ray spectrum of figure 7b is made of photons detected in coincidence with the part of the transmitted Xe^{52+} ions which have suffered low energy loss, i.e. with the well channeled ions that have captured one electron. When compared to the spectrum of figure 7a obtained simultaneously without coincidence, the REC lines of these spectrum are narrower and less asymmetric (the width of the K-REC line is 1.05 keV), which is due to the narrower Compton profile of target electrons that can be captured by well channeled ions. The Lyman lines are here entirely due to deexcitation after radiative capture into excited states. In particular the L-REC and Ly_α peaks correspond to the same process and the difference of the peak areas probably shows that the angular distributions of L-REC and Ly_α emissions are different.

VI - ELECTRON IMPACT IONIZATION

An other interesting feature given by energetic heavy ions channeled in a single crystal is the study of Electron Impact Ionization (EII). For well channeled ions the only available mechanisms for charge exchange are Radiative Electron Capture and Electron Impact Ionization induced by close collisions with target electrons. In the case of ions entering the crystal with a charge state much lower than the mean charge at equilibrum, REC cross sections are vanishingly small with respect to EII. Then the latter can be examined in detail and EII cross sections can be deduced from the experimental charge-state distribution recorded in channeling conditions through a thin crystal.
The first measurement has been performed by Claytor et al [17] at Lawrence Berkeley Laboratory BEVALAC, with 405-MeV/u Uranium ions channeled along the < 110 > axis of 0.11 and 0.37 mm Si crystals. In the rest frame of the Uranium ions, the target electrons have an energy of 222 keV, whereas the binding energy of H-like Uranium electrons is about 133 keV. Then 1s to $2s^2$ electrons of U^{91+} to U^{88+} ions can be ionized. The electron density along the path of the channeled ions has been measured by comparing the cross sections for electron capture by the channeled ions with the capture cross sections for ions in random directions. The least-squares fit of the charge state yields versus target thickness was used to determine EII cross sections for Be-like to H-like Uranium ions by 222 keV electrons. The results are compared with avalaible theories (references cited in ref. [17]). For K-shell electrons of U^{90+} and U^{92+} ions a large difference is found between EII theories and the experimental results. And for L-shell of the U^{88+} and U^{89+} Claytor et al conclude that their results are not accurate enough to distinguish between the various calculations.
Another study of EII has been performed at GANIL [19] at lower energies and ion charge state. A beam of 27 MeV/u Xe^{35+} ions has been channeled along the < 110 > axis of a Si crystal (17 μm pathlength). In this case the kinetic energy of Si valence electrons in the rest frame of the ion is equal to 14.7 keV. The binding energies of the 2p electrons and all M-shell electrons of these $19e^-$-Xenon ions are such that ionization by 14.7 keV electrons is possible. The electron loss cross sections for collisions with target electrons have been extracted from the emerging channeled charge state distribution (as reported in figure 5a). This charge state distribution in the [110] geometry has been fitted using a Monte Carlo simulation. This method is explained in detail by L'Hoir et al [20].The result is reported in figure 8. The Donets's results obtained by Electron Beam Ionization Method (EBIM) [21] are also reported in this figure. Experimental results are compared with predictions of the Lotz empirical formula [22]. The figure 8 shows that cross sections obtained by the channeling method are 2 to 4 times larger than predicted [22] and 1 to 2 times greater than those measured by Donets. This difference is probably due to the fact that in the crossed-beams experiment of reference [21] the electron flux is obviously too low to induce multistep ionization ; on the contrary, in the channeling method indirect ionization, via excitation, is not negligible.

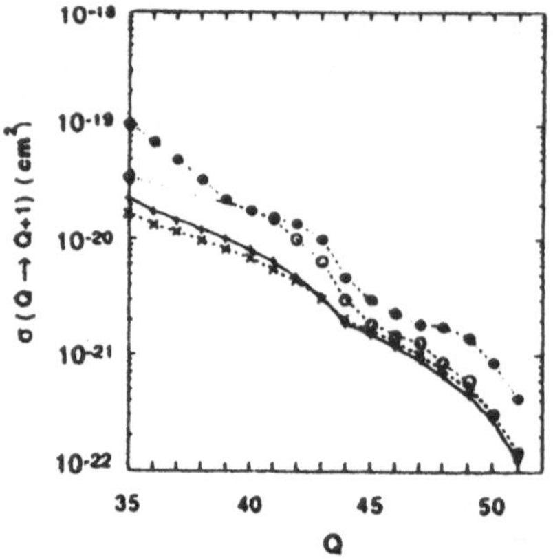

Figure 8: Electron Impact Ionization cross sections $\sigma(Q \rightarrow Q+1)$of Xe^{Q+} ($35 \leq Q \leq 51$) by 14.7 keV electrons. Full circles : results from ref. [20]; open circles : results from Donets [21]; solid curves : calculation from the Lotz formula [22]; dotted curve : calculation from the Thomson formula (see text in ref. [20])

VII - CONVOY ELECTRON PRODUCTION

In the foregoing paragraphs the different aspects of electron capture and loss processes under channeling conditions have been discussed. The methods at disposal to study this phenomena are the measurements of the charge state distributions and the measurement of photons arising from the decay of excited projectile ion states. These methods give access to the behaviour of the manifold of bound states in the projectile ion system, and for the study of primary bremsstrahlung one can also learn about free-free transitions in the continuum [23].

A piece of complementary information in the energy region close to the ionization threshold can be provided by the study of the induced electron emission in these ion-solid interactions.

In the corresponding electron spectrum these electrons, arising from various production processes, are emitted along the beam direction. In the electron velocity spectrum these "convoy electrons" form a cusp shaped peak with a maximum located at the projectile velocity $\vec{v}_e = \vec{v}_p$.

Recent experimental works have shown the close relation between charge exchange and convoy electron production in solids [24]. It has been found also that the two possible processes for convoy electron production in the case of nearly bare ions, the electron loss to the continuum (ELC) and the electron capture to the continuum (ECC) are important. The first observation of convoy electrons in channeling condition is due to the Oak Ridge group [25], with 2.4 MeV/u oxygen ions channeled along the $< 110 >$ and $< 100 >$ axis of a gold crystal. Similar study have been performed very recently at GANIL with Xe^{q+} (q = 37, 44 and 54) ions at two energies 27 and 44 MeV/u. The forward electron emission was observed with a magnetic sector spectrometer [26]. In figure 9 is shown the velocity spectrum of electrons emitted at 0° when 27 MeV/u Xe^{37+} ions are incident on a non-orientated Si crystal. The convoy electron peak, located at $v_e = v_p$ = 32.8 a.u., is on a large background. The origin of these broad electron distribution is the direct ionization of target atoms and the inelastically scattered binary encounter electrons. The spectrum under channeling conditions is also shown in figure 9. The main difference consists in the dramatically changed yield of convoy electrons (a factor 13 less than in the random case), whereas the yield for the binary encounter electrons varies only by a factor of 3.2. For the incident projectile ion Xe^{54+} (44 MeV/u) in figure 10 are shown the electron spectra for aligned and nonaligned situations. Under random incidence the convoy electron peak is clearly observed and it disappears under channeling conditions. This result is not surprising because under channeling condition the charge state of the incoming Xe^{54+} ions is "frozen" (95 %). The only process available for convoy electron production is the radiative electron capture into the continuum or target electron ionization. This figure shows clearly that cross sections for these two processes is very small for 44 MeV/u Xe^{54+} ions channeled in the crystal.

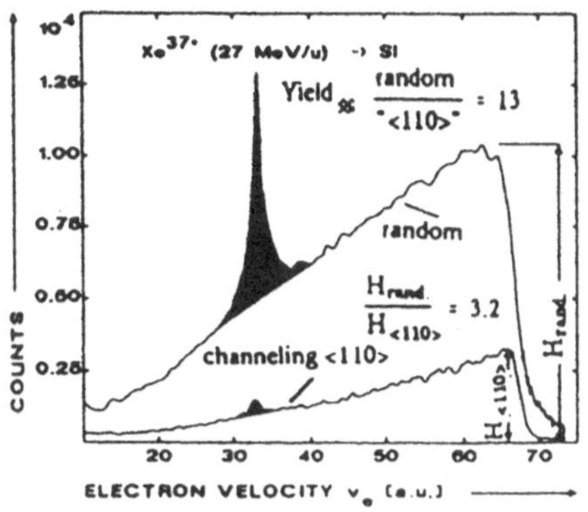

Figure 9: Forward electron spectra obtained with 27 MeV/u Xe^{37+} ions incident on a Silicon crystal (< 110 > axial alignment and random orientation)

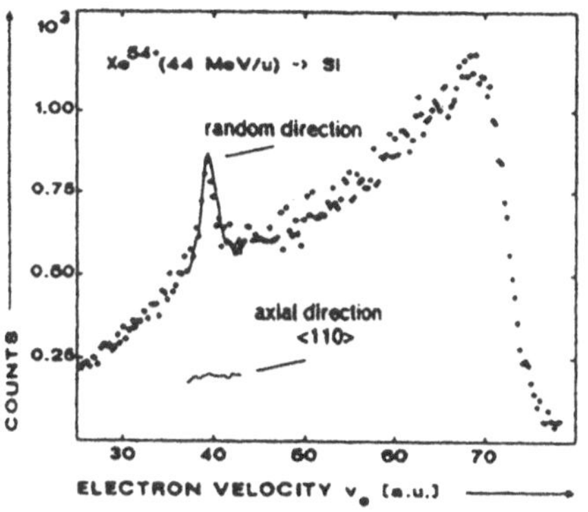

Figure 10: Forward electron spectra obtained with 44 MeV/u Xe^{54+} ions incident on Silicon crystal (< 110 > axial alignment and random orientation)

VIII - RESONANCE PHENOMENA IN CHARGE EXCHANGE PROCESS

In the foregoing paragraphs we have only described non resonant phenomena in charge exchange processes, when energetic heavy ions (Xe and U ions of energy larger than 25 MeV/u) are channeled in a single crystal. It is of course of interest to study resonant phenomena but for that measurement it is necessarly to scan the incident beam energy without changing the optical quality of the beam. Such experiments will be probably possible at GSI - Darmstadt [27] and GANIL - Caen. Of particular interest are the studies of Resonant Transfer and Excitation (RTE) and Resonant Coherent Excitation (RCE) with high energy and high charge state incident ions. Simultaneous electron transfer and excitation of the projectile electron is possible when highly stripped ions collide on atom. This effect can be observed using a multiple coincidence method between emitted projectile X ray and various emerging charge state [8]. For nearly bare nucleus the radiationless electron capture process is accompanied by the excitation of

a projectile inner shell electron and the requirement of energy and momentum conservation is only satisfied at a "resonant" projectile energy [28], [29]. This process is called Dielectronic Recombination (DR) if the electron captured is free and Resonant Transfer and Excitation (RTE) if the electron captured is bound. At Tandem energies, two experiments have been performed very recently by Datz et al [3] and Belkacem et al [4] for the studies of resonance phenomena in the simultaneus electron transfer and excitation of the projectile electron in channeling condition.
Using a 25 MV Tandem accelerator at ORNL Datz et al [3] have observed dielectronic and direct excitation of H-like S^{15+} and Ca^{19+} and He-like Ti^{20+} ions channeled along the $< 110 >$ axis of a 1.2-μm-thick silicon crystal. The same study has been investigated by Belkacem et al [4] at the Argonne Tandem-Linac Accelerator System (ATLAS). Li-like Ti^{19+} and He-like Ti^{20+} having energies between 267 and 320 MeV have been channeled in a thin gold crystal. High-resolution measurements of the energy losses allowed Belkacem et al to show that the measured resonance peak widths for well channeled ions are more than an order of magnitude narrower than any previously observed RTE resonance with hydrogen targets. This result is surprising as it shows that the mean energy of the captured conduction electrons is much lower than one would resonnably expect. Another surprising feature in reference [4] is the large shift in energy between the observed resonance and the predictions. Such a shift can only be interpreted by assuming a very important screening effect of the target electrons on the projectile nucleus.
We present in figure 11, in the projectile rest frame the energy diagram corresponding to the two capture processes that we have already discussed.

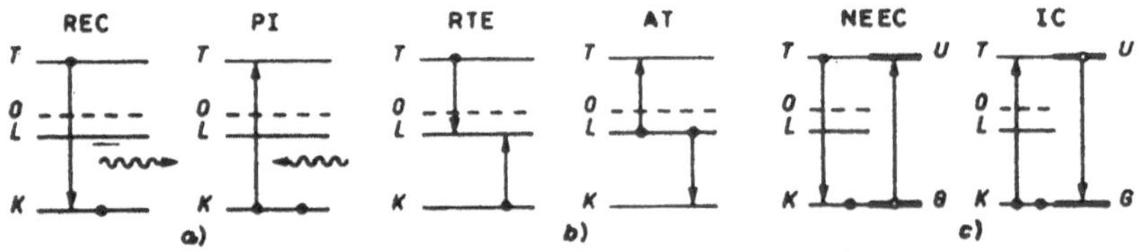

Figure 11 : The processes of target electron capture by a swiftly moving hidrogenic heavy ion (viewed from its rest frame) and their corresponding inverses: T refers to the target electron state, K, L and O to the projectile's atomic orbitals (with O designating the continuum) in the atomic system, while U and G refer to the excited and ground states, respectively, in the nuclear system (from ref. [5]).

One clearly sees that REC is simply the reverse of photo ionization (PI) and that RTE is the reverse of the Auger Transition (AT). We also show in figure 11 the diagram corresponding to another, recently proposed [5], and still never observed, resonant capture process, namely the Nuclear Excitation by Electron Capture (NEEC) which is reverse of Internal Conversion (IC). In RTE the capture leads to excitation of an atomic level while in NEEC a nuclear level is excited. Of course the corresponding cross section is expected to be much smaller than for RTE.

Another resonant process interesting to study with energetic and heavy ion is the Resonant Coherent Excitation (RCE). Channeled ions moving with velocity v along atomic rows experience a coherent periodic perturbation at the string frequency ($\nu = K(v/d)$, $K = 1, 2, 3,\ldots$, where d is the distance between the atoms in the rows). When one of these frequencies coincides with $\nu_r = \Delta E_{ij}/h$, where ΔE_{ij} is the difference between atomic or nuclear states of the ion, a resonant coherent excitation might occur. This effect have been observed for the first time by the Oak Ridge group [30] [31]. Then an important work has been performed with low Z_1 projectiles at Tandem energies. Only resonances corresponding to transitions between $n = 1$ to $n = 2$ (or sometimes $n = 3$), have been observed. The extension of this measurement to high Z_1 ions at higher energies would allow not only to observe the first harmonics of RCE but also stronger resonance effects and resonances for transitions to levels higher than $n = 3$ [32].

In conclusion we have described here some aspects of charge exchange processes for high energy heavy ions channeled in single crystals. Here it is shown that using high energy heavy ions not only can yield quite novel and detailed information on channeling phenomena but also allows to study the interaction of these ions with a dense electron gas. For well channeled ions, the crystal provides a unique source of high density gas of quasi free electrons.

REFERENCES

[1] S. Datz, F.W. Martin, C.D. Moak, B.R. Appleton and L.B. Bridwell, Rad. Eff. 12 (1972) 163

[2] C. Cohen, J. Dural, M.J. Gaillard, R. Genre, J.J. Grob, M. Hage-Ali, R. Kirsch, A. L'Hoir, J. Mory, J.C. Poizat, Y. Quéré, J. Remillieux, D. Schmaus and M. Toulemonde, J. Phys. 46 (1985) 1565

[3] S. Datz, C.R. Vane, P.F. Dittner, J.P. Giese, J. Gomez del Campo, N.L. Jones, H.F. Krause, P.D. Miller, M. Schulz, H. Schöne and T.M. Rosseel, Phys. Rev. Lett. 63 (1989) 742

[4] A. Belkacem, E.P. Kanter, K.E. Rehm, E.M. Bernstein, M.W. Clark, S.M. Ferguson, J.A. Tanis, K.H. Berkner and D. Schneider, Phys. Rev. Lett. 64 (1990) 380

[5] N. Cue, J.C. Poizat and J. Remillieux, Europhys. Lett. 8 (1989) 19

[6] D.S. Gemmell, Rev. Mod. Phys. 46 (1974) 129

[7] S. Datz and C.D. Moak, in Treatise on Heavy Ion Science, ed by D.A. Bromley (Plenum, New-York, 1984) Vol. 6, p. 169

[8] S. Andriamonje, J.F. Chemin, J. Roturier, B. Saboya, J.N. Scheurer, R. Gayet, A. Salin, H. Laurent, P. Aguer and J.P. Thibaud, Z. Phys. A318 (1984) 251

[9] S. Andriamonje, M. Chevallier, C. Cohen, J. Dural, M.J. Gaillard, R. Genre, M. Hage-Ali, R. Kirsch, A. L'Hoir, B. Mazuy, J. Mory, J. Moulin, J.C. Poizat, J. Remillieux, D. Schmaus and M. Toulemonde, Phys. Rev. Lett. 59 (1987) 2271

[10] S. Andriamonje, M. Chevallier, C. Cohen, J. Dural, M.J. Gaillard, R. Genre, M. Hage-Ali, R. Kirsch, A. L'Hoir, B. Mazuy, J. Mory, J. Moulin, J.C. Poizat, J. Remillieux, D. Schmaus and M. Toulemonde, J. Phys. (Paris) 50 (1989) C1-285

[11] Dz. Belkic, R. Gayet and A. Salin, Phys. Rep. 56 (1979) 279

[12] H.D. Betz, Rev. Mod. Phys. 44(1972) 465

[13] J.S. Briggs and K. Dettman, Phys. Rev. Lett. 33 (1974) 1123

[14] J.E. Miraglia, R. Gayet and A. Salin, Europhys. Lett. 6 (1988) 397
[15] R. Anholt, S. Andriamonje, E. Morenzoni, Ch. Stöller, J.D. Molitoris, W.E. Meyerhof, H. Bowman, J.S. Xu, Z.Z. Xu, J.O. Rasmussen and D.H.H. Hoffmann, Phys. Rev. Lett. 53 (1984) 234
[16] B.R. Appleton, R.H. Ritchie, J.A. Biggerstaff, T.S. Noggle, S. Datz, C.D. Moak, H. Verbeek and N. Neelavathi, Phys. Rev. B19 (1979) 4347
[17] N. Claytor, B. Feinberg, H. Gould, C.E. Bemis, Jr., J. Gomez del Campo, C.A. Ludemann and C.R. Vane, Phys. Rev. Lett. 61 (1988) 2081
[18] J.C. Poizat, S. Andriamonje, R. Anne, N.V. de Castro Faria, M. Chevallier, C. Cohen, J. Dural, B. Farizon-Mazuy, M.J. Gaillard, R. Genre, M. Hage-Ali, R. Kirsch, A. L'Hoir, J. Mory, J. Moulin, Y. Quéré, J. Remillieux, D. Schmaus and M. Toulemonde, Invited Paper at XV ICPEAC - New-York (1989)
[19] S. Andriamonje, R. Anne, N.V. de Castro Faria, M. Chevallier, C. Cohen, J. Dural, M.J. Gaillard, R. Genre, M. Hage-Ali, R. Kirsch, A. L'Hoir. B. Farizon-Mazuy, J. Mory, J. Moulin, J.C. Poizat, Y. Quéré, J. Remillieux, D. Schmaus and M. Toulemonde, Phys. Rev. Lett. 63 (1989) 1930
[20] A. L'Hoir, S. Andriamonje, R. Anne, N.V. de Castro-Faria, M. Chevallier, C. Cohen, J. Dural, M.J. Gaillard, R. Genre, M. Hage-Ali, R. Kirsch, B. Farizon-Mazuy, J. Mory, J. Moulin, J.C. Poizat, Y. Quéré, J. Remillieux, D. Schmaus and M. Toulemonde, Nucl. Instr. and Meth. B48 (1990) 145
[21] E.D. Donets, Physica Scripta, T3 (1983) 11
[22] W. Lotz, Z. Physik 216 (1968) 241
[23] J. Remillieux et al, to be published (1990)
[24] J. Kemmler, P. Koschar, O. Heil, C. Biedermann, H. Rothard, K. Kromeberger, S. Leneinas, K.O. Groeneveld and I.A. Sellin, Nucl. Instr. Meth. B33 (1988) 281
[25] S.B. Elston, I.A. Sellin, M. Breinig, S. Huldt, L. Liljeby, R.S. Thoe, S. Datz, S. Overbury and R. Laubert, Phys. Rev. Lett. 46 (1981) 321
[26] J. Kemmler, O. Heil, C. Biedermann, P. Koschar, H. Rothard, K. Kronenberger, K.O. Groeneveld, A. Kvver, G. Szabo, L. Gulyas, D. Berenyi, P. Focke and W. Meckbach, 3rd Workshop on Ion-Atom Collisions in Debrecen/Hungary, Lecture Notes in Physics 294, Springer Verlag (1988) p. 362
[27] F. Bosch, Physica Scripta 36 (1987) 730
[28] See for instance : J.A. Tanis, Nucl. Inst. Meth. A262 (1987) 52
[29] See for instance : M. Schulz, E. Justiniano, R. Schuch, P.H. Mokler and S. Reusch, Phys. Rev. Lett. 58 (1987) 1734
[30] S. Datz, C.D. Moak, O.H. Crawford, H.F. Krause, P.F. Dittner, J. Gomez del Campo, J.A. Biggerstaff, P.D. Miller, P. Hvelplund and H. Knudsen, Phys. Rev. Lett. 40 (1978) 843
[31] C.D. Moak, S. Datz, O.H. Crawford, H.F. Krause, P.F. Dittner, J. Gomez del Campo, J.A. Biggerstaff, P.D. Miller, P. Hvelplund and H. Knudsen, Phys. Rev. A19 (1979) 977
[32] J.C. Cohen, J. Dural, M.J. Gaillard, R. Genre, J.J. Grob, M. Hage-Ali, R. Kirsch, A. L'Hoir, J. Mory, J.C. Poizat, Y. Quéré, J. Remillieux, D. Schmaus and M. Toulemonde, in Relativistic Channeling NATO ASI Series, Ed. by R.A. Carrigan, Jr and J.A. Ellison (Plenum Press, New-York 1987) Vol. 165, p. 493

The development of projectile excitation during ion-solid collisions observed through convoy electron emission

Stuart B. Elston
Department of Physics and Astronomy,
University of Tennessee, Knoxville, TN 37996
and
Oak Ridge National Laboratory, Oak Ridge, TN 37831, USA

Abstract

Preliminary results and interpretation are presented for a recent experiment designed to explore the evolution of electronic excitation of a projectile ion during flight through a solid target by monitoring the development of high-order multipole moments of the convoy electron angular distribution as a function of target thickness.

I. Introduction

A number of studies over recent years has drawn increasing attention to the importance of understanding the detailed nature of and role played by electronic excited states of projectile ions traversing solid media, and the mechanisms by which this excitation develops. This problem is a long- standing one, dating back at least to the early paper of Bohr and Lindhard.[1] Experiments observing the production of photons by projectiles with velocities corresponding to the order of one to a few MeV/u have suggested the existence of states of high angular momentum in bulk solid media, analogous to Rydberg states of isolated atoms.[2] Complementary studies of convoy electron production (from oxygen ions) reveal multipole moments of the emission angular distribution up to order at least 10, consistent with the population of states describable in a hydrogenic basis by principal quantum numbers up to at least 5.[3] In this latter case ($n = 5$), one considers electronic orbitals comparable in size to the typical interatomic target spacing, where dynamic screening of the projectile charge should be significant and the use of such a simple basis as a hydrogenic one should be highly suspect. By means of contrast, however, earlier work aimed at describing the approach of an ion to charge-state equilibrium,[4] which pointed out the need to consider excitation of the projectile, found satisfactory agreement with the exit charge distribution by including only a single, low-lying excited state. While this list is by no means complete, it is seen that information is and has been accumulating which is observable by a variety of techniques whose combined sensitivity spans the full range of states of the projectile.

It is widely recognized that the complex sequence of collisional excitation, ionization, and capture events that occur during the passage of a typical projectile through a solid present a formidable challenge to theory.[5] Central to this challenge is a need to understand the nature of both bound and continuum electronic states of a Coulomb projectile whose trajectory is immersed in the 'sea' of electrons and ion cores of an amorphous or crystalline solid, which sea is strongly disturbed by the projectile's passage. Given even a reasonable approximation to the electronic projectile states, one must then attempt to model the evolution of an initial state during successive collisions which may often be severe enough to make perturbation methods useless. Of particular interest to the analysis of results presented in this paper is that it has lately become clear that one cannot assume that an electron 'pumped up' via the 'density effect'[1] beyond an approximate screening distance becomes forever ionized, but instead must be followed as long as it is transported within the nearby phase space of the projectile Coulomb center.[6]

This work presents preliminary results and interpretation of a recent experiment designed to probe the evolution of excited states of projectile ions while traversing thin amorphous solid targets, as revealed by the development of high-order multipole moments describing the angular distribution of convoy electron emission. The experiment employed zero-, one-, and three-electron Ar projectiles, to permit simple assumptions to be made about initial projectile states. The relatively large projectile velocity (37 a.u.) and the use of thin, amorphous carbon targets were chosen to reduce the above-mentioned transport effects.

II. Method

The basic apparatus used in the present experiment, and the methods used to analyze the angular distributions observed, have been described in some detail previously.[8] Briefly, beams of 36 MeV/u Ar^{q+}, with $q=$ 15, 17, and 18 provided at the LISE facility of the Grand Accélérateur National d'Ions Lourds (GANIL) were directed onto self-supporting carbon foil targets with nominal thicknesses (quoted by the manufacturer[9]) in the range from 0.5 to 21 μg/cm^2. The projectile beam diameter was typically 1.0 mm and the beam emittance was limited to about 1π mm-mrad. Electrons emitted within a forward cone of about 4° were analyzed by an electrostatic spherical sector electron spectrometer equipped with a position-sensitive detector (PSD) to provide simultaneous resolution of energy ($\delta E/E \approx 1\%$) and emission angles ($\delta\theta \approx 0.5^o$). The PSD output, decoded by the ratiometric charge-division method, along with the spectrometer pass energy completely determine, within limits imposed by the resolution, the velocity components of the detected electron. Thus by scanning the spectrometer deflection field we map the entire $\vec{v}$-distribution over a region of interest, in this case around the projectile vector velocity. Multipole moments of the angular distribution were extracted from the emission distributions by using a non-linear least-squares fitting routine as described below. Auxiliary measurement of the exit charge state fractions for 37 a.u. Ar^{15+} incident on both thin ($\leq 15\mu g/cm^2$) and thick ($\geq 100\mu/cm^2$) carbon foils were performed on-line but with separate targets floated from the same stock as those used for the emission distributions. Target thicknesses were calibrated in-situ before and after the main experiment using electron energy loss measurements obtained with an electron gun and the emission spectrometer.

III. Results and Discussion

Two typical observed emission distributions are displayed in Figure 1 as contours of equal emission intensity. Those shown are for Ar^{17+} with projectile velocity $v_p = 37$ a.u. incident on carbon foil targets of nominally 1 and 8 μ gm/cm^2 thicknesses. Considerable evolution of the emission distribution, presumably reflecting projectile excitation as a result of multiple collisions, is evident.

Extraction of multipole moments from the emission distributions was done to quantify the observed evolution and to make contact with theories of the doubly-differential cross-section through the multipole expansion

$$\frac{d\sigma}{d\vec{v}} = \frac{\sigma_0}{v}\sum_{k=0}^{\infty}\beta_k P_k(cos\theta) \quad (\beta_0 \equiv 0),$$

where $\vec{v} = \vec{v}_e - \vec{v}_p$ is the projectile-frame emission velocity, the β_k are the multipole strengths, the P_k are Legendre polynomials, and σ_0 is the total emission cross-section. Non-linear least-squares fitting is used to compare this expansion, transformed into the laboratory frame with first-order relativistic corrections and convoluted with the spectrometer resolution function, to the observed distributions; the β_k are determined as the fitting process converges.

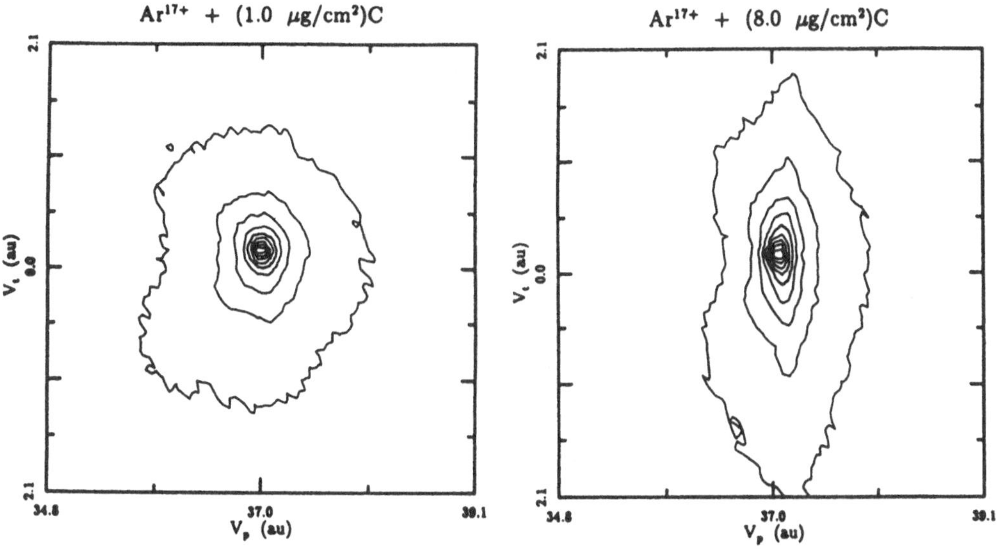

Figure 1. Contours of equal convoy electron emission intensity at intervals of 10% of the peak value for 37 a.u. velocity one-electron argon ions incident on 1 and 8 $\mu g/cm^2$ amorphous carbon targets. The horizontal and vertical axes indicate components of emission velocity parallel (v_p) and transverse (v_t) to the incident projectile directions.

Two particularly interesting multipole moments, β_1 and β_2, are shown as a function of target thickness and incident projectile charge state in Figure 2. We note that in addition to the moments shown in Figure 2, the observations support significant odd order moments up to order 5 and even moments up to order (at least) 10. Efforts are presently underway to model in some detail the multiple collision processes leading to the excitation implied by these multipoles; the present paper will be limited to a discussion of how useful quantitative results can in principle be obtained from multipole strength evolution data like that of the figure.

From the auxiliary exit-charge-state distribution measurements, one can determine total cross-sections for electron capture and loss for each of the relevant projectile charge states, in much the same manner as has been done by Rozet[10] for similar collision systems. The corresponding mean-free-paths (mfp's) for K- and L-shell ionization and K-shell excitation for 36 MeV/u Ar ions in carbon are consistent with theoretical predictions of Gillespie[11] and are of the order of 30000, 10000, and 32000 a.u., respectively. With respect to these and similar processes, then, the present experiment can be considered to be under single collision conditions, since the targets employed have thicknesses in the range 100 to 2000 a.u. However, the excitation cross-sections for shells L- and higher can be expected to be larger, producing smaller mfp's, so that given the population of an L-shell level by excitation or capture, further excitation can be expected to occur rapidly. Of equal importance, inelastic free-electron mean-free-paths for equal-velocity electrons (i.e., $\approx$ 37 a.u.) are typically 600 a.u.[12] Consequently the thicker targets in this experiment should exhibit the effect of multiple collisional excitation and convoy electron transport. The multiple collision aspect should be particularly evident in the case of incident Ar^{15+}, which carries a loosely-bound L-electron into the target, perhaps even at intermediate thicknesses.

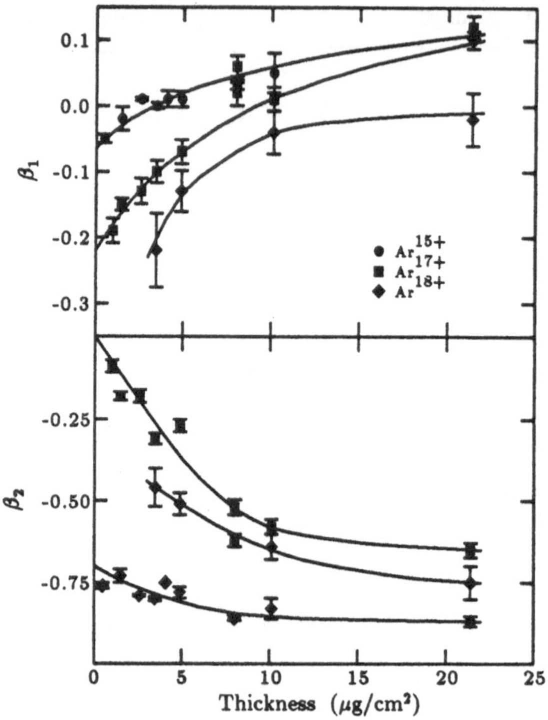

Figure 2. The dependence of dipole (β_1) and quadripole (β_2) moments extracted from the emission angular distributions on target thickness; the solid lines are drawn to guide the eye. Error bars are based on statistical uncertainties associated with the fitting process and on uncertainty in foil thickness.

Considering that the three incident charge states shown in Figure 2 represent three relatively unique initial projectile states (no electron, one 1s electron, and 3 electrons of which one is the more-loosely-bound 2s and is likely to dominate loss and excitation processes), one can take the zero-target-thickness limit of the set of β_k for each given incident charge state to be characteristic of ELC from the appropriate shell or of ECC in the case of the bare projectile. This is possible because the observed thin-target yield of convoy electrons from one-electron Ar^{17+}, i.e. ECC to a charge-17 center plus K-shell ELC, exceeds that from bare Ar^{18+} (ECC) by a factor of 50, so that at worst the one-electron data represents K-shell ELC contaminated by 2% ECC. Further, the Ar^{15+} (three electron) convoy yield is a factor of $\approx$ 20 greater yet, suggesting that the distribution of β_k values for incident Ar^{15+} is dominated by ELC from the L-shell. The specific β_k values obtained in this way from the zero-thickness limit are in quite reasonable agreement with previous experimental data[3,13] and with theory from several sources[14].

The effect of plural- or multiple-collision processes on convoy electron emission is to 'mix' the three characteristic β_k distributions introduced above, and to add other distributions characteristic of higher excited projectile states, in an evolving linear superposition. This mixing occurs with targets of increasing thickness, and the coefficients of this superposition of β_k distributions are related to the cross-sections which connect

Convoy and Rydberg Electrons in Ion-Solid Collisions*

Hans D. Betz, Ralf Schramm, Willi Oswald
Sektion Physik, Universität München
D-8046 Garching

The passage of heavy ions through solid targets is a topic which continues to puzzle investigators in many aspects. While single ion-atom collisions can be treated very accurately, charge exchange and excitation in ion-solid interaction remain partly unclarified. One of the unsolved questions concerns the so-called density effect[1-4]: no final explanation has been given for the 40-years old observation that heavy ions exhibit higher charge states when they emerge from a thin solid as compared to a gas target with equivalent thickness, although considerations of the specific energy loss suggest that the effective charges must be quite similar in both cases. Thus, the higher charges observed behind a foil must be screened while the ions move inside the solid target. An effect of this kind could arise from highly excited states around the projectile continuum, which are transformed to either convoy or Rydberg states upon emergence from the exit surface of the solid.

In order to shed more light on this open question, we performed a series of measurements to determine the absolute number of both convoy and Rydberg electrons formed after passage of heavy ions through thin carbon foils. Preliminary results are given for 125-MeV bromine ions. We are led to conclude that Auger-, convoy- and field-ionized Rydberg electrons do not seem to be present in sufficient number to help explain the charge state density effect.

Quantitative measurement of Rydberg yields turned out to be complicated due to unexpectedly long lifetimes of these highly excited states in the ionizing field. In a recent paper, calculated lifetimes of the order of 10^{-11} s have been reported[5] for quantum numbers n=180 in Ne^{7+}; our data for different Rydberg levels suggests decay times of the order of 10^{-9} s. Theoretical estimates of these times are not yet available.

the various states. For example, the thin-target Ar^{18+} bare incident ion data display the strong, negative dipole moment which is typical of the ECC process at high collision velocities and which is observed in singly-differential as a skewing toward emission energies below the central, 'cusp' energy. This large, negative β_1 value gives way at larger thicknesses to a small dipole moment and an increasingly strong negative quadrupole moment (β_2), which is characteristic of the ELC process as can be seen in the thin-target Ar^{17+} distribution. We interpret this particular behavior as evidence of so-called capture-loss[7] in which an electron is captured into a bound orbital and then, in a subsequent collision, is lost to the continuum in an electron capture-loss to the continuum process (ECLC), and which, to our knowledge, has never previously been resolvable from ECC. The measurements and analysis technique outlined here should provide a quantitative measure of the relative size of these two mechanisms.

Other information of interest, such as the K- to L-shell excitation cross-section which underlies the initial growth of a β_4 moment and the rate of approach to projectile excitation equilibrium, is potentially available, pending analysis, in the data of Figure 2 and corresponding plots of other measurable multipole strengths; space, however, does not permit an exhaustive discussion here.

Acknowledgements

This experiment was conducted at the GANIL accelerator in Caen, France; we are very grateful for the friendly support we received there. This experiment forms part of the doctoral work of John Gibbons of the University of Tennessee. Other collaborators in this work include C. Biedermann, R. DeSerio, N. Keller, and I. A. Sellin (Tennessee); J. P. Grandin, A. Cassimi, and X. Husson (CIRIL/GANIL); M. Druetta (LSIM, Lyon); and L. Liljeby (Manne Siegbahn Institute). This work was also supported in part by the National Science Foundation and by the U. S. Department of Energy under Contract No. DE-AC05-84OR21400 with Martin Marietta Energy Systems, Inc.

References

1. N. Bohr and J. Lindhard, K. Dan Vidensk. Selsk. Mat. Fys. Medd. 28, No. 7 (1954).
2. H.-D. Betz, D. Röschenthaler, and J. Rothermel, Phys. Rev. Lett. 50, 34 (1983); H.-D. Betz, in Forward Electron Ejection in Ion Collisions, edited by K.O. Groeneveld et al. *Lecture Notes in Physics* Vol. 213 (Springer-Verlag, Berlin, 1984), p. 115.
3. S.D. Berry et al., J. Phys. B.19, L149 (1985).
4. C.J. Woods et al., J. Phys. B.18, 4113 (1985).
5. J. Burgdörfer, in proceedings of this conference.
6. J. Burgdörfer, in Proceedings of the 3rd Workshop on High-Energy Ion-Atom Collisions, edited by D. Berenyi and G. Hock, *Lecture Notes in Physics* Vol. 294 (Springer-Verlag, Berlin, 1988) p. 344, and J. Burgdörfer and J. Gibbons, Phys. Rev. A42, 1206 (1990).
7. J. Kemmler et al. Nucl. Instr. and Meth. B33, 281 (1988).
8. S.B. Elston, Nucl. Instr. and Meth. B24/25, 214 (1987).
9. Arizona Carbon Foil Co., Inc., Tuscon, AZ, USA.
10. J.P. Rozet et al., J. Phys. B22, 33 (1989).
11. G. Gillespie, Phys. Rev. A18, 1967 (1978); G. Gillespie, Phys. Rev. A22, 454 (1980).
12. J. Ashley et al., Thin Solid Films 60, 361 (1979).
13. S.B. Elston et al., Phys. Rev. Lett.55, 2281 (1985).
14. J. Burgdörfer, Phys. Rev. A33, 1578 (1985); J. Burgdörfer et al., Phys. Rev. A28 3227 (1983), D.H. Jakubassa-Amundsen, *Forward Electron Ejection in Ion Collisions*, Vol. 213 of *Lecture Notes in Physics*, edited by K.O. Groeneveld et al. (Springer, Berlin 1984), p. 17.

Experimental Technique

The Munich Tandem accelerator was used to prepare 125 MeV bromine ions, directed onto thin carbon foils with thicknesses ranging from 10 to 180 µg/cm² (the latter value allows to provide charge state equilibration). The beam was collimated to a diameter smaller than 1 mm and a divergence of less than 0.1°. A magnetic sector analyzer focusing in two planes was employed to deflect free electrons into a channeltron detector. By rotating the analyzer, electrons emitted into the entire forward cone could be measured step by step. The semi-half-angle of the spectrometer was variable from 0.2° to 2.5° and amounted to 1.2° in this experiment; momentum resolution could be set from 0.1 to 0.8% and was chosen to be 0.2%. These values were determined from the geometry by utilizing ray-tracing techniques and were experimentally verified by resolving a series of known KLL Auger lines from O^{5+} beams[6].

Much care was taken to pin down the magnetic field strength along the various electron trajectories in the spectrograph, especially the fringing field in the entrance and exit region (see Fig. 1 and 4). In particular, transmission factors and effective solid angles have been obtained with good accuracy. As a result, positions, shifts and widths of electron peaks in the velocity spectra could be verified numerically as a function of geometry parameters and enabled us to extract absolute electron yields.

Rydberg electrons have been measured by many groups[5,7-11] and we pursue similar techniques, though our magnetic spectrograph is different from the ones used by others and had to be studied in many details. Excellent separation of convoy and Rydberg peaks was achieved for two reasons. First, a small (≈ 1 V/cm) longitudinal electric field was applied behind the target foil which decelerated free convoy electrons, but did not affect the Rydberg electrons to be ionized later on. Second, the spatially extended fringing field causes ionization of Rydberg electrons to take place in a region behind the one where convoy electrons become deflected. Thus, focusing into the detector is not possible until a higher deflection field is applied (Fig. 1), simulating a higher velocity. The resulting peak shifts are clearly visible in Fig. 2 and 3. Variation of the observation angle allowed to view different parts of the ion trajectory from which electrons can originate. Although the Rydberg ridge in Fig. 2 is spread out in the v-axis, it represents only electrons with exact beam velocity.

Results

Convoy electron spectra have been obtained for observation angles ranging from 0° to 90°. Appropriate integration over the entire solid angle leads to an absolute yield of 1 ± .4 electrons/ion. This number contains no appreciable contribution of electrons from other sources, such as Auger-, Rydberg- or secondary electrons. The number of Rydberg electrons was determined from the data shown in Fig. 2: the observation angle corresponds to a certain emission region and, thus, to a certain magnetic field strength and a minimum quantum number, n_{min}, from which electrons can be ionized. The maximum field was ≈ 350 V/cm and allowed liberation of Rydberg electrons from levels with n > 325. Fig. 4 shows the specific emission of Rydberg electrons along the beam trajectory. It becomes obvious that delay times of emission are of the order of the dwell time of the beam in the fringing field (≈ ns). Preliminary evaluation of the absolute yield was based on the assumption of an initial population a/n^3 of Rydberg levels and yielded $a \approx 50$. The error of this number a is still large, perhaps a factor 3, and we hope to be able to reduce it in refined subsequent experiments.

For the system studied the average equilibrium charge is 23^+ and 17^+ in solid and gaseous targets, respectively. Only one out of the difference of 6 electrons can be traced to convoy electrons. Thus, the nature of the density effect remains essentially unclarified.

References

1. N. Bohr, Mat.-Fys. Medd-K. Dan. Vidensk. Selsk. 18, No.8 (1848) 1.
2. N. Bohr and J. Lindhard, Mat.-Fys. Medd-K. Dan. Vidensk. Selsk. 28, No.7 (1954) 1.
3. H.-D. Betz and L. Grodzins, Phys. Rev. Lett. 25 (1970) 903.
4. M.D. Brown and C.D. Moak, Phys. Rev. B6 (1972) 90.
5. W.-D. Zeitz, R. Kowallik and D. Schneider, Phys. Rev. A39 (1989) 43.
6. N. Stolterfoht, in *Structure and Collisions of Ions and Atoms*, edited by I. A. Sellin, Topics in Current Physics Vol. 5 (Springer-Verlag, Berlin, 1978), p. 155-199 (see Fig. 5.4 for O^{5+} on Ne).
7. Z. Vager, B. J. Zabransky, D. Schneider, E. P. Kanter, G. Yuan- Zhuang, and D. S. Gemmel, Phys. Rev. Lett. 48 (1982) 592.
8. E. P. Kanter, D. Schneider, and Z. Vager, Phys. Rev. A 28 (1983) 1193.
9. E. P. Kanter, D. Schneider, Z. Vager, D. S. Gemmel, B. J. Zabransky, G. Yuan-Zhuang, P. Arcuni, P. M. Koch, D. R. Mariani, and W. van de Water, Phys. Rev. A 29 (1984) 583.
10. Y. Yamazaki and N. Oda, Phys. Rev. A 32 (1985) 1260.
11. K. P. Müller, H. Kuiper, M. Schöberl, and R. Schmelzer, J. Phys. B: At. Mol. Phys. 20 (1987) 2803.

* Partially supported by BMFT (Bonn) under contract number 6ML177I.

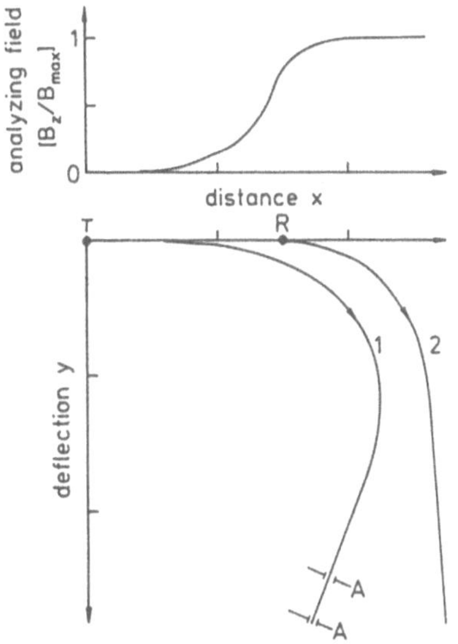

Fig. 1. Experimental separation of convoy and Rydberg electrons in a magnetic spectrograph (not to scale). For a certain deflection field convoy electrons travel along path 1 through apertures A and reach the detector, while Rydberg electrons travel along path 2 and remain undetected. They can be observed, though, at a different angle and at a higher magnetic field.

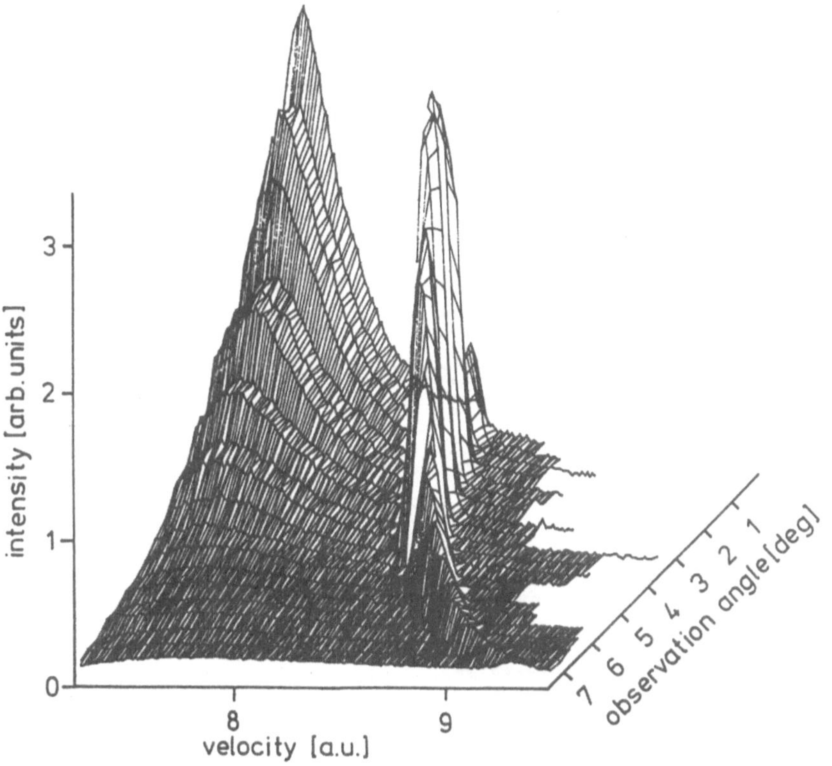

Fig. 2. Velocity distribution of electrons emitted in the forward direction when 125-MeV bromine ions pass through 20 μg/cm² carbon, plotted as a function of the observation angle relative to the beam direction (.6° to 7°; zero degree not included). A small electric field behind the target decelerates convoy electrons and shifts the peak to the left, while field-ionized Rydberg electrons follow a different trajectory and appear at a position shifted to the right.

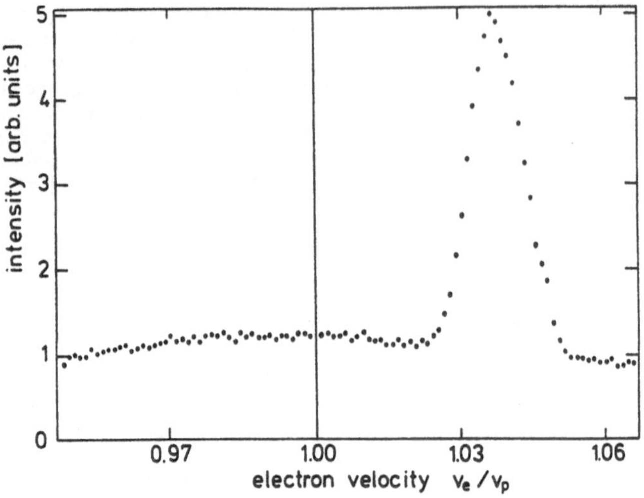

Fig. 3. Electron velocity spectrum for 125-MeV bromine ions in carbon (as in Fig. 2), observed at 4° with respect to the beam direction. Convoy electrons are no longer visible at this relatively large angle. Rydberg electrons still show up at a position shifted to the right; they become ionized deeper inside the spectrometer and travel along a different orbit (see Fig. 1 and 2).

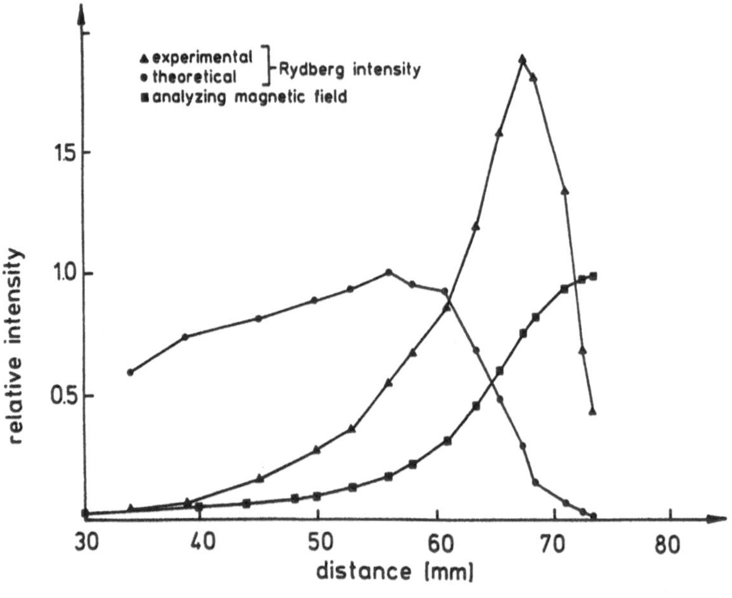

Fig. 4. Lifetime-effect of Rydberg ionization: ▲ observed Rydberg electron intensity; ● intensity expected from the assumption of spontaneous field ionization; ■ fringing field of the magnetic spectrograph, as function of the distance from the target position. The pronounced difference between the former two curves indicates delayed field-ionization (lifetimes in the ns-regime).

FAST-ION RYDBERG STATE PRODUCTION IN CARBON FOILS

J.Wagner, R.Harder, M.Lotter, M.Tempel
H.-G.Toews, B.Hertel, J.Maeritz and H.Kuiper

Physikalisches Institut der Universität Erlangen-Nürnberg
Erwin-Rommel-Str.1, D 8520 Erlangen, F.R.Germany

Abstract: Experimental results of measurements of the O^{7+} Rydberg yield after passage of 13–24 MeV oxygen ions through carbon foils of different thicknesses are presented. The data have been obtained by observing Rydberg states in the $58 \leq n \leq 75$ range of main quantum numbers by means of RYDFISS, a method using field ionization and spatial separation techniques. For thin foils a significant increase of the O^{7+} Rydberg population with increasing thickness is observed, although this population was normalized to the O^{7+} exiting charge fraction. The equilibrium Rydberg yield A_7 shows a pronounced dependence on the projectile velocity.

1 Introduction

When fast ionic projectiles traverse a thin solid foil, a measurable fraction of those projectile ions will be found in high Rydberg states after exiting from the foil into vacuum.

While this has been demonstrated experimentally in various ways [1-8], a complete theoretical understanding of the interplay of a number of different processes leading to the final formation of projectile Rydberg states exiting from the foil has not been achieved yet. In fact, this is not so surprising because many different effects play their rôles in a very complex game: as soon as the fast ion has entered the solid, its charge state will begin to fluctuate stochastically due to electronic capture and loss processes, and "near threshold states" of the projectile, namely convoy electrons and Rydberg states are produced in close succession. At the same time the moving projectile ion will experience excitation and de-excitation processes of its electronic core. The picture of a "random walk in phase space" as suggested by Burgdörfer [9,10] for an electron accompanying an ion moving through a solid either as a *convoy electron* or as a *Rydberg electron* appears to be quite appropriate.

Much theoretical effort has been devoted to special problems, e.g. looking for an explanation for the enhancement of high angular momentum states [11] which is suggested by experimental evidence [12]. Also the problem of capture of secondary electrons into Rydberg states of the moving ion just upon exiting from the solid into vacuum has been attacked from the theoretical side [13].

However, no complete theoretical description exists today, able to predict in analytical form the ionic Rydberg yield after foil passage as a function of the following parameters: projectile nuclear charge Z_p, initial projectile charge state q_i before the beam-foil interaction, projectile exit charge state q_e, projectile velocity v, foil material and foil thickness.

2 Description of Measurements

Experimental Rydberg yields are presented here which have been obtained by collimating a 1 mm diameter beam of O^{q+} ions ($q = 3, 4, 5$) from the Erlangen EN tandem accelerator on thin carbon foils. We use RYDFISS, a method introduced by our group recently [8], which is particularly well suited to determine the absolute ionic Rydberg yield. The Rydberg yield is characterized by the population constant A as shown in ref. [8]. We have measured this population constant for O^{7+} ionic Rydberg states after the foil as a function of foil thickness in the range 4–38 $\mu g/cm^2$ at four projectile energies (13, 16, 19.5 and 24 MeV).

The experimental arrangement was essentially the same as that explained in detail in ref. [8]. The absolute number N_7 of O^{7+} ions that exited from the foil during a run was obtained by current integration of the electrostatically separated O^{7+} charge fraction. The total ion beam was monitored by means of a solid state detector (A) registering the ions backscattered from the stainless steel blades of the rotating chopper [8] placed about 50 cm upstream with respect to the foil. Two solid state detectors (B and C) positioned about 28 cm downstream were used to register the ions scattered from the carbon foil into angles of 4.5° and 6.5°, respectively. The counts of detectors B and C during a run were normalized to the total beam signal from detector A, thus giving a relative measure of the foil thickness. Absolute thickness values were obtained by comparison with foils calibrated by alpha particle energy loss measurements.

It is interesting to note that RYDFISS is equivalent to a coincidence method, e.g. like that used by Gaither III et al. [7]: in both cases the charge state of the detected Rydberg ion is well known. However, as compared to the coincidence method, RYDFISS has the additional advantage that the absolute normalization problem is solved readily.

In describing our results we refer to eqs.(1) and (2) of ref. [8] and to the notation used therein. As explained there, the Rydberg yield parameter A describing the population of O^{7+} Rydberg states was extracted from the slope of the measured ratios $R(E) = N_D/N_7$ plotted as a function of $\sqrt{E}$; here E denotes the electric field strength in the ionizing region, while N_D and N_7 are the symbols used for the number of Rydberg ions and the number of O^{7+} ions, respectively, collected during the run. To be more specific, we use the symbol A_7 here (instead of A as in ref. [8]), in order to stress that in this experiment we are observing exclusively O^{7+} Rydberg states.

3 Results and Discussion

The diagrams of fig.1 show the measured Rydberg population parameter A_7 as a function of foil thickness ρx and projectile energy E_p; the initial charge state q_i of the incoming O^{q_i+} oxygen ion is also indicated in each case. From each of the four diagrams it can be seen, that the O^{7+} Rydberg yield is rising steeply with increasing foil thickness for thin foils of about 5 $\mu g/cm^2$. At a foil thickness of about 15 $\mu g/cm^2$ saturation of the Rydberg yield is reached.

In an attempt to extract more information from these results a least squares fit to the data was made using the functional form

$$A_7(\rho x) = A_{7\,sat}\left(1 - exp(\frac{\rho x}{t_0})\right) \quad (1)$$

and yielding the solid curves in the diagrams fig.1. The four values for t_0 extracted from the four fit curves coincide within about 30 %, which is not unreasonable with a view to the relatively scarce data for thin foils. Their weighted mean is $t_{0\,av} = (2.9 \pm 0.1)\mu g/cm^2$. The question arises what could be the physical meaning of the quantity t_0. According to its definition, A_7 as a function of foil thickness ρx should look like a step function with the step at $\rho x = 0$, if the O^{7+} Rydberg states were produced *exclusively* by excitation of O^{7+} ions. If however, other and particularly capture processes are responsible for the production of Rydberg states, then also the other charge fractions are important. We thus believe that the increase at low foil thicknesses somehow mirrors the development of the ratio of the charge fractions $q = 8$ and $q = 7$. Further investigations in this direction are certainly needed, but due to the nonzero value of t_0 it seems likely that electron capture essentially contributes to the O^{7+} Rydberg state production.

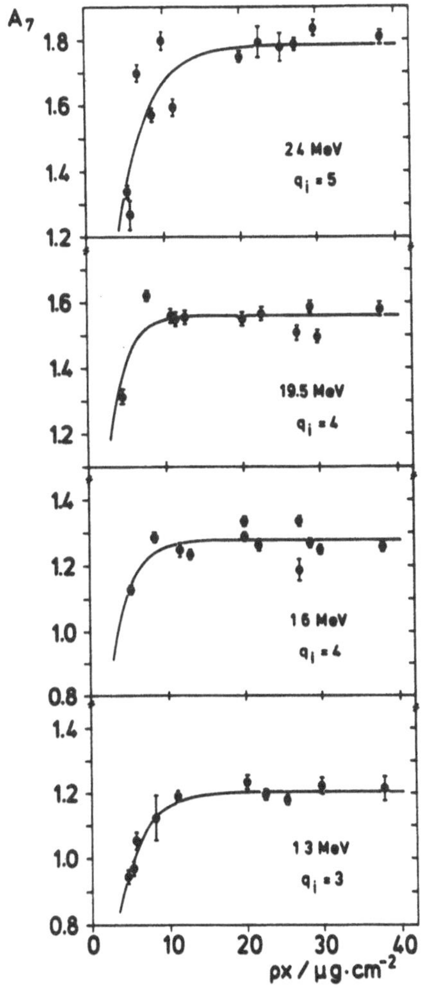

Fig.1 Experimental results at different energies as a function of foil thickness ρx. The initial projectile charge state is denoted by q_i. The solid lines are least square fits according to eq.(1)

The diagrams of fig.2 show the dependence of the saturation Rydberg yield on the projectile velocity v (in atomic units). The yield constant A_7 increases with increasing projectile velocity in an almost linear fashion; a least squares fit using the expression $A_7 = c\,v^r$ yields $c = 0.128 \pm 0.055$ and $r = 1.27 \pm 0.23$. A more interesting result is found by the following renormalization procedure. We denote by ϕ_q the equilibrium charge fraction of the beam after the foil with charge q. Then multiplying A_7 by the factor ϕ_7/ϕ_8 yields the ratio of the number of O^{7+} Rydberg states to the number of O^{8+} ions in the beam. This ratio is of significance if the production mechanism of the observed Rydberg states is predominantly electron capture by the exiting O^{8+} ions. The process of electron production, transport and capture by the exiting ion has been treated theoretically by Schröder [13], who obtained a v^{-6} velocity dependence for the Rydberg yield. Our renormalized data are displayed in the lower diagram of fig.2 showing a decrease of the Rydberg yield with increasing projectile velocity. By means of a least squares fit using a power law $b\,v^s$ we obtain $b = 710 \pm 230$ and $s = -2.47 \pm 0.17$. This differs considerably from Schröder's exponent −6. Schröder himself discusses the possibility of a smaller absolute value of the exponent in question. It should also be noted that Dybdal et al. [4] extracted a v^{-4} dependence from their measurements on Si and F ions, although in the 8–13 range of main quantum numbers.

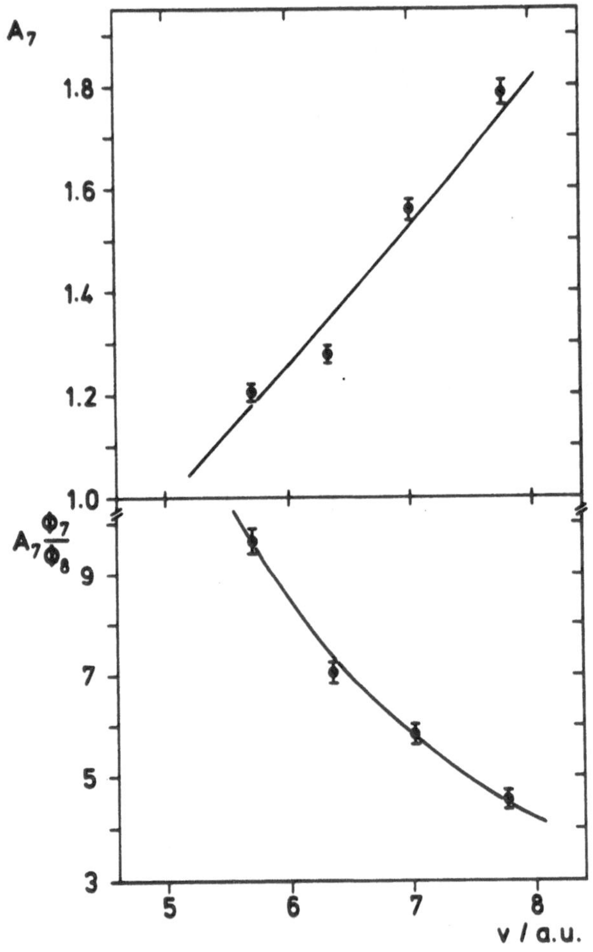

Fig 2 *Upper diagram:* Saturation value $A_{7\,sat}$ of the Rydberg yield constant as defined by eq.(1) as a function of the projectile velocity v in atomic units. *Lower diagram:* The same quantity renormalized as explained in the text.

It is obvious that intense further work, on the experimental as well as on the theoretical side will be necessary until the question of Rydberg state production by fast ions moving through solids can be answered quantitatively.

References

[1] H.-D.Betz, J.Rothermel, D.Röschenthaler, F.Bell, R.Schuch, G.Nolte, Phys. Lett. **A91** (1982) 12

[2] H.-D.Betz, D.Röschenthaler, J.Rothermel, Phys. Rev. Lett. **50** (1983) 34

[3] E.P.Kanter, D.Schneider, Z.Vager, D.S.Gemmell, B.J.Zabransky, Gu Yuan-zhuang, P.Arcuni, P.M.Koch, D.R.Mariani, W.van de Water, Phys. Rev. **A29** (1984) 583

[4] K.Dybdal, J.Sørensen, P.Hvelplund, H.Knudsen, Nucl. Instr. and Meth. **B13** (1986) 581

[5] G.Schiwietz, D.Schneider, J.Tanis, Phys. Rev. Lett. **59** (1987) 1561

[6] K.P.Müller, H.Kuiper, M.Schöberl, R.Schmelzer, J. Phys. B: At. Mol. Phys. **20** (1987) 2803

[7] C.C.Gaither III, M.Breinig, J.Freyou, T.A.Underwood, Nucl. Instr. Meth. in Physics Res. **B40/41** (1989) 56

[8] R.Harder, R.Knauss, K.Kretzschmar, M.Lotter, H.Rauh, J.Wagner, H.Kuiper, Nucl. Instr. Meth. in Physics Res. **B48** (1990) 111

[9] J.Burgdörfer, Lecture Notes in Physics **294** (1988) 344, (Proceedings of the 3rd Workshop on High-Energy Ion-Atom Collisions, Debrecen 1987, D.Berényi, G.Hock eds., Springer-Verlag)

[10] J.Burgdörfer, Nucl. Instr. Meth. in Physics Res. **B42** (1989) 500

[11] J.Burgdörfer, C.Bottcher, Phys. Rev. Lett. **61** (1988) 2917

[12] Y.Yamazaki, N.Stolterfoht, P.D.Miller, H.F.Krause, P.L.Pepmiller, S.Datz, I.A.Sellin, J.N.Scheurer, S.Andriamonje, D.Bertault, J.F.Chemin, Phys. Rev. Lett. **61** (1988) 2913

[13] H.Schröder, Z. Phys. D–Atoms, Molecules and Clusters **7** (1987) 65

G Multi-Electron Processes

Chairman: T. KAMBARA

G.1 Invited Surveys

C.L. COCKE
Review of Some Aspects of Multiply Ionizing Collisions Involving Heavy Ion Projectiles

H. SCHMIDT-BÖCKING, R. DÖRNER, J. ULLRICH, J. EULER, H. BERG, E. FORBERICH, S. LENCINAS, O. JAGUTZKI, A. GENSMANTEL, K. ULLMANN, R.D. DuBOIS, FENG JIAZHEN, R.E. OLSON, A. GONZALES and S. HAGMANN
Multiple Ionisation in Ion–Atom Collisions Investigated by Recoil Ion Momentum Spectroscopy

G.2 Invited Contributions

A.S. SCHLACHTER
Multiple Electron Capture in Fast Ion–Atom Collisions

J. ULLRICH, R. DÖRNER and H. SCHMIDT-BÖCKING
Multielectron Processes

G.3 Contributions (Posters)

B. SULIK, V.J. MONTEMAYOR, N. STOLTERFOHT and I. KÁDÁR
Dynamic Screening in Ion–Atom Collisions

E.M. BERNSTEIN, M.W. CLARK, J.A. TANIS, W.G. GRAHAM, T.J. MORGAN, M.P. STOCKLI, K.H. BERKNER, A.S. SCHLACHTER and J.W. STEARNS
Enhanced Radiative Auger Emission from Lithiumlike Ions

I. KÁDÁR, H. ALTEVOGT, R. KÖHRBRÜCK, V. MONTEMAYOR, A. MATTIS, G. SCHIWIETZ, B. SKOGVALL, K. SOMMER and N. STOLTERFOHT
High-Energy Excitation of Sodium-Like Argon Ions

REVIEW OF SOME ASPECTS OF MULTIPLY IONIZING COLLISIONS INVOLVING HEAVY ION PROJECTILES

C.L. Cocke
J.R. Macdonald Laboratory, Physics Department
Kansas State University, Manhattan, KS 66506-2601 USA

It has been apparent for more than two decades that a violent collision between an energetic heavy ion and a neutral target can remove many electrons from the collision system. Early collisions involved singly charged projectiles scattered at large angles, for which the source of energy for mulitple ionization was the projectile kinetic energy. More recent experiments have involved very highly charged projectiles scattered at modest to very small angles, for which the potential energy carried by the projectile dominates the collision. Multiple ionization of the target for this case does not generally occur at the expense of loss of electrons from the collision system. Early measurements of charge state distributions and electron and x-ray spectra have now been supplemented by measurements differential in both energy and angle of both projectiles and recoils. The complexity of the theoretical treatment of this many-electron problem has invited the building of models based on statistical treatments, independent electron models, and n-body classical calculations. The last of these has had the most success in dealing with the recent data. This talk will survey selected aspects of multiply ionizing ion-atom collisions at intermediate to low velocities, and will attempt to establish some connections between recent results and early work in the field.

I. INTRODUCTION

When a highly charged ion comes into collision with a multi-electron neutral target, many electrons are excited and removed from the target. Much of ion-atom collision work over the past two decades has concentrated on the evolution of the state of the inner shell electrons, whose motion is governed mainly by the large Coulomb potentials of the nuclear charges and which move in a sufficiently high momentum region of phase space that single electron processes of electron ionization, excitation and capture can be isolated and studied. While the inner shell activity is going on, the outer shell electrons are being blown about like leaves in the wind, and the target fragment left after the collision is multiply ionized. The evolution of this incidental fluff escaped attention for some time due to the concentration of effort on inner shell processes, but in recent years mulitple ionization of outer shells has received renewed attention. In this talk I will try to review some pertinent experiments on this subject carried out over many years.

A schematic of the collision system is shown in fig. 1. The target electrons are removed through their interaction with the projectile nuclear potential and with the electrons of the projectile. They can also be lost from the target via interaction with other target electrons after the collision, via Auger processes, but the energy needed for this must first be supplied by excitation from interaction with one of the projectile potentials. If the projectile velocity exceeds substantially the orbital velocities of the outer target electrons, the target electrons will be generally lost to the continuum with the kinetic energy of the projectile driving the ionization and I refer to this as kinetic energy dominated ionization. If the projectile velocity is very slow compared to the orbital velocity, the target electrons are pulled from the target by the strong field of the projectile, but are largely captured into excited states on the projectile and are rarely lost from the system, except by Auger processes on the projectile after the collision. I refer to such a collision as a potential energy dominated one, in which the projectile velocity is very umimportant in determining the outcome of the collision. For slow collisions where the projectile shells are already filled with electrons, this picture is modified substantially, both due to screening of the large projectile charge and due to the Pauli exclusion principle which blocks some of the final states into which the target electrons could be captured. In such a situation, the projectile nuclear potential no longer dominates the collision but the target electrons are lost mainly to the continuum, and the projectile kinetic energy again serves as the driving energy source.

Common to all of these mulitply ionizing collisions is the evolution of the charge state fractions with the violence of the collision. What violence means depends on the collision, but it probably most closely related to the electronic energy made available for ionization in the collision, be it kinetic or potential. In order to investigate to what extent this electronic energy is the major factor determining the charge

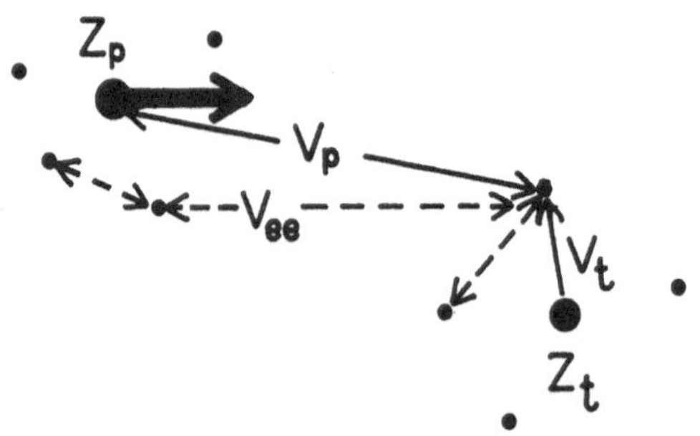

Fig. 1: Schematic of collision.

state distributions, fig. 2 shows plots of charge state fractions of Ne versus energy for some very different physical situations and models. The abscissa in each case is the electronic energy per target electron, but the prescription for assigning this energy is different in each case. For the solar corona fractions, calculated by Jordan[1], the abscissa is the electron temperature. These curves reflect especially high charge state fractions due to the weakness of recombination processes with free electrons. For Ne ions passing through a foil,[2] the abscissa is the equivalent translational

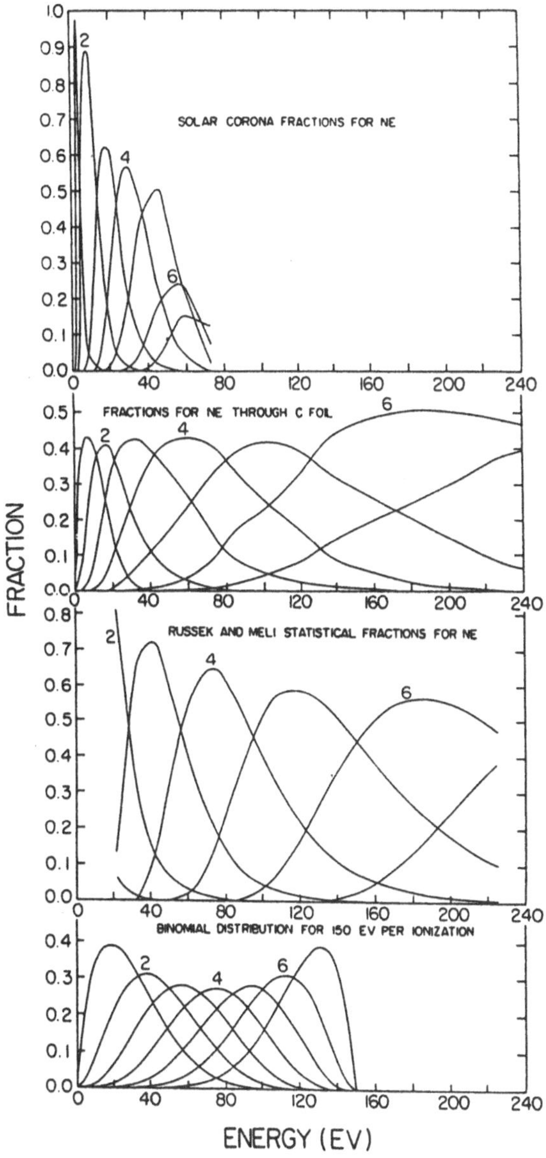

Fig. 2: Charge state fractions versus energy deposited per L shell electron for Ne for different physical processes and models. Charge states 1 to 7 are shown.

energy of the electon travelling at the projectile velocity. The last two curves are model calculations, to be discussed later. While the electronic energy is important, it is clear that this alone is not sufficient to determine the result.

II. EXPERIMENTS

A. Kinetic Energy Dominated Collisions:

1. Impact Parameter Selected Collisions

Early experiments carried out by Everhart and Kessel (KE) and collaborators[3,4] and by Afromisov and Federenko[5] showed that even singly charged projectiles at velocities less than 1 a.u. mulitply ionized both target and projectile when the scattering angle was large, corresponding to small distances of closest approach. The charge state probabilities for projectile and recoil were found to approximately factor into separate and similar probability functions for each fragment, with no evidence that the asymmetry of the indident channel was remembered in the final channel. The charge state distributions were measured as a function of the opening angle between the scattered partners, from which the inelasticity of the collision could be calculated from binary encounter kinematics. From their data for Ar+ on Ar at projectile energies of 50 to 150 keV and θ of 10° to 20°, KE were able to extract charge state fractions as a function of electronic energy deposited in each collision partner (fig. 3).

While data of this type was stimulating the development of the Fano-Lichten model[6] for promotion of inner shell electrons for inner shell vacancy production, the mulitple ionization of the outer shells led Russek

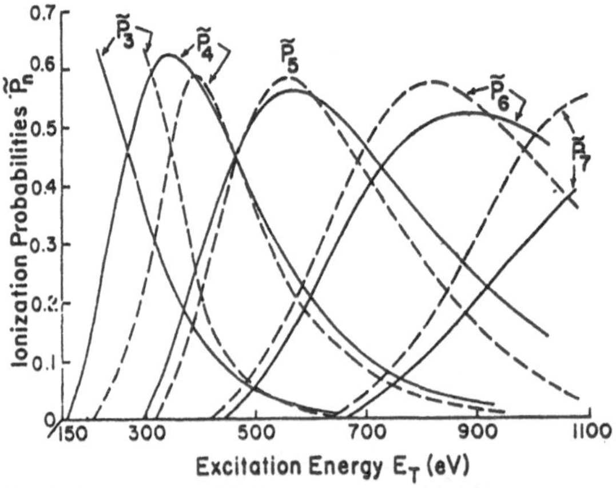

Fig. 3: Charge state fractions versus total excitation energy deposited in the collision for Ar+ on Ar. The solid lines are deduced from the data of Ref. 4, while the dashed lines are from the RM statistical model Eq. (1), Ref. 9.

and collaborators to develop a statistical treatment of this process.[7-9] They described the process in two steps: First, the primary collision deposits energy E_e in the electronic degrees of freedom of the projectile and recoil. Second, each system emits electrons in a statistical manner, populating final charge states according to the phase space available in that ionization state. The model underwent evolution resulting in several prescriptions for calculating the final charge state distributions from the energy deposited. The version used by Russek and Meli (RM) to describe the data of fig. 3 and the curves of fig. 2c gives the fraction of ions in charge state i proportional to

$$F_i \propto \binom{N}{i} S_i(E_K) \tag{1}$$

where $\binom{N}{n}$ is the binomial coefficient, $S_i(E_K)$ is the phase space available for a total electron kinetic energy E_K, as given by Eq.(56) in Ref. 9, and N is the number of active electrons. The F_i must be normalized so that $\sum_i F_i=1$. A comparison of the results of this calculation with experiment is shown also in fig. 3, where good agreement is apparent. The mechanism for the deposition of E_e in the first step was originally taken from a friction model, but later attributed to molecular promotion. Specific promotion mechanisms for the outer shell electrons are difficult to identify due to the large number of orbitals involved.

Later, K-x-ray spectra from heavy ion bombardment of heavy targets showed copious satellite production due to multiple ionization of the L shell accompanying the K vacancy production. If the L electrons are treated as independent , a binomial distribution can be used to describe the probability P_n for ionizing the L shell n times in terms of the probability p for ionizing a single L electron,[10] giving

$$P_n = \binom{N}{n} p^n (1-p)^{N-n} . \tag{2}$$

While p is in principle a function of b, the production of a K vacancy requires b be nearly zero and p(b) becomes p(0). The binomial distribution was quite successful in describing the satellite distributions, although first principles calculations of p from perturbation theories were less so. However, there is a fundamental problem with the use of the binomial distribution. Since the ionization energy for the first electron is very different from that for the last electron, which one do you take in calculating p? It is clear that no single electron removal probability is adequate in principle, and that the agreement between experiment and binomial statistics model is probably fortuitous.

It is possible to select small impact parameter collisions by placing requirements on charge changing transitions on the projectile. Figure 4 shows cross sections for recoil prodution for F^{9+} on Ne for capture of zero, one and two electrons.[11] The recoil charge state distributions shift to larger mean charge as number of projectile captures rises. This shift is partially due to the selection of smaller b by the requirement that multiple capture occur, but partially due to the fact that there is additional correlation of projectile and recoil processes. The importance of the latter effect is still poorly determined.

Kelbch et al.[12] have provided information on this point by measuring projectile angular distributions for charge state resolved collisions. In fig. 5 we show charge state fractions deduced from their data for F^{6+} on Ne at 4.4 MeV for the cases of direct ionization (6+ to 6+) and double capture (6+ to 4+). The distributions are remarkably similar, suggesting that the violence done to the recoil is not stongly dependent on how many electrons

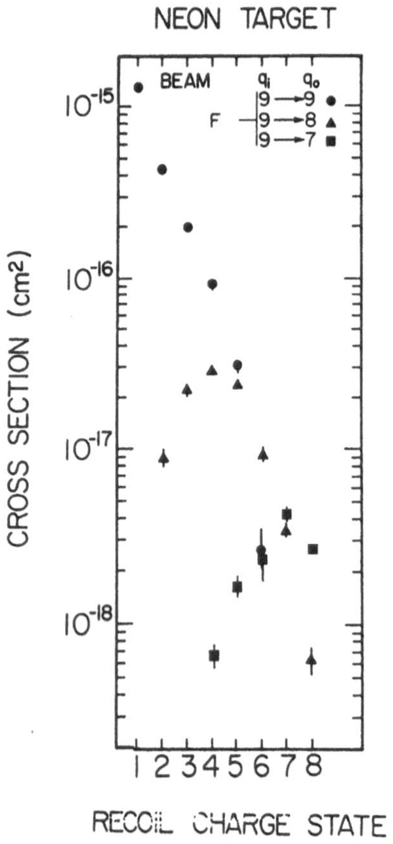

Fig. 4: Cross sections for recoil production in F-Ne collisions as a function of final recoil and projectile charge state.

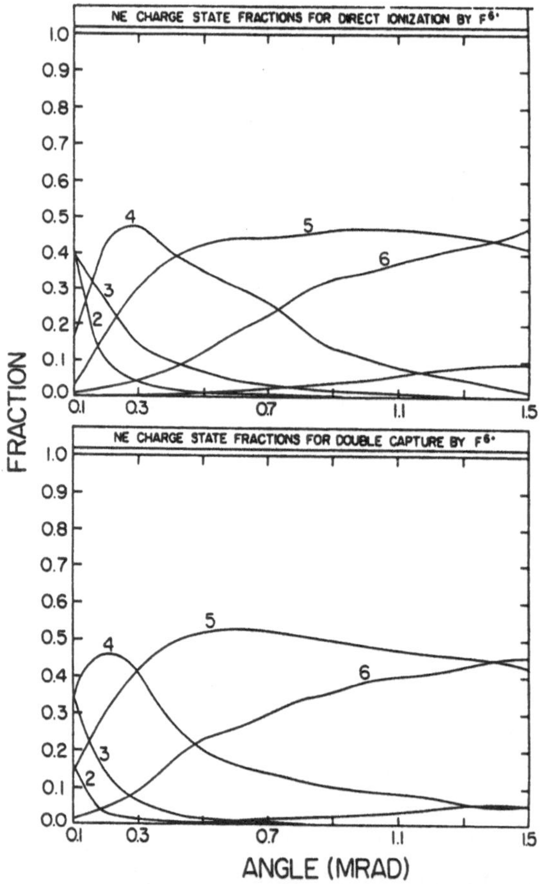

Fig. 5: Charge state fraction for 4.4 MeV F^{6+} on Ne, taken from the data of Kelbch et al. (12).

ended up on the projectle for a given scattering angle θ. One might deduce that the amount of energy deposited in the recoil degrees of freedom is primarily dependent on θ and only weakly dependent on what happens meanwhile to the projectile. A theoretical model for the interpretation of these data was developed by Horbatch,[12,13] who starts from the Vlasov equation to describe the evolution in phase space of the target electron distribution function. In the end, the calculation he actually performs is a classical trajectory Monte Carlo (CTMC) calculation and the results are expressed through the form of single electron transition probabilities converted using multinomial statistics to calculate multielectron transition probabilities. The calculation takes some account of the increase of target binding with ionization state by use of a time dependent nuclear charge of the target. The results are only qualitatively in agreement with experiment for differential cross sections, but do better for total cross sections.

B. Total Cross Sections:

Both the statistical model and the independent electron model reappeared in the early eighties when data began to be available for total ionization cross sections for fast highly charged projectiles. Figure 6

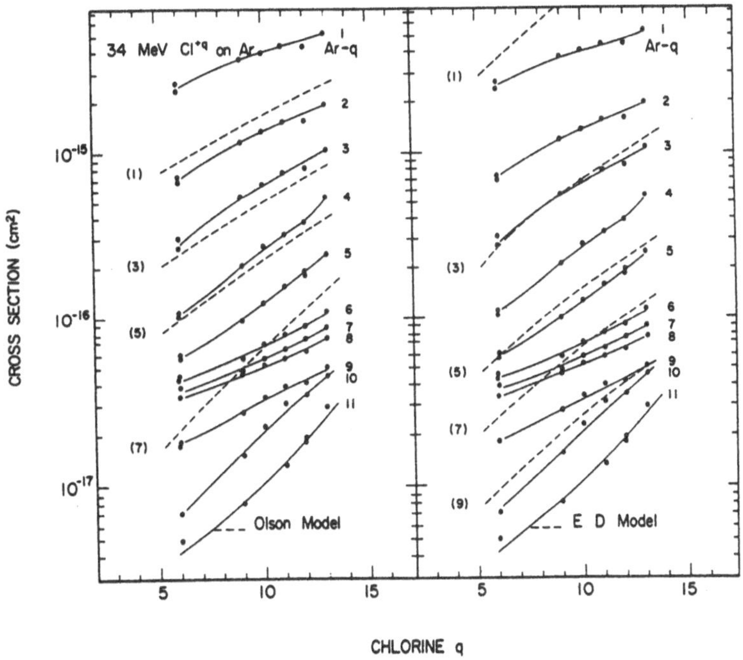

Fig. 6: Total cross sections for recoil production (Ref. 11) compared to two model results. The Olson model uses binomial statistics, and the energy deposition model uses the RM statistical expression (Eq. 1).

shows sample cross sections for Cl projectiles on Ar at 1 MeV/amu where it is seen that the cross sections decrease rapidly with recoil ion charge state. This occurs because the total cross sections integrate over b, and thus the charge state spectrum is not taken at a single value of the violence parameter. These data were compared with both an independent electron model by Olson which used a CTMC calculation for p and the binomial distribution, and with an energy deposition (ED) model which followed Russek's work. The energy deposition calculated for the ED model was taken from the impulse approximation in which the target electrons are treated as free. A comparison of the results from these two models shows that the CTMC is better for the low charge states but overestimates the higher charge states because of the binding energy problem discussed above. One can see how this comes about from fig. 2 (c) and (d), where typical charge state fractions are shown for the RM model (c) and the binomial statistics model (d). For the latter case, the average energy per target electron is taken as the product of p with some average energy per ionization event. What to take? If one takes twice the first ionization energy, roughly correct for single ionization, the multiple ionization is much too high for larger energy depositions. The plot shown takes 150 eV, chosen to make (c) and (d) agree for charge state 4. The average ionization energy for all 8 electrons is 119 eV, and some kinetic energy goes to the continuum, so this value is reasonable. Even with the specially chosen 150 eV, the two models give very different charge state distributions. It is clear that it is difficult to get the relationship between energy deposition and final charge state to be realistic from binomial statistics, and it is not clear what it means to get agreement with experiment for a model based on single electron removal probabilities.

C. Potential Energy Dominated Collisions:

A slow mulitply charged projectile ionizes the target mainly by capturing electrons from it. There are now numerous models which agree that the electrons go into highly excited states on the projectile and are not lost from the collision system. Classically, even during the collision they circulate within the Coulomb potential well formed by the two nuclear charges but the rate of change of this potential is too slow to pump much kinetic energy into them from the translational motion of the projectile.

Groh et al. carried out measurements of total cross sections as a function of final charge states of both target and projectile for a wide range of projectile charge states at low velocities where the cross sections are not varying with velocity.[14,15] From these data, Müller et al.[15] noticed the remarkable result that the charge state fractions for a given target projectile combination seemed to depend only on the total potential energy available to the system. Figure 7 shows their results for the system Xe^{+q} on Xe. Variation of the total potential energy is accomplished

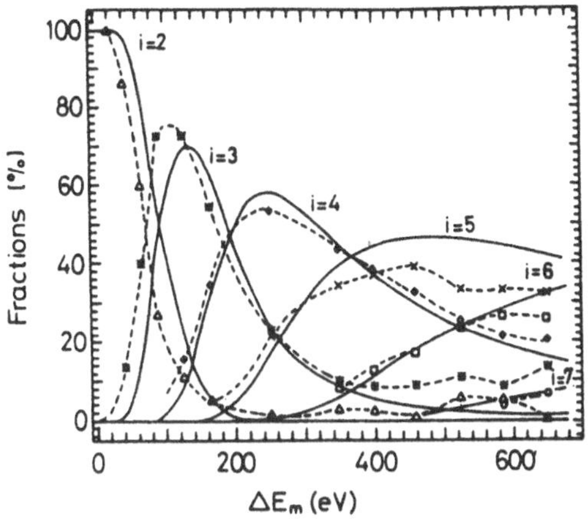

Fig. 7: Charge state fractions versus potential energy available for double capture by Xe^{q+} from Xe. The solid lines are from the statistical model (Müller, et al., Ref. 15).

by varying the projectile charge state and species. This kind of dependence recalls the Russek treatment, and Müller et al. found that indeed their data could be well described by that model (in a slightly different form from that of Eq. 1). Although the agreement of this model with the experiment is remarkable, the physical meaning is not immediately transparent. The total potential energy available is calculated from the electonic energy which would be released if the final ions are in their ground states. Yet physically one might expect that electrons are captured from the target into highly excited states on the projectile, which may later rid itself of this excess electronic energy by Auger processes, but which is far from the target when this occurs. How did the target know how much electronic potential energy was available on the projectile? The actual mechanisms by which electrons are removed from the target must certainly involve capture, Auger emission both during and after the collision and even perhaps some direct ejection to the continuum, and are far from understood. The success of the statistical model in this case is a remarkable solution to a complex problem, but one which does not remove the question of how it really works. (Åberg et al.[16] have presented an alternative maximum entropy description of these data.)

In a low-energy version of the Kelbch experiments, Schmidt-Böcking et al.[17] measured differential cross sections for charge state selected recoils and projectiles in 70 keV Ne^{+7} ions on Ne. This collision system is very reminiscient of the KE experiments,[3,4] but differs from those in two important ways. First, the angular range covered in the Schmidt-Böcking et al. experiments was such the collision did not, except for the very largest scattering angles, attain symmetry between projectile and target in the final charge state distributions. Second, because the projecile was so highly charged in the Ne^{+7} case, few electrons could escape the collision. In fig. 8 are shown the average number of electrons missing from the entire system, the total charge of the system, and the

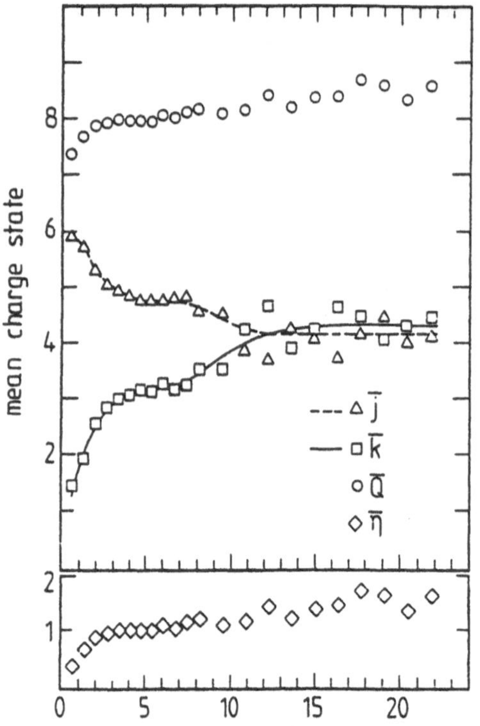

Fig. 8: Mean recoil charge states (k), mean projectile charge states (j), mean total charge (Q), and mean number of electrons missing from the system (η) as a function of θ for 70 keV Ne^{+7} on Ne (Schmidt-Böcking et al., Ref. 17).

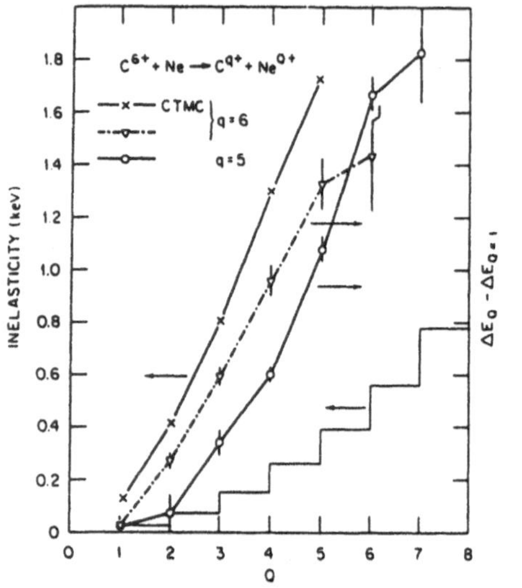

Fig. 9: Measured and calculated projectile inelastic energy loss as a function of recoil charge state for 10 MeV C^{6+} on Ne. The data are the open circles and triangles and the crosses are CTMC calculations (Schuch et al., Ref. 18).

charges of recoil and target as a function of scattering angle for the Ne^{+7} case. The missing electrons are presumably Auger electrons emitted by the projectile following multiple capture to multiply excited states. It is noteworthy that in spite of the heavy ionization of the target so few electrons escape the deep potential wells of the finally highly charged systems. For scattering angles large enough that the L shells of the colliding partners touch for this system, one sees a return to the systematics of the KE experiments where the electronic energy seems to be shared more or less equally between projectile and target, and the charge state asymmetry of the incident channel is lost. Net electron loss to the continuum rises a little only, presumably due to 4fσ promotion and subsequent loss of the promoted electrons.

D. Translational Energy Spectroscopy:

While results abound for translational energy gain spectroscopy for low energy capture by multiply charged ions, with results mainly restricted to capture of one to three electrons, the only experimental results on projectile energy loss for the kinetic energy dominated case are those of Schuch et al. and Schöne et al.[18] These authors used a high resolution single focussing magnetic spectrometer to measure average energy losses in .6 MeV/amu C^{+6} - Ne collisions, and found that the average energy loss of the projectile exceeded by a factor of two or more the sum of the ionization energies needed to produce the measured final charge states of projectile and recoil, as shown in fig. 9. These data were compared to the CTMC calculations of Olson, and good agreement was obtained. Since this measurement provides the Q value of the reaction as a function of projectile scattering angle, it might be expected from the energy deposition picture of the collision that the increase with scattering angle of the recoil charge state should be accompanied, indeed due to, an increase of Q value with angle. No such correlation was seen in this experiment. However, since the experiment measured only the mean Q with a rather broad energy resolution (keV), it is not clear that the experimental resolution would have been adequate to detect a change of Q over the rather restricted angular range for which a given charge state appears. Thus one cannot make a direct comparison between Q and recoil distribution as was possible for the KE experiments.

E. Recoil Ion Momentum Distributions:

For very small impact parameters the main momentum transfer between collision partners is through the nuclear potentials, and the deposition of electronic energy serves only as an energy sink without affecting the momentum balance, as is the case for the KE experiments. From recent experiments, it is now clear that for large impact parameters the net electron momentum carried away in multiply ionizing events is not at all negligible compared to the momentum transfer to the projectile or target. This result is especially important where the loss of electrons to the continuum of the system is important (kinetic energy dominated collisions). The final channel in its most simple description is at least a three body one, the three bodies being the projectile, the recoil and the continuum electrons (taken together).

The experiments which led to this interesting conclusion are direct measurements of the transverse momentum transfer to the recoil ion in collisions of fast heavy ions with neutral gaseous targets. Levin et al.[19] used the broadening of the time of flight peaks of recoils extracted by uniform transverse electric fields to determine the intrinsic transverse energy given to the recoils in such collisions. These measurements confirmed what was known on the basis of circumstantial evidence, from the large total cross sections for recoil production, that the recoil translational energies were extremely small even though large amounts of electronic energy were deposited in the collisions. Ullrich et al.[20] measured transverse momentum distributions of Ne and Ar ions for 1.4 MeV/amu U^{32+} ions on Ne using a field free region in the extraction. These experiments are a real technical feat, as it is very difficult to eliminate electric fields in the collision region to the point that they do not intefere with the nearly thermal energy recoils for lower charge states. His results are shown in fig. 10.

These experiments have now received detailled theoretical attention from Olson, who has used an n-CTMC calculation to describe the U^{+32} on Ne system(20). This calculation integrates the classical equations of motion for all bodies involved, and includes the interaction of the projectile with all of the electrons throughout the collision, thus allowing the calculation of the projectile scattering angle and transverse momentum transfer. The initial conditions are chosen statistically to represent the distribution in phase space in the initial wave functions. The effect of different binding energies for different ionization levels of the recoil is taken into account to some extent by providing, in the initial state, momentum distributions for different effective nuclear target charges for different electrons. The calculation does not include at all the electron--electron potential acting during the collision, since inclusion of this potential would lead to the well known classical instability of the atom which would autoionize all by itself.

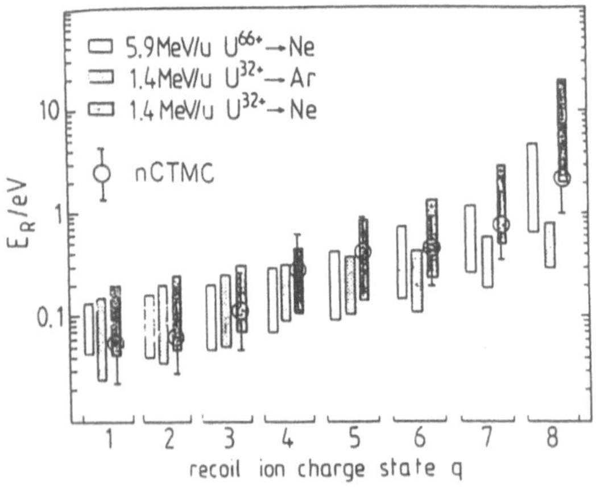

Fig. 10: Transverse recoil energy versus charge state for three systems (Ullrich et al., Ref. 20). The CTMC calculations are by Olson, for Ar.

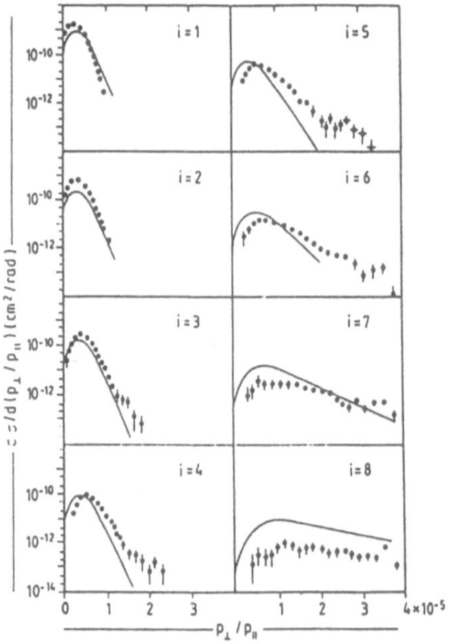

Fig. 11: Distributions of transverse recoil momentum, divided by the projectile longitudinal momentum, versus scattering angle for 1.4 MeV/amu U^{32+} on Ne. The data are solid circles and the solid lines are the CTMC calculation (Olson et al., Ref. 21).

In spite of this, the calculations are in remarkable agreement with the data and reveal several very interesting and unexpected aspects of the momentum balance among the system constituents. Figure 11 shows a comparison between the data of Ullrich et al. and the CTMC results for recoil energies of Ne and Ar ions, where good agreement is obtained. [Note: the CTMC calculation has recently been found to be slightly in error for low momentum transfer. This could lead to slightly too broad distributions for low charge states.] The recoil and projectile transverse momentum distributions are quite different, showing that the electrons must be carrying away significant amounts of net transverse momentum. The experiment measures only the recoils, since the corresponding projectile scattering angles are too tiny to measure directly, and are in excellent agreement with calculation. From the CTMC calculations, Olson concludes that for large scattering angles, the projectiles scatter mainly from the electron cloud, leaving the recoil with almost no transverse momentum transfer at all. Only for hard collisions penetrating well inside the target electron clouds, where high recoil charge states are produced and charge transfer is probable, do the collisions begin to resemble binary encounters between projectile and target.

There are very few experiments which reveal information on the angular distributions of the recoils in such collisions, and this seems to be the next logical area to investigate in such collisions. Recently González et al.[23] have investigated recoil angular distributions from F^{+9} on Ne at .5 MeV/amu. The geometry of the time of flight spectrometer is such that recoils directed toward and away from the detector are registered, but those at other angles are not. The recoils are detected in coincidence with projectiles whose polar and azimuthal scattering angles are measured. The results for double capture show a very remarkable and unexpected feature, as shown in fig. 12. One would expect that on the average the projectile and recoil transverse momenta would lie in a common plane, with the electron momentum affecting the extent to which these momenta do or do not balance. A plot of scattering yield versus the azimuthal angle between recoil and projectile shows, for the higher recoil charges, a double peaked structure, indicating that the electron momentum distribution has peaks above and below the scattering plane. Since the average is still zero, this does not violate the reflection symmetry of the collision, but is nevertheless a very surprising result and one for which no explanation has yet been devised. The CTMC calculations do not reproduce this behavior.

Frohne et al.[24] at KSU have used a geometry which allows the measurement of the whole velocity space distribution of the recoils. Their geometry extracts the recoil with a very weak but very uniform electric field at right angles to the beam and projects them onto the face of a position sensitive detector. The time of flight of the recoil and the position at which it strikes the detector gives all three components of the recoil momentum. The data for 19 MeV F^{9+} on Ne show a correlation between recoil time of flight and projectile scattering angle due to quasi-conservation of transverse momentum in the collision. For the hard collisions, the data confirm that the electron momentum is negligible, but for softer collisions this conservation law clearly breaks down. It is interesting that the charge state of the recoil alone is not sufficient to indicate the hardness of the collision. For example, a Ne^{4+} created in a double capture event obeys the transverse momentum conservation much better than the same ion created in a pure ionization event. Projections onto the transverse plane of the momentum transfer (fig. 13) shows that the velocity space distribution of the higher charged ions is a disk centered on the beam axis, with the recoils leaving near 90°. For the charge states 1^+ and 2^+ no such disk is apparent, and indeed the transverse recoil energy received by charge state one ions is less than thermal. Further reduction of these data are in progress to extract the velocity space distributions for all charge states.

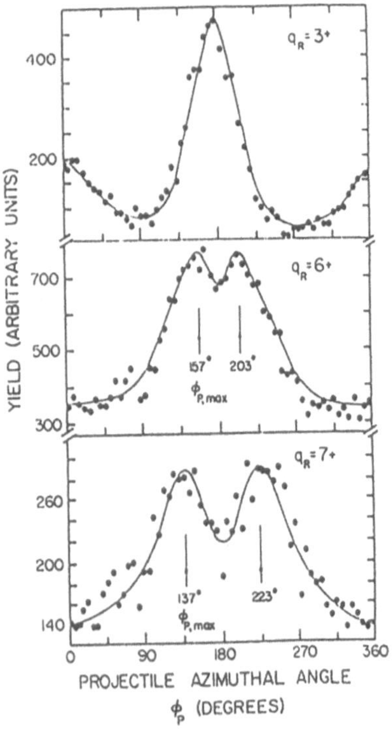

Fig. 12: Azimuthal distributions of recoil transverse momenta for a mean projectile scattering angle of 0.28 mrad in .53 MeV/amu F^{9+} on Ne, double capture (González et al., Ref. 23).

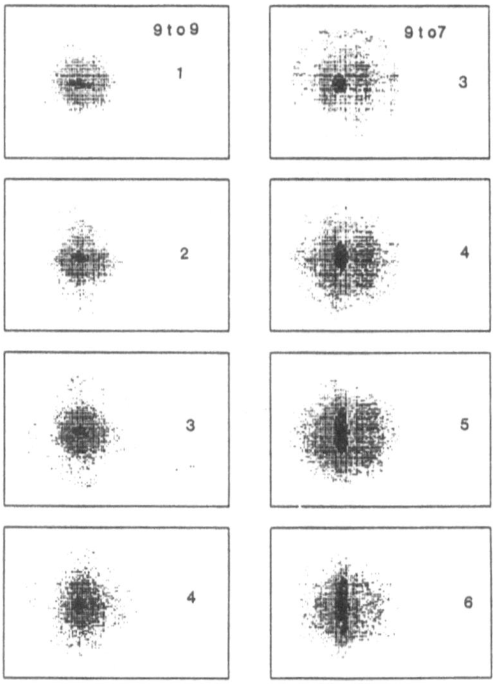

Fig. 13: Projections onto a plane parallel to the beam of velocity distributions of recoils from 1 MeV/amu F^{9+} on Ne. The recoil charge state is indicated. The left hand figures are for direct ionization and the right hand one for double capture (Frohne et al., Ref. 24). The beam moves from left to right in the figure.

Multiple Ionisation in Ion-Atom Collisions

Investigated by Recoil Ion Momentum Spectroscopy

H. Schmidt-Böcking*, R. Dörner*, J. Ullrich§, J. Euler*, H. Berg*, E. Forberich*, S. Lencinas*, O. Jagutzki*, A. Gensmantel*, K. Ullmann*, R. D. DuBois+, Feng Jiazhen°, R. E. Olson■, A. Gonzales××, S. Hagmann×

* Institut für Kernphysik, Johann-Wolfgang-Goethe Universität, D-6000 Frankfurt, FRG

§ GSI Darmstadt, D-6100 Darmstadt, FRG

\+ Pacific Northwest Laboratory, Richland WA 99352, USA

° Department of Physics, University of Lanzhau, P. R. of China

■ Department of Physics, University of Missouri-Rolla, Rolla MO 65401, USA

× J. R. Macdonald Laboratory, Kansas State University, Manhattan KS 66506, USA

×× present adress: Manne Siegbahninstitutet för Fysik, S-10405 Stockholm, Sweden

Abstract:

The technique of recoil momentum spectroscopy as a tool to investigate the details of the kinematics of a many particle ionisation process is discussed. Measuring the recoil ion transverse momentum by time-of-flight technique and the recoil ion emission angle in coincidence with the scattered projectile final transverse momentum precise information on the sum momentum of all ejected electrons can be obtained. Furthermore it is discussed how Q-values and electron mass transfer affect the recoil longitudinal momentum. Recent data are presented.

III. CONCLUSION

Multiply ionizing collisions involving highly charged projectiles are intrinsically complex , experimentally because of the large number of final channels into which the collision can go , and theoretically because of the many bodies involved. There appears to be no successful theoretical substitute for following the detailed evolution of each body in the collision. The CTMC calculations are providing new insight into the mechanical workings of the collisions, but do not yet incorporate the effect of the electron-electron potential during the collision. Such calculations reveal what is happening, but leave room, indeed invite, the development of conceptual generalizations and the identification of patterned behavior in the collision.

ACKNOWLEDGMENT

This work was supported by the Chemical Sciences Division, Office of Basic Energy Research, US Department of Energy.

REFERENCES:

1. C. Jordan, Mon. Notices Roy. Astron. Soc. 142, 501 (1969).
2. C. Zaidins, Ph.D. thesis, Cal. Tech. (1967).
3. E. Everhart and Q.C. Kessel, Phys. Rev. Lett. 14, 247 (1965).
4. Q.C. Kessel and E. Everhart, Phys. Rev. 146, 16 (1966); E. Everhart and Q.C. Kessel, Phys. Rev. 146, 27 (1966).
5. V.V. Afrosimov and N.V. Federenko, Zhur. Tekh. Fiz. 27, 2557 (1957) [JETP 2, Noll, 2378 (1957)].
6. U. Fano and W.L. Lichten, Phys. Rev. Letters 14, 27 (1966).
7. A. Russek and M. Tom Thomas, Phys. Rev. 109, 2015 (1958) and Phys. Rev. 114, 1538 (1959).
8. A. Russek, Phys. Rev. 132, 246 (1963).
9. A. Russek and J. Meli, Physica 46, 222 (1970).
10. J.H. McGuire and O.L. Weaver, Phys. Rev. A16, 41 (1977).
11. T.J. Gray, C.L. Cocke and E. Justiniano, Phys. Rev. A22, 849 (1980).
12. S. Kelbch, et al., J. Phys. B22, 1277 (1989).
13. M. Horbatsch and R.M. Dreizler, Phys. Rev. Lett. A113, 251 (1985); M. Horbatsch, J. Phys. B19, L193 (1986); Z. Phys. D1, 337 (1986); Z. Phys. D2, 183 (1986).
14. Groh et al. J. Phys. B15, L207 (1982); J. Phys. B16, 1997 (1983); Phys. Lett. 85A, 77 (1981).
15. A. Müller, W. Groh and E. Salzborn, Phys. Rev. Lett. 51, 107 (1983).
16. T. Åberg, et al., Phys. Rev. Lett. 52, 1207 (1984).
17. H. Schmidt-Bocking, et al., Phys. Rev. A37, 4640 (1988).
18. R. Schuch, et al., Phys. Rev. Lett. 60, 925 (1988); H. Schöne, et al., Nucl. Inst. Meth. B40/41, 141 (1989).
19. J.C. Levin et al., Phys. Rev. A36, 1649 (1987); J.C. Levin, et al., Nucl. Inst. Meth. A262, 106 (1987).
20. J. Ullrich et al., J. de Physique C1, 29 (1989).
21. R.E. Olson, J. Ullrich and H. Schmidt-Bocking, Phys. Rev. A39, 5572 (1989).
22. C.E.G. Lepera, et al., Nucl. Inst. Meth. B24/25, 316 (1987).
23. A.D. González, S. Hagmann, T.B. Quinteros and B. Krässig, to appear (1990).
24. V. Frohne, et al. private communication (1990).

Introduction:

Multiple ionisation in ion-atom collisions has extensively been studied experimentally [1, 2, 3, 4] and theoretically [5, 6, 7, 8] in recent years. Even antiparticles [9] have been used as projectiles to investigate multiple ionisation cross sections and probabilities. Particular interest have attracted the ratios of single to double ionisation cross sections of He induced by proton and antiproton impact to obtain information on possible electron-electron-correlation effects.

Such correlation effects possibly observable in the double ionisation process might be derived from different measurable quantities. First one can compare multiple ionisation probabilities or cross sections to find deviations from purely statistical expectations or second one can measure the final momenta of the emitted electrons for different degree of ionisation to look directly for correlated electron emission. Using the first method the interpretation of the data may strongly depend on theory, whereas in the second case correlated emission may be a directly measurable quantity. The second method requires in general a multi particle (two nuclei and several electrons) coincidence experiment which is at present not visible. However, as it will be outlined below, a careful measurement of the final recoil-ion momentum $\vec{P}_{Rf}$ and the final tranverse momentum of the projectile $\vec{P}_{Pf}^{\perp}$ already provides all the information needed to deduce information on the correlated multiple electron emission.

Because a historical review on the numerous measurements and theoretical work dealing with total and differential multiple ionisation cross sections or probabilities is given in the paper by Cocke *et al.* [10] and since also in the paper by Ullrich *et al.* [11] recent data on the recoil ion momentum spectroscopy are presented, this paper will mainly discuss the potential of the recoil momentum spectroscopy method in fast ion collisions.

In a typical fast ion-atom collision experiment the following initial parameters can be determined: projectile momentum $\vec{P}_{Pi}$, projectile nuclear charge Z_P and ionic charge state n_i, target atom momentum $\vec{P}_{Ri} \approx 0$ (neglecting thermal motion). Dependent on the degree of the multiparameter coincidence the following final collision parameters can be measured: projectile momentum $\vec{P}_{Pf}$ and projectile charge state n_f, the recoil-ion final momentum $\vec{P}_{Rf}$ and its charge state q_R as well as the momenta of the ejected electrons.

Several coincidence experiments have been performed, where for two or three outgoing particles the momenta have been measured. In nearly all of these experiments the projectile transverse momentum (i.e. the scattering angle) and the outgoing electron momentum has been determined simultaneously [12, 13, 14, 15, 16, 17. 18] using standard position sensitive particle detectors and electron spectrometers. In a number of experiments the multiple ionisation probability of the target has been measured as a function of $\vec{P}_{Pf}^{\perp}$ for the different recoil charge states q_R [19, 20, 21, 22].

The bars on top of v indicate CM-components. $\vec{v}^{\,\parallel}_{Rf}$ is the longitudinal component of the final lab recoil velocity, v_P and E_P are the initial projectile velocity and kinematic energy values, respectively. For $Q < 0$ we obtain a negative $\vec{P}^{\,\parallel}_{Rf}$, i.e. a backward emitted recoil ion ($\vartheta_{Rf} \geq 90°$), if $Q > 0$ we obtain $\vec{P}^{\,\parallel}_{Rf} > 0$, i.e. a forward emitted recoil ion ($\vartheta_{Rf} < 90°$).

In case of mass transfer (e.g. n_e target electrons captured by the projectile and $(q_R - n_e)$-times electrons emitted into the continuum) the initial and final masses are different with

$$M_{Rf} = M_{Ri} - q_R m_e \quad \text{and} \quad M_{Pf} = M_{Pi} + n_e m_e \tag{5}$$

From momentum and energy conversation in the CM-system we obtain

$$M_{Ri} \cdot \bar{v}_{Ri} = -M_{Pi} \cdot \bar{v}_{Pi} \quad \text{and} \quad M_{Rf} \cdot \bar{v}_{Rf} = -M_{Pf} \cdot \bar{v}_{Pf} \tag{6}$$

and

$$M_{Ri} \bar{v}^2_{Ri} + M_{Pi} \bar{v}^2_{Pi} = 2E^{CM}_{kin} \tag{7a}$$

$$M_{Rf} \bar{v}^2_{Rf} + M_{Pf} \bar{v}^2_{Pf} + 2Q = 2E^{CM}_{kin} \tag{7b}$$

From equation 5 to 7 we can derive for $q_R = n_e$ (pure capture) that

$$\bar{v}_{Rf} = \bar{v}_{Ri} \left(1 + \frac{n_e}{2} \cdot \frac{m_e}{M_\mu}\right) \tag{8}$$

where $M_\mu = \dfrac{M_P \cdot M_R}{M_P + M_R}$ is the reduced nuclear mass.

Since $n_e m_e \ll M_P$ or M_R and $Q/E_P \ll 1$ we obtain for the total longitudinal recoil momentum ($Q \neq 0$, n_e electrons captured)

$$M_R \cdot \vec{v}^{\,\parallel}_{Rf} = \vec{P}^{\,\parallel}_{Rf} = \frac{Q}{v_P} - \frac{n_e}{2} m_e \cdot v_P \tag{9}$$

The relation between longitudinal and transverse recoil momentum is

$$\vec{P}^{\,\parallel}_{Rf} = \left|\vec{P}^{\,\perp}_{Rf}\right| \cdot \operatorname{ctg} \vartheta_{Rf} \tag{10}$$

Dependent on possible shell structure effects of the initial multi electron motion in the target atom the momentum $-\sum^{q_R}\vec{P}^{\perp}_{ef} = \vec{P}^{\perp}_{Nf}$ could show a non-monotonic behavior with varying q_R. For a fast heavy ion atom collision at large impact paramaters ($b \gtrsim 1$ a.u.) momentum exchange between projectile and target nucleus is mostly below 1 a.u. and $\vec{P}^{\perp}_{Nf}$ reflects indeed only the multi electron-nuclei interaction. Since the fast ion induced ionisation process occurs typically in time intervalls less than 10^{-16} sec, possible dynamically correlated electron motion could be visible in $\vec{P}^{\perp}_{Nf}$ which is hardly to observe by spectroscopy methods.

Experimental techniques:

In figure 1 the principle of the recoil ion momentum spectrometer is shown [24]. The well collimated heavy ion beam with $|\vec{P}^{\perp}_{Pi}|/|\vec{P}_{Pi}| \lesssim 10^{-5}$ passes a thin gas target, which is located in a cage. The inner surface of this cage is prepared in such a way that possible electrostatic fields in the cage are in the order of a few mV/cm or even smaller. The ionising projectile is detected by a two dimensional position sensitive detector (typical position resolution ≈ 0.2 mm). Using a magnetic or electrostatic deflection system mounted just behind the target the projectiles can easily be separated as function of their final charge state. A careful collimation of the beam allows an overall resolution of the projectile scattering angles of about $2\cdot 10^{-5}$ rad.

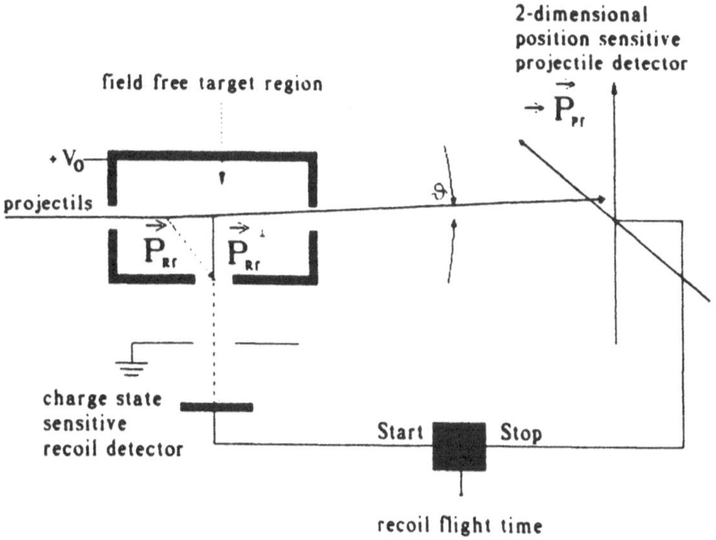

figure 1: experimental set-up

Furthermore, the detection of the scattered projectile provides a clock to measure the start time of the recoiling target ion. After passing a well defined field free region the recoil ion leaves the cage via a high transmission grid, is postaccelerated, magnetically charge state analysed and then detected by a position sensitive channel plate detector. The drift time in the cage yields directly the recoil ion transverse momentum. In figure 2 a typical recoil ion time-of-flight spectrum is shown for 0.5 MeV p on He and a flight path of 3.5 mm.

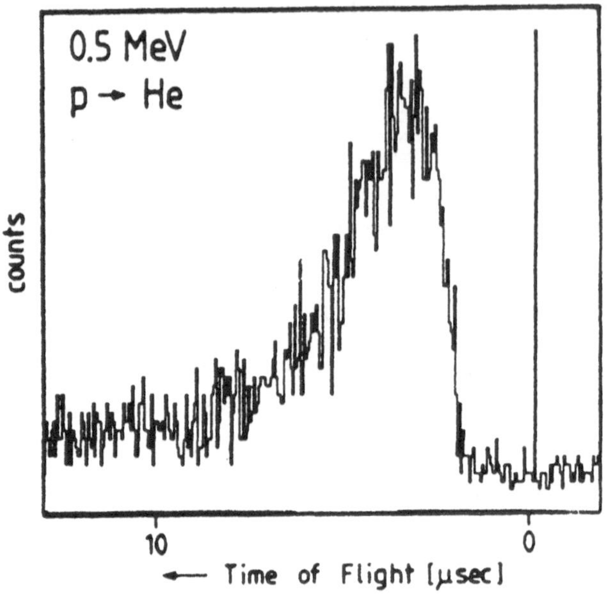

figure 2: recoil ion time-of-flight spectrum

The target cell displayed in figure 1 ensures that practically all recoil ion emission angles are detected with equal efficiency, but this system does not allow the determination of this angle. In order to measure also the recoil-ion emission angle the target area must be localised. This can be achieved by using collimated atomic beams [24, 25] or by shielding the observable target area.

In ref. [24] some recoil momentum spectrometers are described. Also a new type of supersonic jet system is briefly discussed there, which will provide substantial improvements compared with the earlier ones. The target internal velocity distribution width resulting from thermal motion is in all three dimensions very small and corresponds to about 1 K or less yielding a strongly improved resolution for the recoil momentum spectroscopy. Furthermore the high mean velocity of the atomic jet and the recoil ion detection system (see figure 3) ensure strongly increased detection efficiencies. More details about the experimental techniques for recoil momentum spectroscopy are given in ref. [24].

The recoil-ion final charge state is easily determined by time-of-flight technique due to different acceleration of the different q_R in an electrostatic extration field in the target area.

Only very few real triple coincidence experiments have been reported so far in literature, where the projectile transverse momenta, the momentum of one ejected electron and the final recoil-ion charge state have been measured [14, 23]. It is to notice that often pulsed electric field recoil-ion extraction is used to avoid the influence of an electrostatic field in the target area on the electron momentum. However, special care has to be taken here, that the total recoil-ion production rate is small compared to the recoil ion velocity divided by the diameter of the extraction area to avoid uncorrelated coincidence events.

All these coincidence experiments listed above do not yield direct information on correlated multiple electron emission, because so far no electron-electron-particle coincidence experiments have been reported in literature. Improved detection techniques, e.g. position sensitive electron time-of-flight spectrometers (even with simultaneous multielectron detection capability) will allow such experiments in near future.

One important quantity of the final collision parameters is, however, the recoil-ion momentum $\vec{P}_R$. Because the target atom is at rest before the collision in the lab system, $\vec{P}_{Rf}$ directly reflects the transfered momentum, whereas for the projectile the interesting quantity, the transfered momentum $\Delta\vec{P}_P$, is the difference between initial (i) and final (f) momentum. Since $\Delta\vec{P}_P/\vec{P}_{Pi}$ is typically below 10^{-4} in nearly all ionising collisions even a precision measurement of $\delta\vec{P}_{Pf}/\vec{P}_{Pi} \approx 10^{-4}$ would yield only a 100% precision in $\Delta\vec{P}_P$. Particularly the longitudinal component $\Delta\vec{P}_P^{\parallel}$ is hardly to measure with this precision, whereas i.e. the final transverse momentum of the projectile $\vec{P}_{Pf}^{\perp}/\vec{P}_{Pi}$ can be measured down to the 10^{-5} level. From momentum conversation we yield

$$\vec{P}_{Pi} = \vec{P}_{Pf} + \vec{P}_{Rf} + \sum_{\lambda}^{q_R} \vec{P}_{ef}^{\lambda} \qquad (1)$$

where $\vec{P}_{ef}^{\lambda}$ is the final momentum vector of the λth emitted electron. Considering only the transverse components we obtain

$$\sum_{\lambda}^{q_R} \vec{P}_{ef}^{\lambda\perp} = -\left(\vec{P}_{Pf}^{\perp} + \vec{P}_{Rf}^{\perp}\right) = -\vec{P}_{Nf}^{\perp} \qquad (2)$$

i.e., the vector sum $\vec{P}_{Nf}^{\perp}$ of the transverse momenta of both nuclei is a direct measure of the sum momentum of all ejected electrons. It yields therefore comparable information on the ionisation process as obtained in the well known (e,2e) or (e,3e) coincidence studies. For multiple ionisation of degree q_R it represents even an $(e, q_R e)$ experiment.

Remarks to the kinematics of the recoil ion:

In multiple ionising collisions the final recoil-ion momentum depends on the interaction with the projectile and with the participating electrons in their initial and final state. As shown in equation 2 a non-zero nuclear momentum $\vec{P}_{Nf}$ thus reflects the inverse sum momentum of all ejected electrons. In a pure two-body nuclear collision is $\vec{P}_{Nf} \equiv \vec{P}_{Pi}$ with $\vec{P}^{\perp}_{Nf} = 0$ and $\vec{P}^{\parallel}_{Nf} = |\vec{P}_{Pi}|$. Therefore a closer look into the two-body nuclear kinematics for different Q-values and electron mass transfers between the collision partners (and mass transfer to the continuum states) may illustrate the sensitivity of recoil momentum spectroscopy to investigate experimentally the many-particle interaction.

In a pure two-body collision, where the multiple ionisation process consumes energy (Q-value > 0), a backward scattering of the target atom, being initially at rest in the lab frame should not occur. However, considering the electron momenta and in particular the mass transfer (q_R electrons) from the target into the continuum or into the projectile (n_e-electron capture) the recoil ion can have a strongly backward directed emission angle. To illustrate this we will briefly discuss here the two-body nuclear kinematics with a Q-value and n_e times electron capture into the projectile and ($q_R - n_e$) electrons emitted into the continuum. Since nearly all ionisation processes occur at scattering angles below 1 mrad, we have only to discuss the small angle limit and have to estimate the relevant transverse $\vec{P}^{\perp}$ and longitudinal $\vec{P}^{\parallel}$ momentum (velocity) components of recoil-ion and projectile in the lab-system.

In figure 4 (upper part) for three different Q-values (no mass transfer) the final recoil velocity vectors in the lab-system are given as vector sum of the corresponding CM-vectors $\vec{v}_{Rf}(Q)$ plus the vector $\vec{v}_{CM}$ which is the velocity of the CM in the lab-system. In the lower part of figure 4 the decomposition of $\vec{v}_{Rf}(Q_1)$ into the transverse $\vec{v}^{\perp}_{Rf}$ and the longitudinal $\vec{v}^{\parallel}_{Rf}$ components are shown.

For the case of no mass transfer (i.e. pure excitation) and $Q << E_P$ we obtain from

$$-\vec{v}^{\parallel}_{Rf} \approx \vec{\bar{v}}_{Ri} - \vec{\bar{v}}_{Rf} = |\vec{v}_{CM}| \cdot \left(1 - \sqrt{1 - \frac{Q}{E_P} \cdot \frac{M_R + M_P}{M_R}}\right) \qquad (3)$$

that

$$\vec{P}^{\parallel}_{Rf} = M_R \cdot \vec{v}^{\parallel}_{Rf} \approx \frac{Q}{v_P} \qquad (4)$$

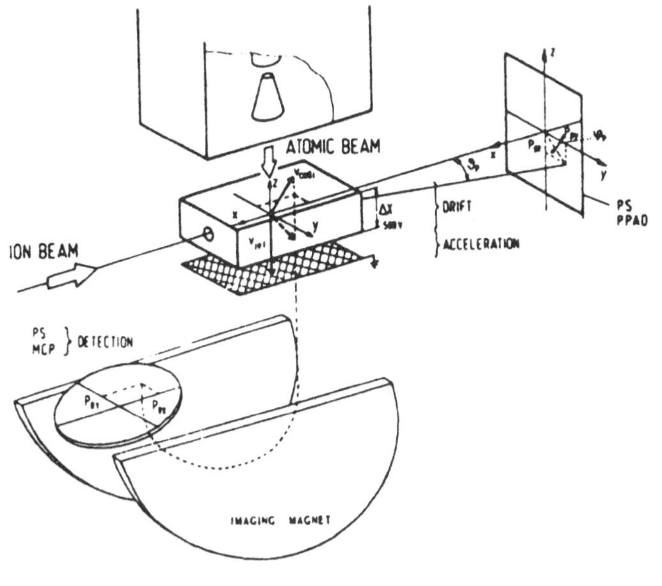

figure 3: atomic gas jet with recoil ion detection system

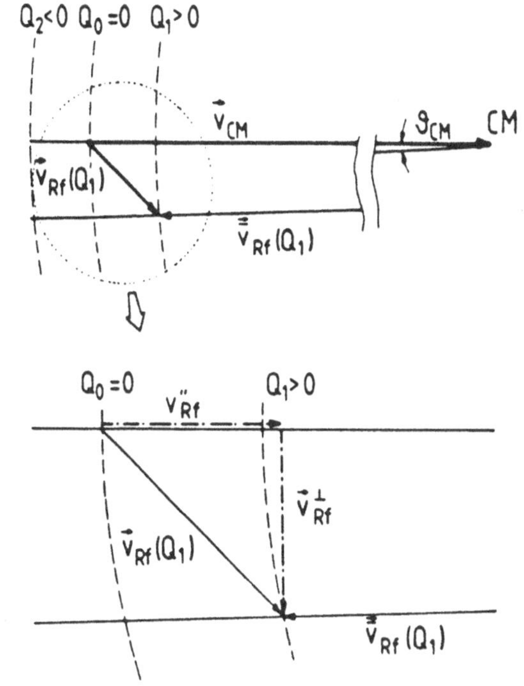

figure 4: kinematics of the collision

Thus not only for a negative Q-value but also, under certain conditions for n_e-fold electron capture $\vec{P}^{\parallel}_{Rf}$ becomes negative and is backward directed. For the pure capture channel (n_e electrons transfered) the recoil longitudinal momentum directly reflects the Q-value corrected for the constant mass transfer contribution. For discrete energy gain values Q in the capture reaction we obtain discrete longitudinal components in the recoil ion momentum. As commonly applied in the method of translational spectroscopy to measure Q-values of low energetic ion-atom capture reactions, even in fast collisions translational spectroscopy seems to be visible detecting the recoil-ion longitudinal momentum. E.g. considering the single capture reaction in 5 MeV/u U^{q+} on He collisions a Q-value of 1 keV energy gain would change the projectile longitudinal momentum by a fraction of less than 10^{-6}. But detecting the He^{+}-recoil ion at a given transverse momentum of about 9 a.u. (which is relatively easily to measure by time-of-flight technique) a Q-value of 1 keV corresponds to about a recoil ion emission angle of $\vartheta_R = 135°$, a Q-value of 0.5 keV corresponds to $\vartheta_R = 129°$ and a Q-value of 3 keV corresponds to $\vartheta_R = 146°$. Recoil ions with Q = 0 are emitted under 108°. Thus the recoil ion angular distribution reflects directly the Q-value distribution of the capture reaction. The mass transfer contribution is for high velocities in the order of many atomic units. For projectiles of 1 MeV/u kinetic energy and double electron capture, the electron transfer results in a backward directed recoil momentum of more than 6 a.u..

Considering also ionisation processes the electron mass transfer of $(q_R - n_e)$ electrons to the continuum contributes by an additional amount of $\frac{q_R - n_e}{4}$ $m_e \cdot v_P$ (backward directed), if we assume that the electrons have zero kinetic energy in the lab-system. Taking the final electron momenta into account too, we have to know exactly whether the final electron momenta are resulting from an electron-projectile (Binary Encounter collision) or from a three-body electron-projectile-recoil ion interaction. In general the momentum exchange between the many particles cannot be predicted easily.

But even if the final electron momenta in the lab-system would be zero, equation 9 just shows that the longitudinal component of the recoil momentum is already strongly influenced by Q-value and mass transfer. It is important to notice that the recoil transverse momentum is not effected by the Q-value and the mass transfer. As derived from equation 2, $\vec{P}^{\perp}_{Nf}$ is indeed the direct measure for the final sum momentum of all ejected electrons. Since $\vec{P}^{\perp}_{Pf}$ is in the order or below 1 a.u. for nearly 99% of all ionising collisions, $\vec{P}^{\perp}_{Rf}$ alone often reflects the interesting electron sum momentum. Thus $\vec{P}^{\perp}_{Rf}$ represents an interesting parameter to study details of the multiple ionisation process.

A corresponding equation 9 can be derived for the projectile longitudinal momentum transfer $(\Delta\vec{P}^{\parallel}_{Pf})$. n_e captured electrons will increase the forward directed momentum components by the same amount of $n_e \cdot \frac{m_e}{2} v_P$. But because of the increased projectile mass this means due to factor of $\frac{1}{2}$, a slightly reduced final lab-velocity. Since $\Delta\vec{P}^{\parallel}_{Pf}/P_{Pi}$ is typically below 10^{-4} a determination of the Q-value by measuring $\Delta\vec{P}^{\parallel}_{Pf}$ directly appears very difficult [21], whereas for recoil ions $\Delta\vec{P}^{\parallel}_{Rf}/P_{Pi} \approx 10^{-7}$ seems to be visible [24].

We can conclude from the discussions above that $\vec{P}^{\perp}_{Nf} = -\sum_{\lambda}^{q_R} \vec{P}^{\perp}_{ef}$ yields direct information on the electron sum vector, whereas the recoil emission angle ϑ_{Rf} is a direct measure of the Q-value and mass transfer.

Thus measuring the recoil and projectile transverse momentum component and the recoil emission angle important information on the momentum, energy and mass transfer in the multiple ionising collision can be obtained.

Data and discussion:

So far only very few experiments have been performed to measure $\vec{P}^{\perp}_{Nf}$ and ϑ_{Rf}. A few groups are just working to start with such experiments. Therefore only very preliminary and speculative results can be presented here.

In figure 5 calculated recoil momentum distributions are presented for 1 MeV/u F^{8+} on Ne → F^{nf} + Ne^{q_R}. The calculations have been performed in the nCTMC approach [26]. Figure 5 clearly shows the pronounced backward scattering of the recoil ions with increasing n_e as expected according to equation 9. This system has been investigated very recently by Gonzales *et al.* [27], Cocke [25] and Forberich [28] using different recoil ion extraction techniques. Since the data analysis of ref. [25] and [28] are presently not completed, only the data of Gonzales *et al.* [27] can be compared to these calculations. They have found for n_e = 2 (double capture) and increasing q_R an increasing electron sum momentum in the azimuthal recoil-projectile coincidence distribution. Since an electrostatic extraction field with rectangular recoil detector aperture was used they can hardly distinguish between a possible non-coplanar azimuthal or backward directed (polar) nucleus-nucleus scattering. Nevertheless they should be able to derive information on the size of the electron sum momenta. Their experimentally derived electron sum momenta are in qualitative agreement with the nCTMC predictions for the backward directed longitudinal recoil momenta. However, the data of ref. [27] do not allow any final conclusion. The techniques applied in ref. [25] and [28] should give much more precise results in very near future.

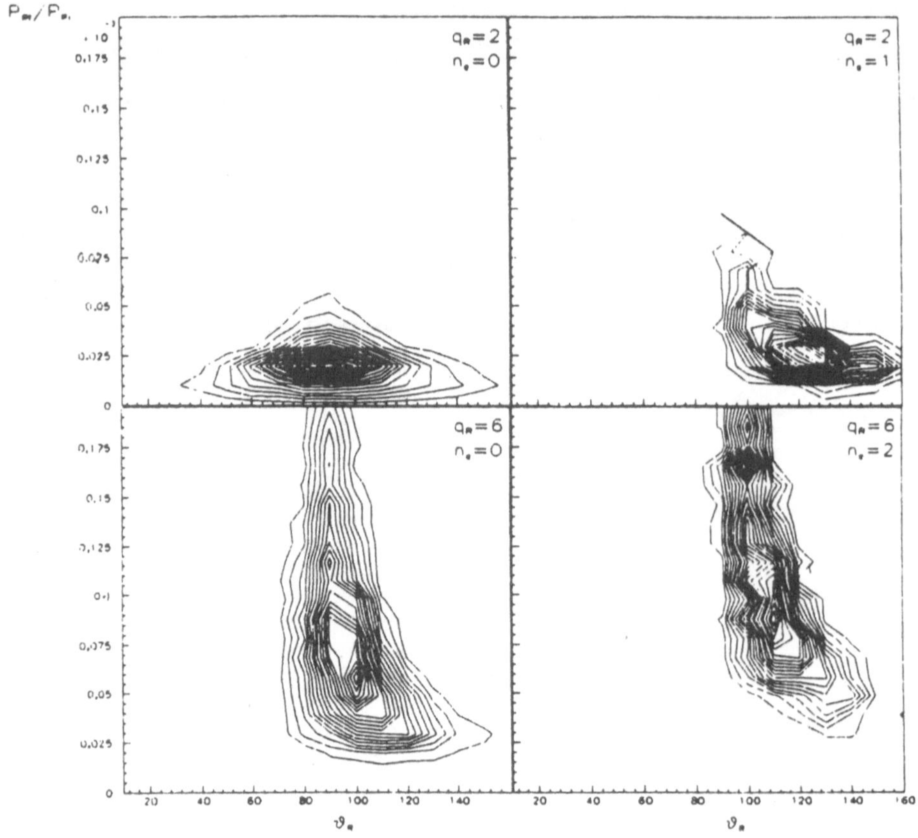

figure 5: calculated recoil momentum distributions for 1 MeV/u F^{8+} on Ne → F^{nf} + Ne^{qR}

Very recently for 1 MeV α on He the angular distributions for different recoil transverse momenta were measured [29]. They clearly show the backward directed scattering of the recoil ions. In these experiments $\vec{P}^{\perp}_{Pf}$ has not been measured simultaneously. For p on He at different proton energies Dörner has measured [30] for He single and double ionisation $\vec{P}^{\perp}_{Rf}$ as a function of $\vec{P}^{\perp}_{Pf}$. From $\vec{P}^{\perp}_{Pf}$ and $\vec{P}^{\perp}_{Rf}$ the resulting azimuthal dependence of the single and double electron emission has been derived. It is shown there that for more than 97% of all ionisation processes the final electron momenta strongly influence the two-body nuclear kinematics. As shown in ref. [30] for very low $\vec{P}^{\perp}_{Pf}/P_{Po} < 5 \cdot 10^{-4}$ rad a clear discrepancy for the measured recoil momenta with existing theories has been found [31, 32]. Due to extremely diffcult experimental procedures (e.g. control of 2...4 meV contact potentials) at present no final conclusions from these discrepancies can be drawn. Experiments with new improved target systems as described above are in preparation to exclude possible unknown systematic errors.

Instead of measuring $\vec{P}^{\parallel}_{Rf}$ Schuch *et al.* [21] have measured for C^{6+} on Ne the projectile final longitudinal momentum $\vec{P}^{\parallel}_{Pf}$ with high resolution ($\gtrsim 10^{-4}$) using a high resolution magnetic spectrometer. They could derive from their data mean Q-values for the different ionisation and capture channels. The experimental data are in fair agreement with nCTMC calculations.

Conclusion:

The recoil-ion momentum spectroscopy in fast ion atom collisions is a new starting activity in atomic collision physics. With improved target and detector techniques it will provide new detailed information on the many particle interaction process. It will yield a momentum resolution which is hardly to obtain by standard projectile detection techniques. We believe that it will in particular provide new basic information on the sum momentum exchange of many electron systems and thus possibly give an experimental tool for the investigation of dynamical electron-electron correlations.

Acknowledgement:

This work was supported by BMFT, DFG, GSI and DOE. Several of us are thankfull for the support they obtained from foundations (R. D. DuBois - Fullbright, F. Jiazhen - Heraeus-Stiftung).

References:

[1] C.L. Cocke, Phys. Rev. **A20**, 749 (1979)

[2] T.J. Gray, C.L. Cocke and E. Justiniano, Phys. Rev. **A22**, 849 (1986)

[3] H. Tawara, T. Tonuma, H. Kumagai and T. Matsuo, Phys. Rev. **A41**, 116 (1990)

[4] J.C. Levin, R.T. Short, C.O.H. Cederquist, S.B. Elston, P. Gibbons, I.A. Sellin and H. Schmidt-Böcking, Phys. Rev. **A37**, 1649 (1987)

[5] R.E. Olson in *Electronic and Atomic Collisions* edited by H.B. Gilbody, W.R. Newell, F.H. Read and A.C.H. Smith (Elsevier Science, New York 1988) pp. 271 - 285

[6] M. Horbatsch, J. Phys. **B22**, L639 (1989)

[7] R. Gayet and A. Salin, J. Phys. **B20**, L571 (1987)

[8] J.F. Reading and A.L. Ford, Phys. Rev. Lett. **58**, 543 (1987)

[9] L.H. Andersen, P. Hvelplund, H. Knudsen, P. Moller, K. Elsener, K-G. Rensfelt and E. Uggerhoj, Phys. Rev. Lett. **57**, 2147 (1986)

[10] C.L. Cocke *et. al.*, see this proceedings

[11] J. Ullrich *et. al.*, see this proceedings

[12] S. Hagmann, S. Kelbch, H. Schmidt-Böcking, C.L. Cocke, P. Richard, R. Schuch, A. Skutlartz, J. Ullrich, B. Johnson, M. Meron, K. Jones, D. Trautmann and F. Rösel, Phys. Rev. **A36**, 2603 (1987)

[13] C.L. Cocke, H. Schmidt-Böcking and R. Schuch, J. Phys. **B15**, 651 (1982)

[14] G. Schiwietz, B. Skogvall, N. Stolterfoth, D. Schneider and V. Montemayor, AIP Conference Proceedings **205**, 358 (ICPEAC 1989)

[15] A. Skutlartz, S. Hagmann, C. Kelbch and H. Schmidt-Böcking, Physica Scripta **T22**, 307 (1988)

[16] O. Jagutzki, R. Koch, A. Skutlartz, C. Kelbch and H. Schmidt-Böcking, (1990) to be published in J. Phys. **B**

[17] H. Schmidt-Böcking, A. Skutlartz and S. Hagmann, Phys. Lett. **A122**, 421 (1987)

[18] C. Kelbch, Diplomarbeit, Universität Frankfurt (1986), unpublished

[19] J. P. Giese and E. Horsdal, Phys. Rev. Lett. **60**, 2018 (1988)

[20] H. Sharabati, K. Bethge, J. Ullrich, R. Dörner, R.E. Olson, V. Dangendorf, R. Koch and H. Schmidt-Böcking, J. Phys. **B23**, 2957 (1990)

[21] R. Schuch, H. Schöne, P.D. Miller, H.F. Krause, P.F. Dittner, S. Datz and R.E. Olson, Phys. Rev. Lett. **60**, 925 (1988)

[22] S. Kelbch, C.L. Cocke, S. Hagmann, M. Horbatsch, C. Kelbch, R. Koch, H. Schmidt-Böcking and J. Ullrich, J. Phys **B22**, 1277 (1989)

[23] J. Freyou, M. Breinig, C.C. Gaither III and T.A. Underwood, Phys. Rev. **A41**, 1315 (1990)

Multiple Electron Capture In Fast Ion-Atom Collisions*

A. S. Schlachter
Accelerator and Fusion Research Division, Lawrence Berkeley Laboratory,1 Cyclotron Road, Berkeley, California, USA

Abstract

Many electrons can be captured by an energetic, highly charged ion in collision with a target atom, demonstrated for 47-MeV Ca^{17+} ions in Ar.

Introduction: Multiple-Electron Processes

Charge transfer, ionization, and excitation are fundamental processes in fast ion-atom collisions. Each of these processes can occur alone in a collision, or more than one process can take place in the same collision. When more than one electron is involved, electron correlation, due to an electron–electron interaction, can be important, as in resonant transfer and excitation (RTE) or in Auger-electron emission. When electron correlation is not important, the process can be described by an independent electron model, as is generally the case, for high-energy ion-atom collisions, in continuum-electron emission and recoil-ion production. A variety of these processes has been studied in experiments at the Lawrence Berkeley Laboratory, results of which have previously been reported [1].

Introduction: Multiple-Electron Capture

Many electrons can be lost by a projectile in a high-energy collision. The number of electrons lost to the continuum is limited only by the number of electrons originally belonging to the projectile. In principle, the number of electrons captured by a projectile in an energetic ion-atom collision is limited only by the number of vacant spaces available on the projectile (the projectile charge state) or by the number of electrons belonging initially to the target (the target atomic number). In fact, capturing electrons is, generally, more difficult than losing them, and capture of more than two electrons is

[24] J. Ullrich, R. Dörner, S. Lencinas, O. Jagutzki and H. Schmidt-Böcking, Proccedings of the 11th International Conference on the Application of Accelerators in Research and Industry, Denton 1990 (Nucl. Instr. Meth.)

[25] C.L. Cocke (1990), private communication

[26] R.E. Olson (1990), private communication

[27] A.D. Gonzales, S. Hagmann, T.B. Quinteros, B. Krässig, H. Schmidt-Böcking, R. Koch and A. Skutlartz, J. Phys. B23, L303 (1990)

[28] E. Forberich, Diplomarbeit, Universität Frankfurt (1991), unpublished

[29] S. Lencinas, Dissertation, Universität Frankfurt (1991), unpublished

[30] R. Dörner, Dissertation, Universität Frankfurt (1991), unpublished

[31] H. Fukuda, I. Shimamura, L. Vegh and T. Watanabe, (1990) to be published in Phys. Rev. A

[32] R. Dörner, J. Ullrich, O. Jagutzki, S. Lencinas, R.E. Olson and H. Schmidt-Böcking, Proceedings of the Vth International Conference on the Physics of Highly-Charged Ions, Gießen 1990, to be published

rarely observed. Electron-capture cross sections generally decrease with increasing energy, and are thus small at high energies, while electron capture at lower energies often goes to excited states, resulting in multiple Auger decay, and few electrons remain attached to the projectile when viewed long after the collision.

Observation of capture of up to four or five electrons has been reported in the literature. The observations fall generally into several categories:

- slow collisions, in which potential energy plays more of a role than kinetic energy;
- intermediate-energy, nearly symmetric collisions;
- fast collisions with a nearly bare highly charged projectile.

An example of multiple-electron capture in slow collisions is found in the work of Salzborn and Müller and their collaborators [2], who found that cross sections for electron capture by low-energy highly charged rare-gas ions in rare-gas targets are large and nearly independent of energy; cross sections were measured for capture of up to four electrons in a single collision. An example is shown in Fig. 1, for xenon ions in krypton, with the projectile energies of the order of 1 keV/u. Examples of the second category, intermediate-energy, nearly symmetric collisions, are found in the work of Macdonald and his collaborators [3] for 0.3–3 MeV/u fluorine ions in nitrogen and argon targets, as shown in Fig. 2, and for oxygen ions in various gases. An example of multiple-electron capture in a fast collision is found in the work of Anholt and his

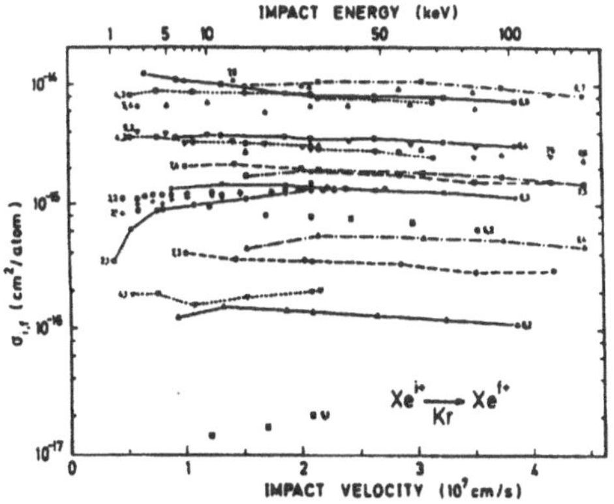

FIG. 1. Electron-capture cross sections $\sigma_{i,f}$ for Xe^{i+} ions incident on krypton, as a function of projectile velocity. From Ref. 2 by Salzborn and Müller.

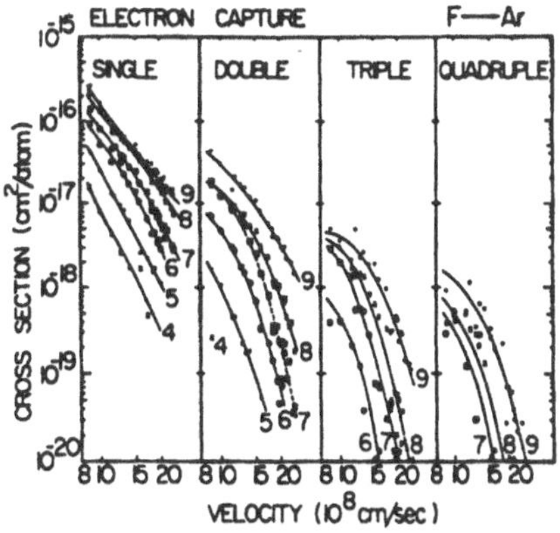

FIG. 2. Electron-capture cross sections for fluorine ions in argon. Curves are labeled with the charge state of the incident ion. From Ref. 3 by Ferguson et al.

collaborators [4] for 82 MeV/u Xe^{54+} ions in various solid targets (capture of up to three electrons observed) and for 105 MeV/u U^{90+} ions in various solid targets (capture of up to 5 electrons observed), shown in Fig. 3. However, in general, the cross section for capturing even two electrons in a single collision is small [5].

Total Electron Capture: 47-MeV Ca^{17+} in Ar

A collision system of particular interest is Ca^{17+} in an argon target. The calcium-ion projectile is lithium-like, with a filled K shell and 7 vacancies in the L shell; higher shells are entirely empty. The target argon atom is neutral, and thus has filled K, L, and M shells. For a projectile velocity corresponding to an energy of 47 MeV, or 1.2 MeV/u, the projectile has nearly the same velocity as that of electrons in the target L shell. It is thus plausible that a large number of electrons be transferred from the target to the projectile, especially from the target L shell to the projectile L shell, although transfer of electrons from or to other shells cannot be excluded. We have reported measurements of the total electron-capture cross section for this system [6], with the observation of capture of up to 8 electrons in a single collision, shown in Fig. 4. It would appear that,

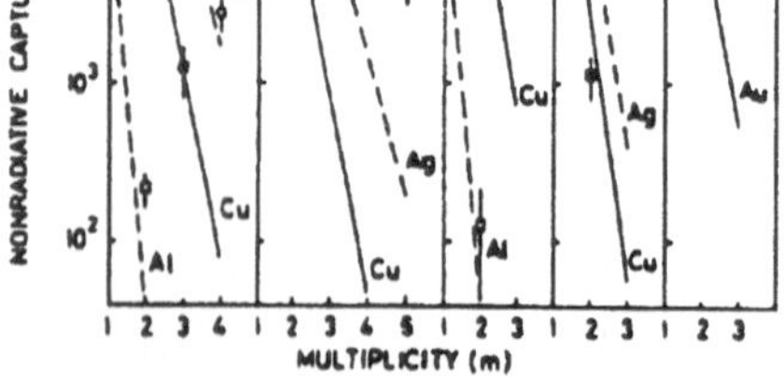

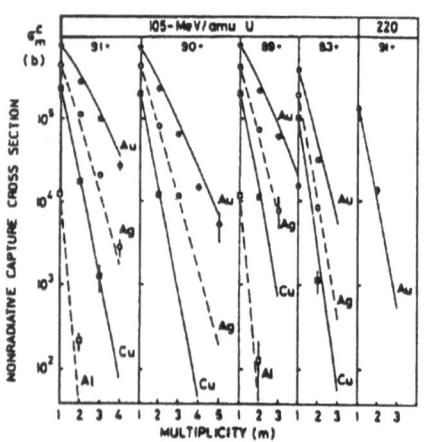

FIG. 3. Electron-capture cross sections for 105- and 220-MeV/U ions passing through Al, Cu, Ag, and Au target foils, as a function of the multiplicity (*m*) of the capture. From Ref. 4 by Anholt et al.

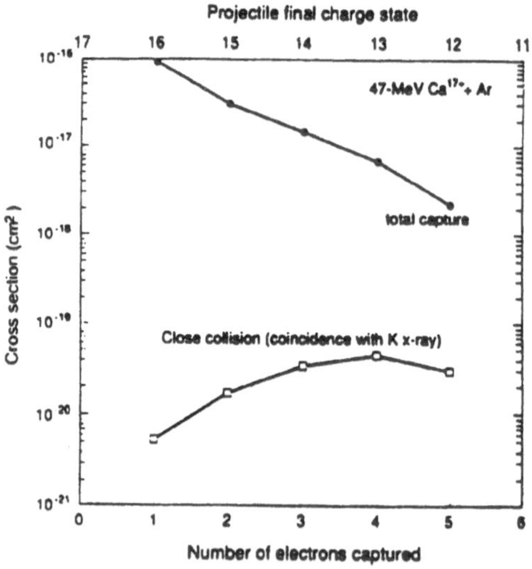

FIG. 4. Electron-capture cross sections for 47-MeV Ca^{17+} in Ar, as a function of the final charge state of the projectile and of the number of electrons captured. Lower curve shows electron capture for a close collision. Upper curve shows total electron capture. From Ref. 6.

in some collisions, an appreciable number of electrons is transferred, and, if they come from the L shell of the target, some appreciable autoionization must occur to stabilize the target after the collision. It is also likely that many of these electrons are transferred to low-lying levels of the projectile, otherwise many would be lost by autoionization of the projectile following the collision. In summary, multiple electron capture is observed because the projectile has a nearly empty shell for electrons to be captured into, and the target has many electrons available, for a projectile having a velocity similar to that of electrons in one shell of the target. The same system has been studied as a function of projectile charge state; similar results have been obtained, which will be reported elsewhere [7].

Electron Capture in Close Collisions: 47-MeV Ca^{17+} in Argon

We have studied close collisions for this system [6] by measuring electron capture in coincidence with emission of a projectile or target K x ray; emission of a K x ray indicates a collision with an impact parameter less than the radius of the K shell. A collision within the K-shell radius ensures that the projectile and target L shells have maximum interpenetration. Again, the projectile has a nearly empty L shell and the projectile velocity approximately matches that of electrons in the target L shell. Results for close collisions of 47-MeV Ca^{17+} ions in an argon target are shown in Fig. 4. Results for coincidence with an argon K x ray have been shifted by one charge state to account for the difference in the production mechanism of the K vacancy [6], and cross sections for Ar and Ca K x-ray coincidence have been summed. It is seen that the cross section for capturing several electrons exceeds that for capturing only one electron in a single close collision; a fit to a binomial distribution indicates an electron-capture probability of the order of 0.5, under appropriate assumptions. The relative electron-capture probabilities are observed to agree with a binomial distribution [6], which is consistent with uncorrelated electron capture.

These results for close collisions are possibly related to previous observations of multiple-electron capture or loss in a single close collision, where close collisions have been selected by scattering of the projectile through a large angle [8]. This connection is being studied at present [7].

***This work was supported by the Director, Office of Energy Research, Office of Basic Energy Sciences, Materials Sciences Division of the U.S. Department of Energy, under Contract No. DE-AC03-76SF00098**

References

[1] A. S. Schlachter, *Nuclear Instruments and Methods* B43, 265 (1989).

[2] E. Salzborn and A. Müller, Electronic and Atomic Collisions, edited by N. Oda and K. Takayanagi (North-Holland Publishing Co., Amsterdam 1980) p. 407.

[3] S. M. Ferguson, J.R. Macdonald, T. Chiao, L.D.Ellsworth, and S.A. Savoy, Phys. *Rev. A* 8, 2417 (1973); J.R. Macdonald, S.M. Ferguson, T. Chiao, L.D. Ellsworth, and S.A. Savoy, *Phys. Rev. A* 5, 1188 (1972).

[4] R. Anholt, W.E. Meyerhof, X.-Y. Xu, H. Gould, B. Feinberg, R.J. McDonald, H.E. Wegner, and P. Thieberger, *Phys. Rev. A* 36, 1586 (1987); W.E. Meyerhof, R. Anholt, J. Eichler, H. Gould, A. Munger, J. Alonso, P. Thieberger, and H.E. Wegner, *Phys. Rev. A* 32, 3291 (1985).

[5] J.A. Tanis, E.M. Bernstein, M.W. Clark, W.G. Graham, R.H. McFarland, T.J. Morgan, A. Müller, M.P. Stockli, K.H. Berkner, P. Gohil, A.S. Schlachter, J.W. Stearns, B.M. Johnson, K.W. Jones, M. Meron, and J. Nason, Proceedings of the Second U.S.–Mexico Symposium on Atomic and Molecular Physics: Two-Electron Phenomena, published in *Notas de Fisica* 10, 467 (1987).

[6] A.S. Schlachter, E.M. Bernstein, M.W. Clark, R.D. DuBois, W.G. Graham, R.H. McFarland, T.J. Morgan, D.W. Mueller, K.R. Stalder, J.W. Stearns, and J.A. Tanis, *J. Phys. B* 21, L291 (1988); A.S. Schlachter, J.W. Stearns, K.H. Berkner, E.M. Bernstein, M.W. Clark, R.D. DuBois, W.G. Graham, T.J. Morgan, D.W. Mueller, M.P. Stockli, J.A. Tanis and W.T. Woodland, *Nuclear Instruments and Methods* B40/41, 21 (1989); A.S. Schlachter, J.W. Stearns, K.H. Berkner, E.M. Bernstein, M.W. Clark, R.D. DuBois, W.G. Graham, T.J. Morgan, D.W. Mueller, M.P. Stockli, J.A. Tanis, and W.T. Woodland, in Proceedings of the 16th International Conference of Physics of Electronic and Atomic Collisions, edited by A. Dalagarno, R.S. Freuknd, P.M. Kock, M.S. Lubell, and T.B. Lucatorto (American Institute of Physics, New York, New York, 1989) pp. 366–371.

[7] A.S. Schlachter, J.W. Stearns, K.H. Berkner, E.M. Bernstein, M.W. Clark, R.D. DuBois, W.G. Graham, T.J. Morgan, D.W. Mueller, M.P. Stockli, J.A. Tanis, and W.T. Woodland, (article in preparation).

[8] B. Rosner and D. Gur, *Phys. Rev.* ,A15, 70 (1977); E.N. Fuls, P.R. Jones, F.P. Ziemba and E. Everhart, *Phys. Rev.*, 107, 704 (1957); E. Everhart and Q.C. Kessel, *Phys. Rev.* 146, 16 (1966); V.V. Afrosimov, Yu. S. Gordeev, M. Panov and N.V. Fedorenko, *Zh. Eksp. Teor. Fiz. Pisma Red.* 2, 291 (1965); Q.C. Kessel, *Phys. Rev.* A2, 1881 (1970).

MULTIELECTRON PROCESSES*

J. Ullrich, R. Dörner, H. Schmidt-Böcking

GSI-Darmstadt

IKF Frankfurt, FRG

Abstract

Recoil-ion momentum spectroscopy (RIMS) has been applied to obtain differential single and multiple ionisation cross sections in energetic, ion-atom collisions. Investigating various collision systems, different aspects of the multiple ionisation reaction have been elucidated: The ionisation collision dynamics, the role of many-body interactions, has been explored for single ionisation of He by 0.3 and 0.5 MeV proton impact measuring the transverse momentum of the recoil-ion in coincidence with that of the projectile; deviations from two-body kinematics have been observed. For the 1.4 MeV/u U^{32+} on Ne collision system multiple ionisation coincident with single electron capture of the projectile has been measured in dependence of the recoil-ion transverse momentum. In a third experiment, 1.4 MeV/u U^{32+} on Ar, the contribution of autoionisation processes to the production of Ar^{6+} ions is investigated: Part of the Ar^{6+} ions were found to be produced with recoil-ion momenta typical for Ar^{5+} production.

1 Introduction

The detailed understanding of the mechanisms of momentum transfer from a fast charged projectile to a many-electron atom in a single ion-atom collision is of basic interest from a fundamental point of view but also of considerable practical importance for applied physics and biophysics. Experimental and theoretical data for single collisions would provide a firm basis for the description of more complex problems like the interaction of fast ions with dense matter (e.g. the implantation of ions into semiconductors) or the damage of biological systems under energetic ion-irradiation. In most of the practical cases the energetic ion losses all of its kinetic energy during its propagation through the dense matter. Thus, experimental data on energy transfer mechanisms are needed over a wide regime of projectile velocities ranging from GeV/u down to eV/u requiring experimental techniques which can be applied in that whole regime. Simultaneously, there is a strong demand for theories capable to treat the many-body Coulomb problem over a wide energy range.

Three basic reaction channels of energy transfer are usually separated if an energetic ion hits a many-electron target: First, one or more electrons of the target atom might be transferred to

*supported by GSI, Darmstadt, BMFT and DFG

continuum states without any change in the projectile's electronic configuration (usually denoted as pure multiple ionisation, MI). Secondly, in addition to target MI, the projectile itself may lose or capture one or more electrons and, third, in addition to the ionisation of the target atom it may be highly excited during the collision and autoionise after the encounter, thus further increasing its ionic charge.

Within the last decade, the first two processes have been explored intensively, using effective time-of-flight (TOF)-techniques [1, 2, 3, 4, 5] to determine the recoil-ion and projectile charge state after the collision. Several attempts have been made to investigate all these reactions differentially in dependence of the final projectile transverse momentum $P_{p\perp}$, i.e. the laboratory scattering angle $\vartheta = P_{p\perp} / P_{\|}$ ($P_{\|}$ being the incoming projectile momentum) [6, 7, 8]. In the medium and high energy regime, however, such measurements can only provide differential data for a very narrow impact parameter regime since the determination of ϑ is technically limited to values above 5×10^{-5} rad.

In order to circumvent this problem, we have developed a method which enables the simultaneous spectroscopy of the transverse recoil-momentum of the target-ion $P_{R\perp}$ and of the recoil-ion charge state q [9]. The recoil-ion momentum spectroscopy (RIMS) for the first time provides differential cross sections for many-electron transitions for energetic and heavy projectiles at those momentum transfers to the nuclei (i.e. impact parameters b) which are typical for the reactions under consideration. The related projectile scattering angles mostly range in the submicrorad regime.

In this paper we report on single ionisation of the fundamental proton-He collision system at 0.3 and 0.5 MeV projectile energy. Measuring the recoil-ion transverse momentum and charge state in coincidence with the projectile transverse momentum, detailed information of the dynamics of pure ionisation (reaction mechanism 1) and the role of the emitted electron momenta could be obtained. We further report on RIM-dependent multiple ionisation cross section for simultaneous electron capture of the projectile in 1.4 MeV/u U^{32+} on Ne collisions (reaction type 2). Moreover, it is demonstrated (1.4 MeV/u U^{32+} on Ar) that RIMS can help to separate autoionisation processes (reaction type 3). Future high precision RIMS might give quantitative information about the electronic configuration of the target atom formed in an encounter with an energetic, highly charged ion.

2 Experimental Method

The 1.4 MeV/u U^{32+} beam was delivered by the UNILAC of GSI, the 0.3 and 0.5 MeV proton beam by the Van-de-Graaff accelerator of the Institut für Kernphysik, Universität Frankfurt. In both cases the beam was tightly collimated to a spot of 0.05 mm^2 at the interaction region which was a triply differentially pumped static gas cell, cylindrically shaped, the axis being collinear with the beam direction. For U^{32+} impact, the projectiles were charge-state analysed after the collision by magnetic deflection and detected by a one-dimensional position-sensitive parallel plate avalanche detector (PPAD) (details of this experiment can be found in Ref. [10]). For proton impact the PPAD was replaced by a two-dimensional position sensitive microchannel-plate detector with a resolution of 130 μm in both directions. This enabled to identify the very small projectile scattering angles of interest in the sub-mrad regime (for complete information see Ref. [11]).

The recoil-ion transverse momentum $P_{R\perp} = m_R \frac{\Delta x}{\Delta t}$ (m_R: recoil-ion mass) is measured by a time-of-flight technique. Recoil-ions produced by the beam along the beam axis inside the target cylinder (Ø 10 mm) drift from the axis to the wall (typical drift length, $\Delta x = 5$ mm) in a time interval Δt inversely proportional to their transverse momentum. The recoil-ions are then accelerated and charge-state analysed; Δt is obtained from a coincidence with the projectiles. (A complete description of the present apparatus as well as on the future scope of the method can be found in Ref. [12]).

The geometry of the RIM-spectrometer defines the scattering plane as illustrated in Fig. 1. The additional two-dimensional detection of the projectile deflection (transverse projectile momenta in x and y direction P_{px}, P_{py}) fully determines the transverse momentum balance for single ionisation events. This enables to investigate the φ angular emission pattern of the ejected particles (φ: azimuthal angle) as well as the azimuthal emission angles relative to each other.

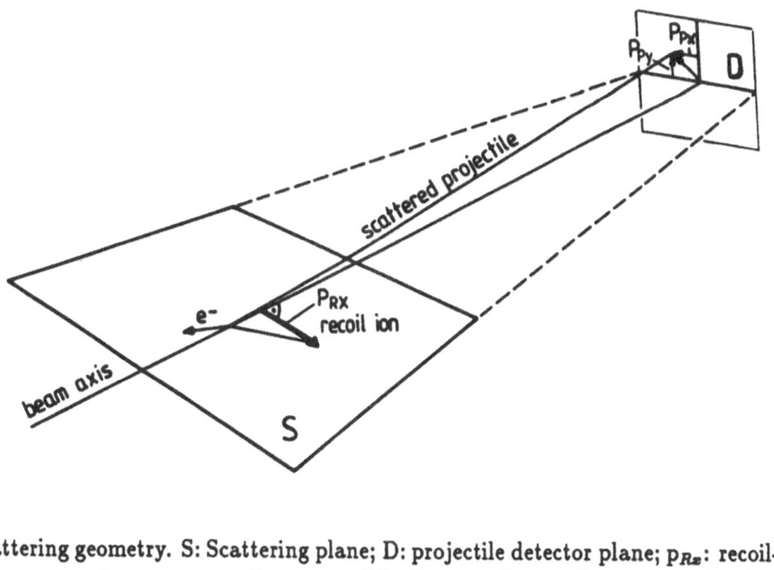

Figure 1: Scattering geometry. S: Scattering plane; D: projectile detector plane; p_{Rx}: recoil-ion transverse momentum component, measured in the experiment; P_{px},P_{py}: projectile transverse momenta.

3 Proton on He Single Ionisation (reaction type 1)

In a previous publication [11], the influence of three-body interactions on the He single ionisation collision dynamics has been demonstrated. This influence manifested itself in observed deviations of the relationship between projectile scattering anlge ϑ and transverse recoil-ion energy $E_{R\perp}$ from the kinematical relationship for a two-body collision $E_{R\perp} = E_{\parallel} \frac{m_P}{m_R} \cdot \vartheta^2$ ($E_{\parallel}$ is the projectile energy; m_P, m_R are the masses of the projectile and the recoil-ion, respectively). Hence, the reason for such a deviation can only lie in well defined relative angular emission patterns between the particles which have been investigated for $\vartheta = 0.5$ mrad for 0.5 MeV p on He, where the recoil-ion energy considerably falls below the two-body value (see Fig. 2a). To fulfill momentum conservation, this means that the ejected electron is more probably scattered to the recoil-ion side: The electron emission is found to be peaked opposite to the projectile (Fig. 3a) $\varphi_{pe} \sim 180^0$ similar to the relative behaviour between the projectile and the recoil ion (Fig. 3b, $\varphi_{PR} \sim 180^0$).

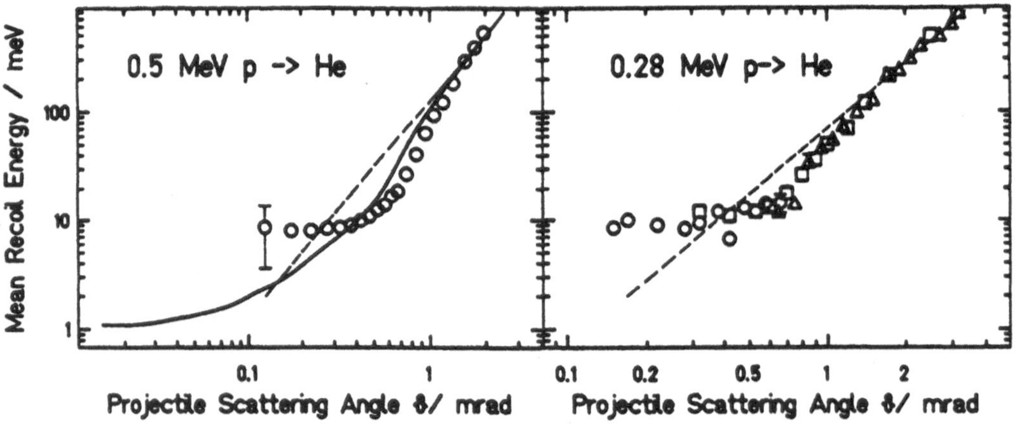

Figure 2: Mean transverse recoil-ion energy $<E_{R\perp}>$ (open symbols) in dependence of the projectile scattering angle ϑ for 0.5 (2a) and 0.3 MeV (2b) singly ionising H^+ – He collisions. Full line: nCTMC calculations [13]. Dashed line: Results for a two-body collision between the proton and the He nucleus.

It would be interesting to investigate this behaviour at very small ϑ below 0.2 mrad which mainly contribute to the total single ionisation cross section. New nCTMC calculations [13] (full line in Fig. 2a) predict a recoil-energy of about 1 meV nearly independent of the projectile scattering angle for $\vartheta \leq 0.05$ mrad. (This theoretical value is lower than the one, 6.5 meV, given in Ref. [11] which was influenced by the approximation that the target nucleus and not the whole atom was at rest before the collision. For a detailed discussion see [14]). Our experimental value for the recoil-ion

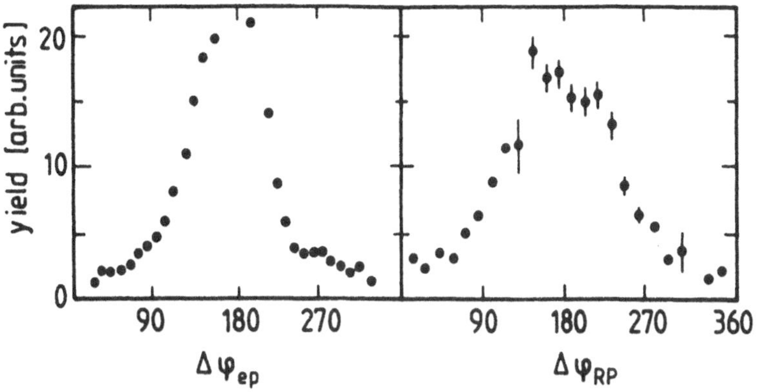

Figure 3: Intensity of the emerging particles (electron, recoil-ion, projectile) in dependence of the relative azimuthal emission angle between projectile and electron $\Delta\varphi_{ep}$ (3a) as well as projectile and recoil-ion $\Delta\varphi_{RP}$ (3b) for 0.5 MeV p-He single ionisation.

energy at $\vartheta = 0.12$ mrad is around 8 meV with a possible systematical error of $\pm$ 5 meV which is in disagreement with the theoretical value of 2.5 meV.

In order to check this result we measured the $E_{R\perp}$-ϑ relationships for 0.28 MeV proton impact where the ϑ-determination is less critical. In addition, the recoil-ion drift path was decreased to $\Delta x = 2.5$ mm compared to $\Delta x = 3.5$ mm for 0.5 MeV proton impact. If the measured saturation value would be due to an electric potential distribution over the drift path, it is expected to change for different Δx. The results are shown in Fig. 2b: Again the expected two-body behaviour is observed for $\vartheta \geq 1$ mrad, then $E_{R\perp}$ drops below the two-body values at $0.5 \leq \vartheta \leq 1$ mrad and again, a saturation value of about 8 meV is observed. (Different symbols in the figure denote measurements using different cup diameters to dump the undeflected beam.)

Apart from the observed discrepancies with theory at very small momentum transfers, these results underline the previous statement, that the ionisation collision dynamics, the motion of the heavy particles is strongly influenced by their interaction with the ionised electrons showing that any calculations based on a central deflection potential is unable to follow the facets of the dynamics. It should be noted that this holds for more than 95 % of all ionising collisions, which are known to be related to projectile deflections of less than 0.5 mrad for the collision systems investigated.

The results show that it is necessary to further improve the resolution of RIMS by a factor of 10 to become sensitive on the sub-meV level. (A possible approach to achieve this is discussed in Ref. [12]). It is expected that this resolution will be sufficient to investigate the ionisation process even for GeV/u U-He collision at all impact parameters of interest. On the other hand, our results demonstrate that RIMS provides identical data as an angular differential measurement for "close" collisions, where the momentum transfer is dominated by the Coulomb-interaction between the two nuclei ($\vartheta \geq 1$ mrad for 0.28 and 0.5 MeV proton on He collisions). Since the resolution of RIMS is independent of the projectile energy the present setup is capable to provide transverse momentum dependent data at high projectile velocities and masses where a ϑ-determination is impossible which is illustrated in the next chapter.

4 Differential MI Cross Sections Coincident to Projectile Electron Capture (reaction type 2)

Projectile electron capture takes place at impact parameters considerably smaller than those contributing to pure ionisation. For the 1.4 MeV/u U^{32+} on Ne collisions this results in typical transverse recoil-ion energies of more than 100 meV and, hence, the related transverse momenta of $P_{R\perp}/P_{\parallel} \geq 4 \times 10^{-6}$ rad might be identical with the projectile scattering angle $\vartheta = P_{P\perp}/P_{\parallel}$. However, due to the large number of electrons involved — 5 additional target electrons are emitted in the average coincident to single electron capture — their sum momentum might still disturb the two-body kinematical relationship as has been observed for double electron capture plus MI in the F on Ne collision system [15].

RIM-dependent MI cross sections coincident to single electron capture are shown in Fig. 4 (full circles) together with previously published data for pure MI (full line). The maximum in the charge state distribution (integration over d($P_{R\perp}$ / $P_{\parallel}$) for all q) is found for the Ne charge state q = 5, which is in slight disagreement to experimental results of Müller et al. [16] who found the maximum intensity for q = 6. A significant peak in the RIM distributions is observed for recoil-ion charge states q = 2 to q = 7 shifting from $P_{R\perp}$ / $P_{\parallel}$ = 4×10^{-6} (q = 2) to about 1.3×10^{-5} for q = 7. Simultaneously, the peak structure gets less pronounced and a nearly flat dependence is obtained for q = 8. The comparison with the experimental data for pure MI (full line in Fig. 4) shows that MI plus capture occurs at smaller impact parameter for the same degree of target ionisation: The maximum is at higher recoil-ion momenta and the distributions extend to larger momentum transfers (smaller b).

5 Contributions of Autoionisation to MI (reaction type 3)

It was observed in numerous studies of the multiple ionisation reaction that the total cross section for the production of certain recoil-ion charge states significantly exceeds the values calculated by theories using the Independent Particle Model (IPM, Ref. [17]). This behaviour which was found to be more pronounced for Ar than for Ne targets usually was attributed to the contribution of autoionisation cascades [2]. Clearly, due to the nature of autoionisation as an electron-electron

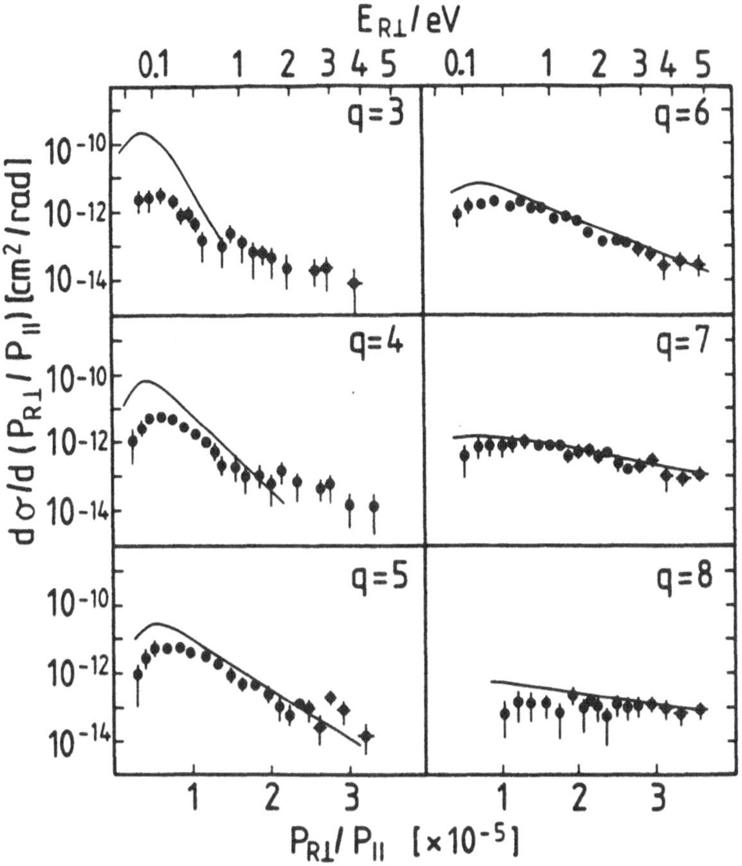

Figure 4: Differential multiple ionisation cross sections (1.4 MeV/u U^{32+} on Ne) for pure MI (full line) and MI plus single electron capture by the projectile (full circles) in dependence of the relative transverse momentum transfer to the recoil-ion $P_{R\perp}/P_{\parallel}$. Upper scale: Transverse recoil-ion energy.

interaction between well defined initial and final electronic configurations, such constributions will be in disagreement with the simple binomial statistical behaviour resulting from the assumption that the electrons do not interact with each other (IPM).

Along with the results of Müller et al. [16] who provided systematical data sets on Ar and Ne, total MI cross sections at 1.4 MeV/u projectile velocity, we observe an enhanced production of Ar^{6+} recoil-ions for 1.4 MeV/u U^{32+} on Ar collisions in disagreement with a simple binomial statistical behaviour. In addition to this total cross section information we measure the recoil-ion transverse energy dependence (b-dependence) of the creation of the various Ar charge states, shown in Fig. 5a,b for Ar charge states q = 1 to q = 7. With increasing q the recoil-ion energies increase steadily up to q = 5, the TOF distributions show systematically, with each single charge state, a shift to lower TOF, i.e. higher energies (Fig. 5a). This systematic behaviour vanishes for q = 6: The low energy tail extends to values which are typical for Ar^{5+} (Fig. 5b). This indicates, that part of the Ar^{6+} ions are created at impact parameters typical for Ar^{5+} production, probably as Ar^{5+} but in excited states resulting in additional autoionisation and a final charge state of q = 6.

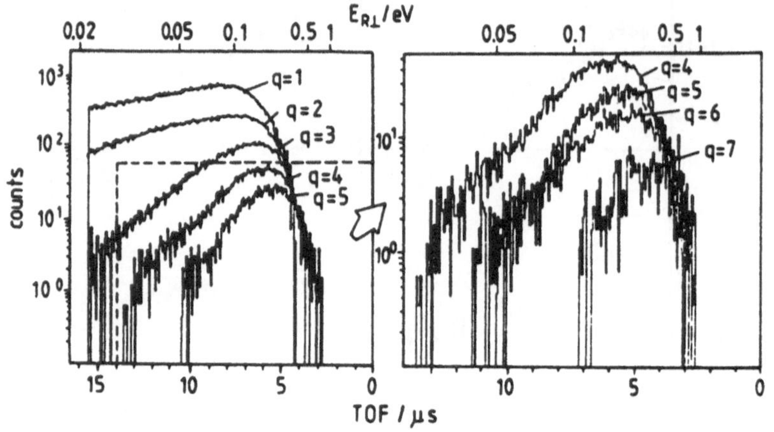

Figure 5: TOF spectra for the production of q-times charged recoil-ions in 1.4 MeV/u U^{32+} on Ar collisions. Fig. 5a: q = 1 to 5. Fig. 5b: q = 4 to 7. Upper scale: Transverse recoil-ion energy. The charge state q is denoted in the figure.

Due to the limited energy resolution induced by the target thermal motion no further information on the involved levels and magnitude of autoionisation contributions can be obtained. However, these results illustrate that RIMS is capable to elucidate much more details of MI and autoionisation reactions compared to a simple total cross section measurement. Using a future high resolution apparatus, which is presently being machined [12], it is expected to obtain a sub-meV recoil-ion energy resolution which is comparable to 10 eV electron energy determination. Under certain kinematical conditions, we then might be able to identify even distinct autoionising states due to the recoil-momentum transfered to the target nucleus during the Auger decay.

6 Conclusion

Investigating several collision systems using recoil-ion momentum spectroscopy (RIMS) different aspects of the multiple ionisation reaction were elucidated.

The coincident measurement of the recoil-ion and projectile momentum in a plane transverse to the beam direction for proton on He collisions yielded detailed information on the ionisation collision dynamics. For single ionisation, the ejected electron is shown to considerably influence the momentum exchange between the heavy particles. The results demonstrate that three-body interactions (recoil-ion, projectile, emitted electron) have explicitly to be accounted for to follow the facets of the ionisation collision dynamics. The experimental results were found to be in good agreement with nCTMC calculations for closer collisions $\vartheta \geq 0.5$ mrad. For the smaller $\vartheta \leq 0.5$ mrad the present experiments yield a saturation value of about 8 meV in contrast to new nCTMC results of about 2 meV.

The coincident determination of the recoil-ion and final projectile charge state in dependence of the recoil-ion transverse momentum provided differential MI plus capture cross sections for the 1.4 MeV/u U^{32+} on Ne collision system. Using RIMS, we were able to investigate this process at those impact parameters which mainly contribute to the total MI plus single capture cross section.

In a third study, 1.4 MeV/u U^{32+} on Ar, the potential of RIMS to provide details on the contribution of autoionisation to total MI cross sections is indicated. The enhancement of Ar^{6+} production for highly charged ion impact, also observed in previous experiments is shown to be due to contributions from impact parameters typical for Ar^{5+} production. Future high precision RIMS will not only deliver information on the momentum transfers during a MI reaction which is equivalent to the measurement of angular differential cross sections but in addition might be sensitive enough to identify even the exact autoionising states contributing to MI.

References

[1] C.L. Cocke, Phys. Rev. A20, 749 (1979)

[2] S. Kelbch, J. Ullrich, R. Mann, P. Richard, and H. Schmidt-Böcking, J. Phys. B18, 3423 (1985)

[3] A. Müller, B. Schuch, W. Groh, E. Salzborn, H.F. Beyer, P.H. Mokler, and R.E. Olson, Phys. Rev. A33, 3010 (1986)

[4] T.J. Gray, C.L. Cocke, and E. Justiniano, Phys. Rev. A22, 849 (1986)

[5] H. Tawara, T. Tonuma, H. Kumagai, T. Matsuo, Phys. Rev. A41, 116 (1990)

[6] R. Schuch, H. Schöne, P.D. Miller, H.F. Krause, P.F. Dittner, S. Datz, and R.E. Olson, Phys. Rev. Lett. 60, 925 (1988)

[7] S. Kelbch, C.L. Cocke, S. Hagmann, M. Horbatsch, C. Kelbch, R. Koch, H. Schmidt-Böcking, and J. Ullrich, J. Phys. B22, 1277 (1989)

[8] E.Y. Kamber, C.L. Cocke, S. Cheng, S.L. Varghese, Phys. Rev. Lett. 60, 2026 (1989)

[9] J. Ullrich, H. Schmidt-Böcking, and C. Kelbch, Nucl. Instr. Meth. A268, 216 (1988)

[10] J. Ullrich, M. Horbatsch, V. Dangendorf, S. Kelbch, and H. Schmidt-Böcking, J. Phys. B21, 611 (1988)

[11] R. Dörner, J. Ullrich, H. Schmidt-Böcking, and R.E. Olson, Phys. Rev. Lett. 63, 147 (1989)

[12] J. Ullrich et al., to be published in Nucl. Inst. Meth.

[13] R.E. Olson in Electronic and Atomic Collisions, edited by H.B. Gilbody, W.R. Newell, F.H. Read, and A.C.H. Smith (Elsevier Science, New York, 1988), pp. 271 – 285

[14] R. Dörner et al., Proc. of the Vth International Conference on the Physics of Highly-Charged Ions, Gießen 1990, to be published

[15] A.D. Gonzales, S. Hagmann, T.B. Quinteros, B. Krässig, H. Schmidt-Böcking, R. Koch, and A. Skutlartz, J. Phys. B23, L303 (1990)

[16] A. Müller, B. Schuch, W. Groh, and E. Salzborn, Z. Phys. D7, 251 (1987)

[17] J.H. McGuire and L. Weaver, Phys. Rev. A16, 41 (1977)

H Alignment in Atomic Collisions; Laser Assisted Collision Processes

Chairman: W. MEHLHORN

H.1 Invited Surveys

R. HIPPLER
Coherence Studies of Atomic Collisions

T. ÅBERG
Collisions and Half Collisions in Strong Electromagnetic Fields

H.2 Invited Contributions

A. BORDENAVE-MONTESQUIEU, P. MORETTO-CAPELLE, P. BENOIT-CATTIN and A. GLEIZES
Search for an Alignment in Double Capture Processes between Multicharged Ions and Helium below 8 keV/amu

I.C. LEGRAND, R. DÖRNER, H. SCHMIDT-BÖCKING and V. ZORAN
Influence of Rotating Electron Wave Functions on L-Shell Ionization and Alignment

H.3 Contributions (Posters)

S.M. BLOKHIN and V.P. PETUKHOV
The Polarization of Satellites of the X-Ray Lines of Titanium Atoms in Compounds under Proton Ionization

A. GUTENKUNST, S. ZUCCATTI and W. MEHLHORN
Alignment of $Mg^+(2p_{3/2}^{-1}\ ^2P_{3/2})$ Ions after Electron Capture by Protons

T. PAPP and I. TÖRÖK
Angular Distribution Measurements of Th M X-Ray Lines

Y. AWAYA, A. HITACHI, T. KAMBARA, Y. KANAI, T. MIZOGAWA and T. PAPP
Angular Distribution of the X-Rays of 1 MeV/amu Argon Beams Passing through Carbon Foil

E. TAKÁCS, S. RICZ, J. VÉGH, B. SULIK, T. PAPP, J. PÁLINKÁS, B. TÓTH and D. BERÉNYI
Systematic Study of the Anisotropy of the Ne K Auger Satellites in Ion-Atom Collisions

Coherence Studies of Atomic Collisions.

Rainer Hippler

Fakultät für Physik, Universität Bielefeld, D-4800 Bielefeld 1, Fed. Rep. Germany

I. Introduction

Electronic processes in simply-structured atomic collision complexes are of fundamental importance for our understanding of collision dynamics and also of practical relevance for a variety of experimental fields including fusion technology and astrophysics. In continuation of our previous work [1–2] we concentrate on the coherent excitation of H(n=2) states either by direct or by charge changing processes at incident energies of 1–25 keV. This energy region is of particular interest as it covers the transition regime from small velocities, where the collision partners transiently form a quasi-molecule [3], to intermediate velocites, where the projectile velocity v_p equals the orbiting velocity v_e of bound electrons. The excitation of an hydrogen atom to the H(n=2) states is a prototype case for the study of coherent excitation mechanisms operating during atomic collision processes [4]. This is because of the near-degeneracy among the different H(n=2) levels, for which due to the small energy splitting $\Delta E \lesssim 10^{-5}$ eV significant differences in the time-development occur after $\tau = \frac{\hbar}{\Delta E} \approx 10^{-11}$s ($\hbar$ is Planck's constant); this time is considerably larger than typical collision times $\tau_{col} \approx 10^{-16} \ldots 10^{-15}$s. The wavefunction Ψ of a coherently excited H(n=2) state at the instant of the collision ($t = 0$) may be expressed as

$$\Psi = f_{00}|2s_0\rangle + f_{10}|2p_0\rangle + f_{11}|2p_1\rangle + f_{1-1}|2p_{-1}\rangle\,, \tag{1}$$

where $|2s_0\rangle$ and $|2p_m\rangle$ ($m = 0, \pm 1$) are the magnetic substates and f_{lm} are the corresponding (complex) excitation amplitudes. For symmetry reasons (conservation of reflection symmetry with respect to a given scattering plane), $f_{11} = -f_{1-1}$ is required [5] which leaves a total of 3 independent excitation amplitudes. In a properly chosen 'complete' experiment, the above wavefunction (Eq. (1)) can be fully determined. So far, such an experiment has not been performed yet. Instead, in many experiments carried out thus far certain average quantities have been determined. This is particularly true for 'integral' experiments in which the direction of the outgoing projectile is not specified; in this case the excited state is best described in terms of a (integral) density matrix [6] $\rho_{n=2}$ at the instant of the collision ($t = 0$),

$$\rho_{n=2} = \begin{pmatrix} \sigma_{2s_0} & 0 & \sigma_{sp} & 0 \\ 0 & \sigma_{2p_{-1}} & 0 & 0 \\ \sigma^*_{sp} & 0 & \sigma_{2p_0} & 0 \\ 0 & 0 & 0 & \sigma_{2p_{+1}} \end{pmatrix} \tag{2}$$

where the $\sigma_{lm} =<| f_{lm} |^2>$ are the partial cross sections for excitation into the H(2s) and H(2p$_m$) ($m = 0, \pm 1$) substates, f_{lm} is the corresponding excitation amplitude, $\sigma_{sp} =< f_{00} f^*_{10} >$ is the so-called s–p coherence, and where $< \ldots >$ denotes an integration over all scattering angles. The real part of the s–p coherence, $\Re(\sigma_{sp})$, is proportional to a collision-induced electric dipole moment, i.e., to a shift of the center-of-charge with respect to the center-of-mass (for example, an electron trailing behind a proton); it manifests itself in a forward-backward asymmetry with regard to an external electric field [7–8]. Such collision-induced dipole moments $< D >_z$,

$$< D >_z = \frac{6\,\Re(\sigma_{sp})}{\sigma_t}, \tag{3}$$

where $\sigma_t = \sum \sigma_{lm}$ is the total cross section for H(n=2) excitation, have, for example, been observed before in charge exchange excitation of H(n=2) and H(n=3) in H^+–He collisions at incident energies above 20 keV [9–10] and in 'electron-capture-to-the-continuum (ECC)' processes [11]. The imaginary part $\Im(\sigma_{sp})$ is related to $< \vec{L} \times \vec{A} >_z$,

$$< \vec{L} \times \vec{A} >_z = -\frac{2\,\Im(\sigma_{sp})}{\sigma_t}, \tag{4}$$

where $\vec{L}$ and $\vec{A}$ are the angular momentum vector and the Runge-Lenz vector, respectively; classically, the vector $\vec{L} \times \vec{A}$ points along the orbital electron velocity at the perihelion of its Kepler orbit [8,9]. Other quantities of interest here are the alignment A_{20} and the orientation O_1 [12]. The alignment A_{20} provides detailed information about the relative population of the H(2p_m) magnetic substates; it is defined as [13]

$$A_{20} = \frac{\sigma_{2p_1} - \sigma_{2p_0}}{\sigma\,(2p)} \tag{5}$$

where $\sigma(2p) = \sigma_{2p_0} + 2\,\sigma_{2p_1}$. In order to obtain all the relevant elements of the $\rho_{n=2}$ density matrix, we utilize Stark mixing of the H(2s) and H(2p) states in an external electric field and measure the linear and circular polarizations of Lyman–α radiation (λ = 121.6 nm) emitted during the decay of the excited H(n=2) state to the H(1s) ground state. These measurements were performed with the help of a Brewster-type linear polarizer; i.e., we used a lithium-fluoride (LiF) plate inclined at $\sim 60^o$ with respect to the optical axis resulting in a polarization sensitivity of 90 %. For circular polarization measurements, a quarter-wave-plate made from magnesium fluoride (MgF_2) and with a polarization sensitivity of $\sim$ 98 % was inserted in front of the linear polarizer. Details of the apparatus have been described previously [13] and shall not be repeated here. To extract the density matrix elements $\rho_{lml'm'} = f_{lm} f^*_{l'm'}$ from the measured degree of polarization one (i) has to expand the initial state Ψ in terms of Stark states $|k\rangle$, (ii) take the time evolution of the Stark states into account, and (iii) calculate matrix elements for the radiative decay of these states to the H(1s) ground state. The corresponding intensities $I_{\vec{r}}(t)$ of the emitted light with polarization vector $\vec{r}$ are calculated using

$$I_{\vec{r}}(t) = \Big| \sum_{k=1}^{4} \langle 1s_{1/2} \,|\, e\vec{r} \,|\, k \rangle \, e^{-(i\omega_k - \frac{\gamma_k}{2})t} \, \langle \mathrm{k} \,|\, \Psi \rangle \Big|^2, \tag{6}$$

where $\hbar\omega_k$ and γ_k are the electronic binding energy and the radiative width, respectively, of the $|k\rangle$ Stark state. Following Lüders [14], the (relatively weak) external field causes a mixing of the unperturbed hydrogenic states $\Phi_i = |jm_j\rangle$; the perturbed H(n=2) eigenstates (Stark states) $|k\rangle$ are then given as linear combinations $|k\rangle = \sum a_{ki}\Phi_i$, where the a_{ki} (k=1...4) are expansion coefficients which depend on the electric field strength F [15]. The time-dependent intensities $I(t)$ have been integrated over the finite observation intervall over which production and decay of the excited H(n=2) atoms occurs to yield the integrated intensitities; this smears out the rapid intensity oscillations (Stark beats [16–17]) which where not observable in the present experiment. The measured polarizations were fitted according to Eq. (6) and the non-zero density matrix elements were extracted [18]. This procedure was previously tested for the adiabatic [15] and the non-adiabatic [19] electric-field-induced Lyman–α radiation from metastable H(2s) atoms.

II. One-electron systems

For the by far simplest atomic collision system, H^+–H, excitation of H(2p) in

$$H^+ + H \rightarrow \begin{cases} H^+ + H\,(2p) \\ \\ H\,(2p) + H^+ \end{cases} \tag{7}$$

collisions is predicted to proceed at low incident velocities through a united-atom 2pσ–2pπ rotational coupling in the transiently formed quasi-molecule (Fig. 1), giving rise to the population of H(2p$_{\pm 1}$) magnetic substates only. As is indicated in Fig. 2, the measured integral alignment [20] is rather large at small projectile energies and close to the maximum value of +50% permitted by Eq. (5). These measurements are thus in good agrement with theoretical calculations which properly account for this rotational coupling mechanism [21–25]. Other calculations, for example coupled-state calculations employing separated-atom electronic wavefunctions only [26–28], are at lower velocities at variance with our experimental results.

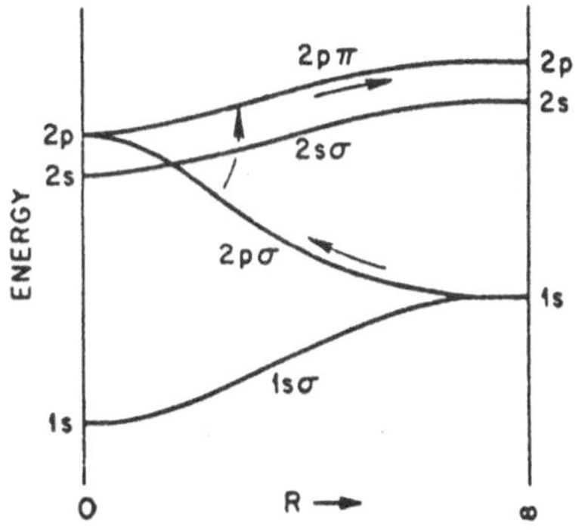

Fig. 1: Schematic energy-level diagram for H^+–H. The binding energy of an electron moving in the field of two approaching protons is shown as a function of the internuclear separation R. For clarity, the l–degeneracy has been removed.

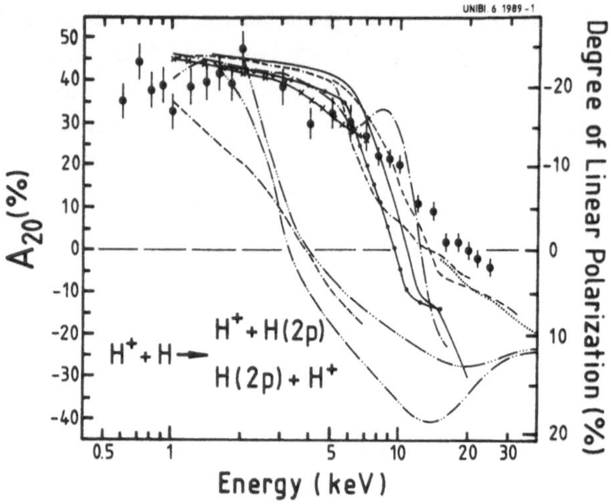

Fig. 2: Integral alignment A_{20} (left-hand scale) and degree of linear polarization P_1 (right-hand scale) for H(2p) production in (•) H^+–H collisions versus incident energy [20]. Also shown are the predictions of various theoretical calulations (see text).

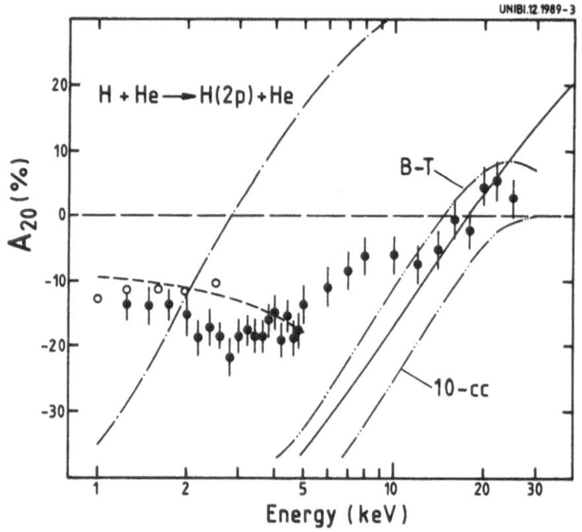

Fig. 3: Integral alignment A_{20} for H(2p) production in H–He collisions (•, Ref. 31; ∘, Ref. 32) versus incident energy. Also shown are the molecular orbital (MO) calulations of Kimura and Lane [33] (dashed line), the optical potential calculations of Ast et al. [34] (dash-dotted line), the 10-state close-coupling (10 *cc*) calculations and the first Born calculations accounting for 'double transitions' ($B-T$) of Bell et al. [29], and model calculations according to Nielsen and Dahler [30] (solid line).

III. Quasi-one-electron systems

A different picture emerges for the (asymmetric) quasi-one electron system

$$H + He \rightarrow H\,(2p) + He \tag{8}$$

which was investigated at incident energies of 1–25 keV. Fig. 3 displays the measured integral alignment for this collisions system [31–32], which differs significantly from our results for H^+–H collisions [20]. Again, as becomes obvious from a corresponding correlation diagram [33], a united-atom 2pσ–2pπ rotational coupling provides an important mechanism for H(2p) excitation in H–He collisions at low incident velocities. In addition, radial couplings occur among the 2pσ and the 2sσ and 3dσ orbitals, resulting in a negative A_{20} for small incident velocities. Differential measurements show that the radial couplings are most pronounced for small impact parameters (Fig. 4), whereas the 2pσ–2pπ rotational coupling dominates at larger impact parameters [2].

Towards larger velocities the integral alignment becomes positive. This result is an indication of a direct 1s–2p excitation process which at medium velocities favours the H(2p$_{\pm1}$) magnetic substates. Above $E \gtrsim 10$ keV our results are in reasonable agreement with the 10-state close-coupling calculations and the first Born calculations accounting for 'double transitions' of Bell et al. [29], and with model calculations according to Nielsen and Dahler [30], but not with optical potential calculations of Ast et al. [34].

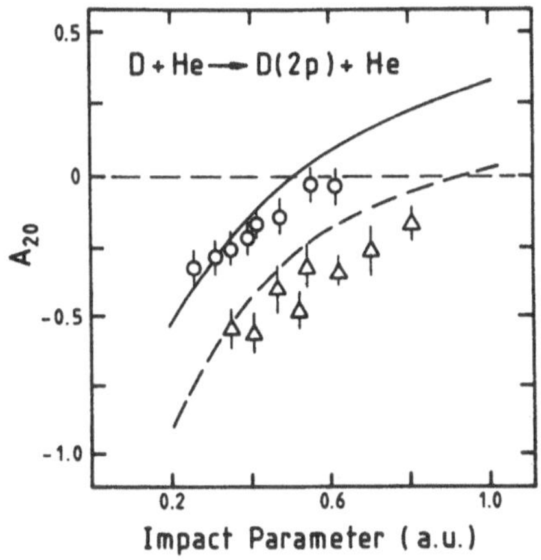

Fig. 4: Differential alignment A_{20} versus impact parameter for D(2p) production in (Δ) 2 keV and (o) 3 keV D–He collisions [2]. Also shown are corresponding calulations of Kimura and Lane [33] (dashed and solid lines).

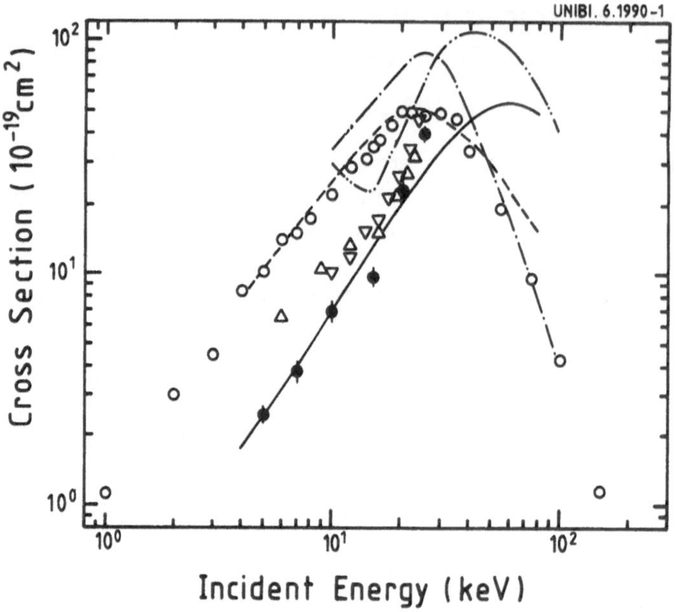

Fig. 5: Total cross sections σ_s and σ_p for H(2s) and H(2p) excitation, respectively, in H^+–He collisions versus incident energy. H(2s): Ref. 18 (•), Ref. 36 (▽), Ref. 37 (△), AO-MO calculations [33] (—), AO+ calculations [8] (– · · – · · –); H(2p): Ref. 35 and Ref. 18 (o), AO-MO calculations [33] (– – –), AO+ calculations [8] (– · – · –).

IV. Two-electron systems

Two-electron systems, for example H–H or $(H–He)^+$, are of considerable interest as they allow an investigation of spin and, hence, of electron-electron interaction effects during the collision. Experimentally, one of the best studied collisions is H^+–He; for this system we have studied charge exchange excitation to H(2s) and H(2p) (Fig. 5). Up till recently there has been a puzzling discrepancy between theoretical predictions of total cross sections for H(2p) excitation and experimental investigations by different authors. It now appears settled that the former experiments were at low velocities significantly in error, probably due to either target gas or ion beam impurities, whereas more recent experiments are in satisfactory agreement with theory [35].

The experimental results for the neutral

$$H\,(1s) + H\,(1s) \rightarrow H\,(2p) + H\,(nl) \tag{9}$$

collision system (Fig. 6) indicate that $H(2p_0)$ and $H(2p_{\pm 1})$ states are populated with about equal probability below $E \ < \ 10$ keV [38]. This results is at variance with recent theoretical calculations [42], where the electron wavefunction has been expanded in travelling atomic orbitals (AO) on both centers. Molecular orbitals are not included in these calculations and, hence, the above mentioned $2p\sigma$–$2p\pi$ united atom rotational coupling mechanism is not properly accounted for. In fact, when comparing these calculations for H–H with previous atomic orbital calculations for H^+–H collisions (e.g., Fig. 2), we note that even on a quantitative basis these calculations predict a rather similar energy dependence of A_{20}. This points to the necessity of including united-atom wavefunctions in the calculations. The inclusion of united-atom couplings is more complicated in H–H collisions, however, where above $E > 1$ keV both the ionization and the capture ($H + H \to H^+ + H^-$) channel become comparable or even larger than the H(2p) excitation channel [39–40]. Recent calculations of Borondo et al. [41] have shown that population of H(2s) and H(2p) states in H–H collisions is in fact influenced by a long-range radial coupling with the $\to H^+ + H^-$ charge exchange exit channel.

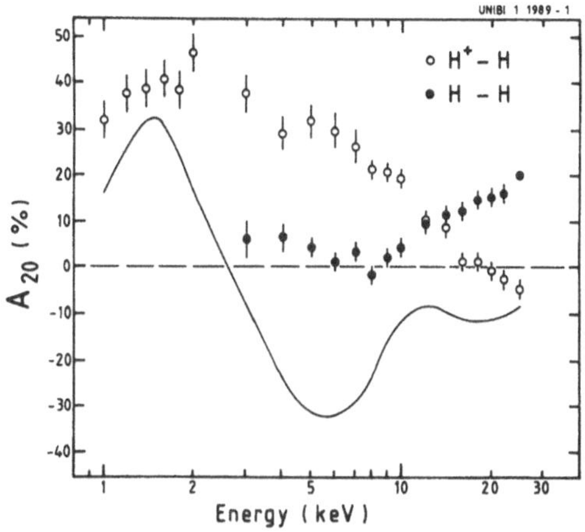

Fig. 6: Integral alignment A_{20} for H(2p) production in (o) H^+–H and (•) H–H collisions versus incident energy [38]. Also shown are theoretical calulations of Singal et al. [42] for H–H (solid line) collisions.

V. Spin effects

The $(H\text{–}He)^+$ complex in a simple way permits an investigation of spin effects in atomic collisions. For this system we have investigated the following two reaction channels

$$\left.\begin{array}{r} He^+ + H \\ \\ H^+ + He \end{array}\right\} \to H\,(2p) + He^+ \tag{10}$$

for which, in an independent-electron picture and at low velocities, we could expect the same alignment A_{20}. As is illustrated in Fig. 7, excitation to the H(2p) state in He^+–H collisions occurs dominantly at small internuclear separations via couplings of the $2p\sigma$ MO with other near-degenerate MOs (for example, $2p\pi$ or $2s\sigma$). In H^+–He collisions, on the other hand, H(2p) excitation is thought to arise from a two-step mechanism. In a first step a long-range radial coupling populates the charge exchange channel He^+–H; then the second step, as before, involves couplings of the 2pσ MO with other high-lying MOs leading to H(2p) excitation. Thus, although the two incident channels are different they appear to proceed along the same intermediate states and through the same couplings; in a one-electron picture, these states and couplings are identical for H^+–He and He^+–H collisions. If this picture were correct, the similarity of H^+–He and He^+–H would be reflected in certain characteristic signatures of the collision. For example, the alignment depends sensitively on the contributions of σ and π states to the excitation process. In a one-electron picture there is little reason why the two collision systems would produce a different alignment. In a two-electron picture, on the other hand, the different spin channels have to be considered: In H^+–He collisions there exists only one (singlet) spin channel, whereas in He^+–H collisions both singlet and triplet channels can contribute. The spin symmetry, however, decides the symmetry of the spatial part of the electron wave function and, hence, its energy.

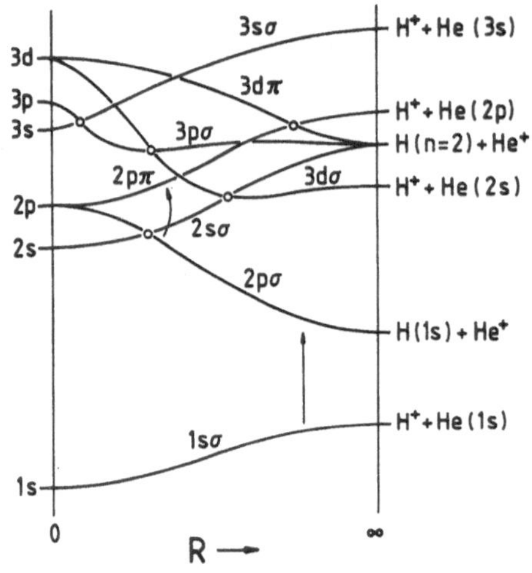

Fig. 7: Schematic (*diabatic*) energy level diagram for $(H\text{–}He)^+$ collisions (singlet case).

In fact, the measured A_{20} differs significantly [43–45] for the two reaction channels (Eq. (10)). In the following we shall discuss the origin of this difference in more detail. This is facilitated by the fact that the $(H\text{–}He)^+$ collision complex has been studied quite thoroughly. In particular, recent differential measurements [13] have demonstrated that for incident H^+–He collisions the 2pσ–2pπ rotational coupling dominates at impact parameters $b \approx 1$ a.u.; this excitation mechanism populates the H($2p_{\pm 1}$) magnetic substates. At smaller b a double-rotational coupling 2pσ–2pπ–2sσ leads to H($2p_0$) excitation [46]. This 2sσ state population is produced by the radial coupling among the 2sσ and 2pσ states around $R \approx 0.4$ a.u.. At low incident energies of a few keV, the integral populations caused by these two couplings are roughly equal; as a consequence, the integral alignment A_{20} is approximately zero. This does not hold for the reverse system He^+–H, where we observe a significantly larger (positive) alignment. It indicates that the above double-rotational coupling mechanism is now of a lesser importance.

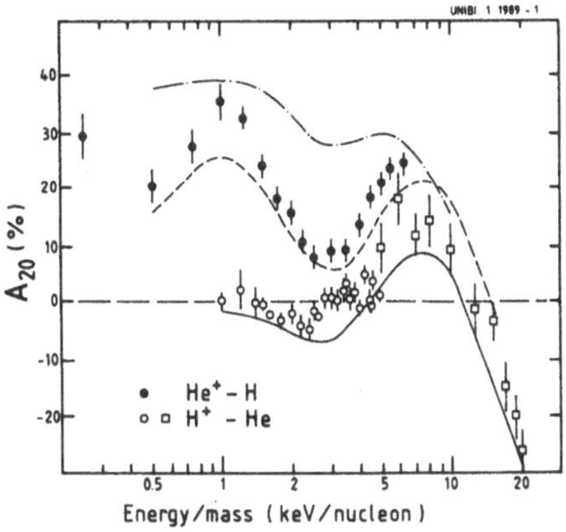

Fig. 8: Integral alignment A_{20} for H(2p) production in H^+–He and He^+–H collisions versus incident energy. Experimental results for He^+–H collisions (•, Ref. 45) and for H^+–He collisions (o, Ref. 43; ▫, Ref. 44) are compared with corresponding theoretical calculations (dashed and solid lines) and with recent calculations for He^+–H by Errera et al. [47] (dash-dotted line).

This experimental result is confirmed by theoretical calculations, in which the molecular states for both singlet and triplet manifolds were obtained by the full configuration interaction method [45]. The most notable difference is that, for the triplet manifold, the separate atom level H^++He(2^3P) lies lower than the H(n=2)+He^+ levels, while the situation is just the opposite for the singlet manifold. In the singlet case, H($2p_{\pm 1}$) population is produced by a 2pσ–2pπ rotational coupling, whereas H($2p_0$) excitation is produced by the 2pσ–2pπ–2sσ coupling. At low energies, these contributions produce σ and π states with comparable strength, and the integral alignment in H^+–He collisions (pure singlet case) is small. The same couplings also operate in the triplet manifold; however, the important difference is that for the triplet channel the 2pσ state does no longer correlate with the He^++H(n=2) outgoing channel. This leads to the result that H(2p) excitation in He^+–H collisions at low incident energies is dominated by a rotational coupling mechanism leading to population of outgoing π-states. It has the consequence that predominantly the H($2p_{\pm 1}$) magnetic substates become populated, which is reflected in the large and positive alignment.

VI. s–p coherence

The measured degree of linear polarization P_l in H^+–He collisions shows a pronounced forward-backward asymmetry with respect to an external electric field which is due to the collision-induced electric dipole moment (Fig. 9). In general, $< D >_z$ is small around incident energies of 5–15 keV and rises towards higher energies. The sign of the measured dipole moment indicates that the captured electron is lagging behind the proton. This result is intuitively acceptable, as it correspond to an electron which is attracted by both the incident proton and by the product helium ion and, therefore, tends to stay in between. This argument seems to be further supported by measurements for the neutral H–He collision system, where significantly smaller dipole moments and of opposite sign are observed (Fig. 9).

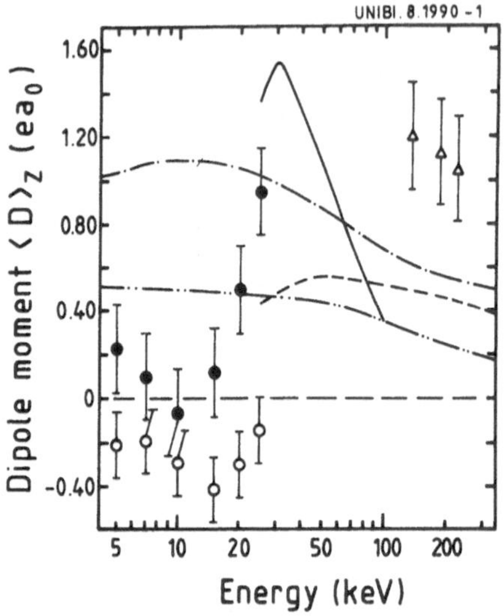

Fig. 9: Dipole moment $< D >_z$ for H(n=2) excitation in (a) H^+–He and (b) H–He collisions versus incident energy. (a): •, Ref. 18; △, Ref. 9; AO+ calculations (solid line, Ref. 8); CDW-PCI calculations (dashed line, Ref. 9); model calculations (dash-dotted line); (b): o, Ref. 18; model calculations (dash-two-dotted line) (see text).

In order to illustrate the origin of the collision-induced dipole moment $< D >_z$ on a more quantitative basis, we first express the coherently excited H(n=2) wavefunction Ψ at $t = 0$ as

$$\Psi = f_+|+\rangle + f_-|-\rangle + f_1|2p_1\rangle + f_{-1}|2p_{-1}\rangle \tag{11}$$

where $|\pm\rangle = \mp\frac{1}{\sqrt{2}}(|2s_0\rangle \pm |2p_0\rangle)$ are the (large-field) Stark states, and where $f_\pm = \mp\frac{1}{\sqrt{2}}(f_{00} \pm f_{10})$ are the corresponding excitation amplitudes. Rewriting Eqs. (3) and (4) we obtain

$$< D >_z = \frac{3\,(\sigma_+ - \sigma_-)}{\sigma_t} \tag{12a}$$

$$< \vec{L} \times \vec{A} >_z = -\frac{2\sqrt{\sigma_+\sigma_-}}{\sigma_t}\sin\bar{\chi}\,, \tag{12b}$$

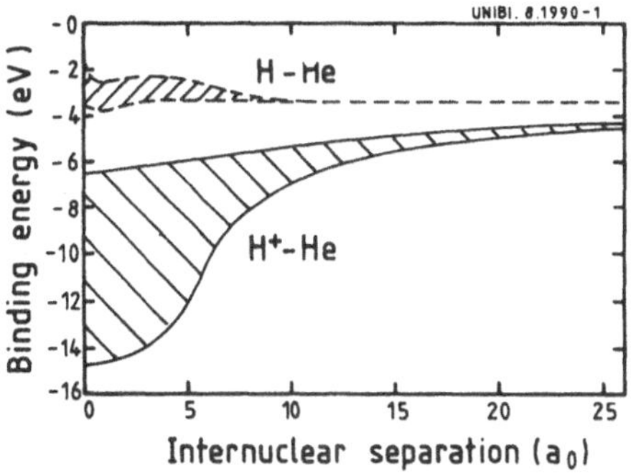

Fig. 10: Binding energies of the relevant quasi-molecular states correlating with the $|+\rangle$ and $|-\rangle$ Stark states in H^+–He (solid lines) and H–He (dashed lines) collisions estimated from calculated molecular energies [33,48].

with $\bar{\chi}$ the average phase between f_+ and f_-. The two Stark states $|\pm\rangle$ possess an electric dipole moment $d = \pm 3ea_0$ (e is the electron charge, and a_0 is the Bohr radius); a net dipole moment hence results, if the two excitation cross sections $\sigma_\pm = <| f_\pm |^2>$ are not equal. Unequal cross sections may result, for example, if the two excitation energies E_+ and E_- are different. Since during the collision the atomic states are considerably perturbed by the presence of the other collision partner, significant energy splittings among formerly degenerate electronic states may arise. At low incident energies, the two collision partners form a transient quasi-molecule [3]; for the singlet $(H\text{–}He)^+$ system we use calculated molecular energies [48] to estimate the energy splitting among the diabatic $2s\sigma$ and $3p\sigma$ molecular states [45] which, for large internuclear separations $R \rightarrow \infty$, correlate with the $|+\rangle$ and $|-\rangle$ Stark states, respectively (Fig. 10). Qualitatively, due to the largely different excitation energies in the $(H\text{–}He)^+$ system, we may expect significantly different cross sections σ_+ and σ_-. As estimates of σ_+ and σ_- we use the Oppenheimer-Brinkman-Kramers (OBK) approximation [49]. At low velocities we take the molecular $2s\sigma$ and $3p\sigma$ energies at $R = 0$ [48]. At large incident velocities, due to the short interaction time, the atomic states experience little perturbation and we have to take the energies of the unperturbed states. In between, we average the molecular energies over the typical response time $\tau_s = \frac{\hbar}{E_s}$ of the atomic system to obtain average energies $\bar{E}_s = \frac{1}{\tau_s} \int_{-\tau_s/2}^{\tau_s/2} E_s(R(t))\, dt$. The estimated dipole moment $< D >_z$ is displayed in Fig. 9 (dash-dotted line) and shows the expected behaviour: it is large at small incident velocities and approaches zero for high velocities. Qualitatively, the predicted behaviour is in reasonable agreement with predictions from the continuum-distorted-wave approximation accounting for post-collision-interaction effects (CDW-PCI, Ref. 9) and with AO+ calculations of Jain et al. [8]. The same analysis applied to the H–He system [33] shows that the relevant energy splitting is considerably smaller; the predicted $< D >_z$ is, hence, less pronounced although still significantly larger than zero (Fig. 9) [50]. Our simple model qualtitatively explains the observed differences between $(H\text{–}He)^+$ and H–He collisions; it is, most likely due to the radial couplings which operate in the transient quasi-molecule but have not been considered here [33,48], not able to predict the correct $< D >_z$ at low velocities.

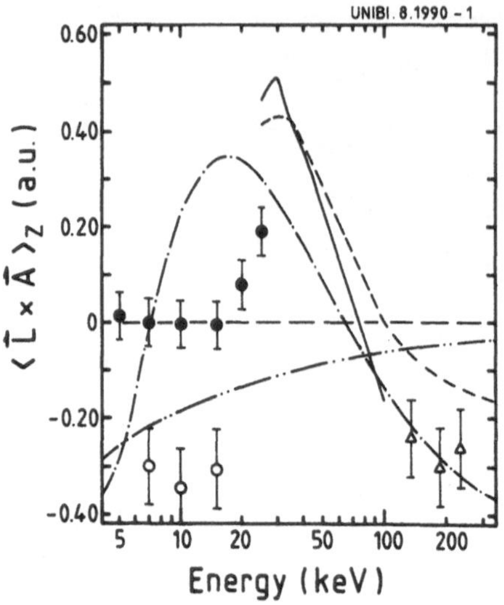

Fig. 11: The moment $< \vec{L} \times \vec{A} >_z$ for H(n=2) excitation in (a) H^+–He and (b) H–He collisions versus incident energy. (a): •, Ref. 18; △, Ref. 9; AO+ calculations (solid line, Ref. 8); CDW-PCI calculations (dashed line, Ref. 9); model calculations (dash-dotted line); (b): ∘, Ref. 18; model calculations (dash-two-dotted line) (see text).

In Fig. 11 we display our results for $< \vec{L} \times \vec{A} >_z$ which were extracted from measurements of the circular polarization P_c in transverse electric fields. As before the dipole moment, $< \vec{L} \times \vec{A} >_z$ is small in H^+–He collisions up till 15 keV and rises towards larger energies. Together with previous measurements of DeSerio et al. [9], these data indicate an oscillatory behaviour whose origin becomes evident if realized that the relative phase $\bar{\chi}$ is at least in part due to the accumulated phase difference $\Delta\phi = \frac{1}{\hbar v} \int_{R_0}^{\infty} \Delta E(R)\, dR$ among the $|+\rangle$ and $|-\rangle$ substates. This quantity is thus sensitive to the very details of the collision process. The fair agreement between experiment and the present model calculations is somewhat fortuitious, however, since a constant phase $\phi_0 = \frac{\pi}{2}$ which possibly arises from the before-mentioned radial couplings was added to the calculated $\Delta\phi$. In H–He collisions, the imaginary part $\Im(\sigma_{sp})$ at the here considered energies is negative and in fair agreement with our calculation.

The author acknowledges helpful discussions with Prof. Dr. H.O. Lutz, Prof. Dr. H. Kleinpoppen, Dr. H. Ast, Dr. D. Dowek, Dr. W. Fritsch, Prof. C.D. Lin, Dr. W. Harbich, Dr. H.J. Lüdde, Dr. M. Kimura, Dipl. Phys. H. Madeheim, Dipl. Phys. O. Plotzke, and Dr. U. Wille. This work was supported by the Deutsche Forschungsgemeinschaft in Sonderforschungsbereich 216 "Polarisation und Korrelation in atomaren Stoßkomplexen".

References

1. R. Hippler, M. Faust, R. Wolf, H. Kleinpoppen, H.O. Lutz, Phys. Rev. A 31, 1399 (1985); Phys. Rev. A 37, 4644 (1987).
2. R. Hippler, W. Harbich, M. Faust, H.O. Lutz, J. Phys. B 21, 103 (1988).
3. U. Fano, W. Lichten, Phys. Rev. Lett. 14, 627 (1965).
4. e.g., *Coherence in Atomic Collision Physics* (H.J. Beyer, K. Blum, R. Hippler, Eds.), New York: Plenum (1988).
5. K. Blum, H. Kleinpoppen, Phys. Reports 52, 203 (1979); Phys. Reports 96, 251 (1983).
6. U. Fano, Rev. Mod. Phys. 29, 74 (1957).
7. T.G. Eck, Phys. Rev. Lett. 31, 270 (1973).
8. A. Jain, C.D. Lin, W. Fritsch, Phys. Rev. A 36, 2041 (1987).
9. R. DeSerio, C. Gonzalez-Lepera, J.P. Gibbons, J. Burgdörfer, I.A. Sellin, Phys. Rev. A 37, 4111 (1988).
10. C.C. Havener, N. Rouze, W.B. Westerveld, J.S. Risley, Phys. Rev. Lett. 53, 1049 (1984); J.R. Ashburn, R.A. Cline, C.D. Stone, P.J.M. van der Burgt, W.B. Westerveld, J.S. Risley, Phys. Rev. A 40, 4885, (1989).
11. e.g., W. Meckbach, R. Vidal, P. Focke, I.B. Nemirowsky, E. Gonzalez-Lepera, Phys. Rev. Lett. 52, 621 (1984).
12. U. Fano, J. Macek, Rev. Mod. Phys. 45, 553 (1973).
13. R. Hippler, In: *Fundamental Processes in Atomic Collision Physics* (H. Kleinpoppen, J.S. Briggs, H.O. Lutz, Eds.), Plenum: New York (1985), p. 181; In: XV ICPEAC Electronic and Atomic Collisions (H.B. Gilbody, W.R. Newell, F.H. Read, A.C.H. Smith, Eds.), North-Holland: Amsterdam (1988), p. 241.
14. G. Lüders, Z. Naturforschung 5 a, 608 (1950); Annalen der Physik, 6. Folge, Band 8, 301 (1951).
15. W. Harbich, R. Hippler, H. Kleinpoppen, H.O. Lutz, Phys. Rev. A 39, 3388 (1989).
16. I.A. Sellin, J.R. Mowat, R.S. Peterson, P.M. Griffin, R. Laubert, H.H. Haselton, Phys. Rev. Lett. 31, 270 (1973).
17. A. Gaupp, H.J. Andrä, J. Macek, Phys. Rev. Lett. 32, 268 (1974).
18. R. Hippler, O. Plotzke, W. Harbich, H. Madeheim, H. Kleinpoppen, H.O. Lutz, Z. Physik D (1991), in press; and to be published.
19. O. Plotzke, U. Wille, R. Hippler, H.O. Lutz, submitted (1990).
20. R. Hippler, H. Madeheim, W. Harbich, H. Kleinpoppen, H.O. Lutz, Phys. Rev. A 38, 1662 (1988).
21. M. Kimura, W.R. Thorson, Phys. Rev. A 24, 1780 (1981).
22. W. Fritsch, C.D. Lin, Phys. Rev. A 26, 762 (1982); Phys. Rev. A 27, 3361 (1983).
23. H.J. Lüdde, R.M. Dreizler, J. Phys. B 15, 2703 (1982).
24. T.G. Winter, C.D. Lin, Phys. Rev. A 29, 567 (1984).
25. R. Shingal, C.D. Lin, private communication (1989).

26. D. Rapp, D. Dinwiddie, J. Chem. Phys. 57, 4919 (1972).
27. C. Gaussorgues, A. Salin, J. Phys. B 4, 503 (1971).
28. R. Shakeshaft, Phys. Rev. A 18, 1930 (1978).
29. K.L. Bell, A.E. Kingston, W.A. McIlveen, J. Phys. B 6, 1237 and 1246 (1973); K.L. Bell, A.E. Kingston, T.G. Winter, J. Phys. B 7, 1369 (1974).
30. S.E. Nielsen, J.S. Dahler, J. Phys. B 12, 2435 (1980); S.E. Nielsen, J.S. Dahler, Phys. Rev. A 28, 3308 (1983).
31. H. Madeheim, R. Hippler, H.O. Lutz, Z. Physik D 15, 327 (1990).
32. B. Van Zyl, M.W. Gealy, Phys. Rev. A 35, 3741 (1987).
33. M. Kimura, N.F. Lane, Phys. Rev. A 37, 2900 (1988).
34. J. Ast, H.J. Lüdde, R. Dreizler, J. Phys. B 21, 4143 (1988).
35. R. Hippler, W. Harbich, H. Madeheim, H. Kleinpoppen, H.O. Lutz, Phys. Rev. A 35, 3139 (1987).
36. D.H. Jaecks, B. Van Zyl, R. Geballe, Phys. Rev. 137, A 340 (1965).
37. E.P. Andreev, V.A. Ankudinov, S.V. Bobashev, Sov.-Phys. JETP 23, 375 (1966).
38. R. Hippler, H. Madeheim, H. Kleinpoppen, H.O. Lutz, J. Phys. B 22, L 257 (1989).
39. G.W. McClure, Phys. Rev. 166, 22 (1968).
40. M.W. Gealy, B. Van Zyl, Phys. Rev. A 36, 3100 (1987).
41. F. Borondo, F. Martin, M. Yanez, Phys. Rev. A 35, 60 (1987); Phys. Rev. A 36, 3630 (1987).
42. R. Shingal, B. H. Bransden, D. R. Flower, J. Phys. B 22, 855 (1989).
43. R. Hippler, W. Harbich, M. Faust, H.O. Lutz, L. Dubé, J. Phys. B 19, 1507 (1986).
44. P.J.O. Teubner, W.E. Kaupilla, W.L. Fite, R.J. Girnius, Phys. Rev. A 2, 1763 (1970).
45. R. Hippler, H. Madeheim, H.O. Lutz, M. Kimura, N.F. Lane, Phys. Rev. A 40, 3446 (1989).
46. J. Macek, C. Wang, Phys. Rev. A 34, 1787 (1986).
47. L. F. Errera, L. Mendez, A. Riera, Z. Physik D, to be published; L. Mendez, private communication.
48. T.A. Green, H.H. Michels, J.C. Browne, M.M. Madsen, J. Chem. Phys. 61, 5186 and 5198 (1974);
A. Macias, A. Riera, M. Yanez, Phys. Rev. A 27, 206 and 213 (1983).
49. H.D. Betz, In: *Methods of Experimental Physics*, Vol 17 (P. Richard, Ed.), New York: Academic (1980), p. 73; the OBK is known to provide reasonable estimates for the relative capture cross sections which are only needed in this case.
50. Here we used $\sigma \propto n^{-3} \propto E^{3/2}$ since $E \sim n^{-2}$, cf., e.g., J. van den Bos, *Ph.D. thesis*, University of Amsterdam (1971).

COLLISIONS AND HALF COLLISONS IN STRONG ELECTROMAGNETIC FIELDS

Teijo Åberg

Laboratory of Physics, Helsinki University of Technology,
02150 Espoo, Finland

The theory of laser-assisted (half)collisions is discussed with emphasis on the relationship between multiphoton ionization and electron capture processes.

1. INTRODUCTION

The most thoroughly studied case among laser-assisted collisions is the electron scattering from atoms in the presence of an electromagnetic (EM) radiation field [1-4]. The essential features of this process is schematically described in Fig. 1 which illustrates elastic scattering within a pulsed laser field. This process can roughly be divided into five stages [1], depending on the temporal and spatial characteristics of the pulse: (i) When an electron with momentum $\mathbf{q}_i$ is impinging on an atom which is exposed to a laser field it may be scattered by the pulse. This is the Kapitza-Dirac effect [5]. (ii) If it enters the spatial region of the laser pulse it starts to exchange photons with the field, i.e. it gets dressed by the photons. In classical terms, it starts to wiggle with an average oscillatory kinetic energy (ponderomotive potential energy Δ) as it proceeds [3].(iii) During the interaction it scatters from a dressed target atom into (iv) a final dressed scattering state. (v) The process is completed when the electron leaves the spatial region of the pulse and enters the detector with momentum $\mathbf{q}_f$.

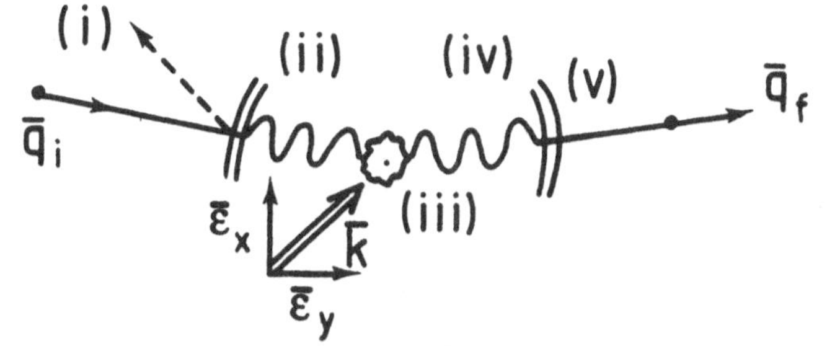

Fig. 1. The five stages (i-v) in electron-atom collisions. The laser pulse propagates in the direction of the wave vector **k**.

Disregarding resonance effects the central result concerning elastic scattering is given by the low-frequency, $\omega \to 0$, approximation of Kroll and Watson [6]. They showed in the case of a spatially uniform and linearly polarized monochromatic field that

$$(d\sigma_j/d\Omega)_{on} = (Q_f/Q_i)|J_j(\alpha_o \mathbf{e}_x \cdot \mathbf{Q})|^2 (d\sigma(\mathbf{Q}_f^*, \mathbf{Q}_i^*)/d\Omega)_{off} \qquad (1)$$

which relates the differential cross section for the laser on to that of laser off. Here j is either the number of absorbed (j<o) or emitted (j>o) photons in $(Q_f^2/2\,m_e) = (Q_i^2/2\,m_e) - j\hbar\omega$. The momentum transfer $\mathbf{Q} = \mathbf{Q}_f - \mathbf{Q}_i$ refers to the dressed states, and $\mathbf{Q}_{i(f)}^* = \mathbf{Q}_{i(f)} + (j\hbar\omega m_e \alpha_o \mathbf{e}_x)/(\alpha_o \mathbf{e}_x \cdot \mathbf{Q})$. The Bessel function J_j is the j th coefficient of the Fourier expansion of the time-dependent part of the overlap between the final and initial dressed scattering states. It depends on the important parameter $\alpha_o = eF/m_e\omega^2$ (e<o). The absolute value $|\alpha_o|$ is the classical radius of oscillation of the electron in a periodic field with field strength F.

Equation (1) holds to all orders in the scattering potential and has thus received considerable attention with generalizations not only to elastic [4,7] but also to inelastic [8] processes. Its validity has been qualitatively demostrated by Weingartshofer et al [9] who measured low-energy electron scattering from argon atoms in the presence of CO_2 laser pulses. They showed that the experimental scattering cross section $(d\sigma/d\Omega)_{on}$ exhibited side bands corresponding to both j>o and j<o in accordance with (1). These results also implicitly demonstrate the possibility of subthreshold excitation of electrons with the assistance of the laser [10].

The situation is less clear concerning collisions in which the impinging electron is replaced by an ion. Exploratory work has been made by Mittleman [2] with emphasis on charge transfer. In the low velocity regime it is usually described to be a consequence of energy-level crossings in the temporary collisional ion-atom complex. It is hence clear that since emission and absorption of photons can both alter existing crossings and induce new ones the electron transfer probabilities may be drastically influenced by the field, especially in the case of multiple electron capture.

If the first two phases in electron scattering are left out there still remains the possibility that instead of scattering the laser pulse ionizes the atom. The electron enters and leaves the laser pulse as described by stage (iv) and (v) in Fig. 1. We have the situation of a half collision or multiphoton ionization (MPI) which has been vividly studied in recent years [11]. Even if MPI appears simpler than laser-assissted electron scattering, experiments have revealed an unexpectedly large body of strong-field phenomena such as above-threshold ionization (ATI) [12] and multiple ionization [13]. In ATI the side bands predicted by (1) are replaced by almost equally spaced photoelectron peaks, corresponding to absorption of photons in excess of the minimum number that is required to ionize the atom.

From the point of view of ion-atom collisions MPI is an interesting process since it resembles electron capture (EC) in many ways. As demonstrated below strong-field MPI may be viewed in the long-pulse limit as a two-step process in which the electron is first captured by the field and then expelled by the field. It is shown that a description [14] of the first step is obtained within the framework of time-independent scattering theory by replacing the projectile in the theory of EC by the photon field. It is indicated that this relationship is not only formal but may have observable consequences.

We proceed by introducing the concept of strong laser fields. Before we consider the relationship between MPI and EC we shall establish the link between the quantum-mechanical time-independent and the semiclassical time-dependent description of the laser field in a scattering situation. It serves as a background to the use of formal time-independent scattering theory in our comparison between MPI and EC. We shall conclude with a brief analysis of the electron-field decoupling mechanisms [15] at the end of the pulse.

2. STRONG FIELDS

Strong fields usually mean peak intensities which are close to I_0=1 a.u.=3.5·10^4 TWcm^{-2}. This is the time-averaged intensity which corresponds to the field strength (F_0=0.51 TVm^{-1}) the electron experiences at the distance of the Bohr radius a_0 from the nucleus in the hydrogen atom. What is strong depends however on the frequency as well as the binding energy of the electron.

In the following "strong" means peak intensities at which the perturbative description of the interaction between the laser and the outermost atomic ground-state electrons ceases to be valid. Signs for the break-down are for example the appearance of photoelectron peaks that correspond to absorption of more than the minimum number of photons (ATI) or the divergence of the perturbation expansion of the ac Stark shift [16]. Combining existing evidence indicates that the perturbative approach becomes questionable when the ratio between the quiver energy or the ponderomotive potential energy

$$\Delta = e^2F^2/4m_e\omega^2 = (1/4)eF\alpha_0 \quad (2)$$

and the photon energy $\hbar\omega$ is larger than or of the order of 0.1. This criterion corresponds to about I=12 TWcm^{-2} for the Nd-YAG wavelength λ=1.064 μm in agreement with a suggestion by Reiss [17]. The solid line in Fig. 2 represents the border line between the perturbative and nonperturbative regime for existing laser frequencies. It corresponds to $|\alpha_0|$ ranging from about 10a_0 to a_0 with increasing ω.

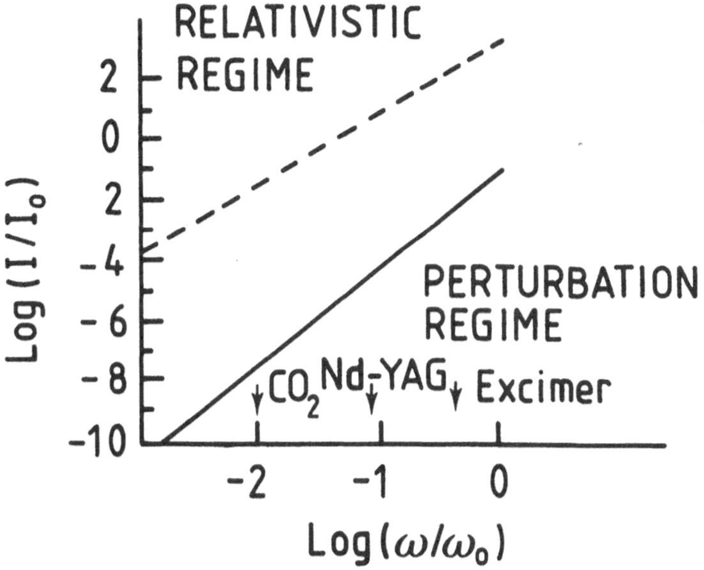

Fig. 2. The strong-field regime is represented by the area above the solid line (I_0=1 a.u., ω_0=0.5 a.u.).

In the strong-field regime relativity including retardation effects may become important. The region below the dashed line in Fig. 2 corresponds to the requirement that the instant oscillatory velocity should not exceed 0.5 c, i.e. the kinetic energy should be less than about 0.1 m_ec^2 in order to have a nonrelativistic situation. This is however not the only constraint since the electron may easily reach velocities close to c in the propagation direction of the pulse [18].

Electric field strengths achieved by present-day lasers often reach or exceed the strength of the transient field by which slowly moving highly-charged ions affect the electrons of the target atom at distances of closest approach. The pulse durations are comparable to typical collisions times which range from a few femtoseconds to picoseconds.

3. THE MANY-PHOTON CORRESPONDENCE PRINCIPLE

In view of the foundation [19] of the time-dependent impact parameter method it is of interest to know under what circumstances the formal time-independent scattering theory [20] and the semiclassical time-dependent S-matrix theory [17,21] yield identical transition rates for laser-assisted (half)collisions [1-3]. We shall demonstrate this connection in the case of a monochromatic elliptically polarized EM field [22].

Suppose that the electron is in a stationary scattering state $\psi=\Psi\exp(-iEt)$ which is a solution of the Schrödinger equation (we use atomic units)

$$[H_\gamma + (1/2\, m_e)\hat{\mathbf{p}}^2 + U(\mathbf{r})]\psi = i\,(\partial\psi/\partial t) \tag{3}$$

with appropriate boundary conditions regarding the scattering by the atomic potential $U(\mathbf{r})$ in the presence of the photon field $H_\gamma = (\omega/2)(a^\dagger a + aa^\dagger)$. In (3) $\hat{\mathbf{p}} = -i\nabla - e\mathbf{A}$ (e<o) is the canonical momentum with

$$\mathbf{A}(-\mathbf{k}\cdot\mathbf{r}) = g[\boldsymbol{\varepsilon}\exp(i\mathbf{k}\cdot\mathbf{r})a + \boldsymbol{\varepsilon}^*\exp(-i\mathbf{k}\cdot\mathbf{r})\,a^\dagger)] \tag{4}$$

where $g = (2V_\gamma\omega)^{-1/2}$ depends on an arbitrary normalization volume V_γ. The polarization properties are described by

$$\boldsymbol{\varepsilon}\cdot\boldsymbol{\varepsilon} = \boldsymbol{\varepsilon}^*\cdot\boldsymbol{\varepsilon}^* = \cos\xi, \quad \boldsymbol{\varepsilon}\cdot\boldsymbol{\varepsilon}^* = 1 \tag{5}$$

such that $\xi=0$ corresponds to linear polarization and $\xi=\pm(\pi/2)$ to circular polarization. It follows from (3) by applying the transformation $\psi_t = \exp(iH_\gamma t)\psi$ and by projecting on the coherent state $|\alpha\rangle = D^\dagger|0\rangle$, where the shift operator $D=\exp[\alpha^* a - \alpha a^\dagger]$ depends on the free parameter α, that

$$\sum_{n=0}^{2} \langle 0\,|\,DH_tD^\dagger|n\rangle\,\langle n|D\psi_t\rangle = i(\partial\langle 0|D\psi_t\rangle/\partial t) \tag{6}$$

Here $H_t = (1/2\, m_e)\mathbf{p}(t)^2 + U(\mathbf{r})$ contains the time-dependent canonical momentum $\mathbf{p}(t)$ which is obtained from $\mathbf{p}$ in (3) by replacing $-\mathbf{k}\cdot\mathbf{r}$ by $kx=\omega t-\mathbf{k}\cdot\mathbf{r}$ in (4). Equation (6) is an exact equation for ψ_t in which

$$\Psi = \sum_{j=-N}^{\infty} c_j(\mathbf{r},\mathbf{k},\boldsymbol{\varepsilon})\,|\,N+j\rangle \tag{7}$$

is expressed as a superposition of the photon number states $|N+j\rangle$.

The semiclasscial Hamiltonian $H_c=H_c(t)$ is defined by $H_c=(1/2\, m_e)\hat{\mathbf{p}}_c^2 + U(\mathbf{r})$, where $\hat{\mathbf{p}}_c$ contains the classical vector potential

$$A_c(kx) = \Lambda\,[\boldsymbol{\varepsilon}\exp(-ikx) + \boldsymbol{\varepsilon}^*\exp(ikx)] \tag{8}$$

It can be shown by explicit evaluation that in (6) $\langle\alpha|H_t|\alpha\rangle=\langle 0|D\,H_tD^\dagger|0\rangle = H_c+(e^2/2\,m_e)g^2$ provided $\Lambda=g\alpha$. The corresponding n=1,2 terms are proportional to g and g^2, respectively [22]. The projected wave function $\psi_c = \langle\alpha|\psi_t\rangle = \langle 0|D\,\psi_t\rangle$ is thus a solution of the semiclassical Schrödinger equation

$$H_c\,\psi_c = i(\partial\psi_c/\partial t) \tag{9}$$

in the limit $g\rightarrow 0$ ($\alpha\rightarrow\infty$, Λ constant).

This limit can be associated with a large N in (7) by putting $\alpha=\sqrt{N}$, i.e. $\langle\alpha|H_\gamma|\alpha\rangle = (N+1/2)\omega$. In this case $\langle\psi|H_\gamma|\,\psi\rangle\cong N$ which means that the average numer of transferred photons in a collision process, described by j, is much smaller than N. The corresponding eigenenergy is $E=(N+1/2)\,\omega + E_e$, where E_e is the interaction energy, and the corresponding semiclassical solution of (9) is

$$\psi_c^\infty = \exp(-iE_e t)\sum_{j=-\infty}^{+\infty} c_j(\mathbf{r},\mathbf{k},\boldsymbol{\varepsilon})\exp(ij\omega t) \tag{10}$$

which fulfils the exact equation (6) to the order of N^{-1}. It is obtained from $\psi_c = \langle\alpha|\psi_t\rangle$ by an application of Stirling's formula to the coefficients $\langle\,\alpha|N+j\,\rangle$ [22].

Our derivation shows that the stationary wave function (7) and the semiclassical wave function (10) have identical expansion coefficients c_j provided that they correspond to the same asymptotic energy and boundary conditions. The formal time-independent scattering theory thus leads in the large-photon-number limit to the same transition rates as the semiclassical approach in the monochromatic limit (8). The multimode case with various photon statistics as well as the relationship to the impact parameter method require further investigations.

To give an example consider phase (ii) and (iv) in Fig. 1 for which U=0 in (3). The Volkov solution ψ_v which approaches the plane wave $\chi(\mathbf{r})=(2\pi)^{-3/2}\exp(i\mathbf{P}\cdot\mathbf{r})$ ($\mathbf{P} = \mathbf{q}_{i(f)}$) in the time-dependent picture of adiabatic switch on and off of V is represented by

$$c_j = \chi(\mathbf{r})\exp[i\,(Z\,\mathbf{k} - j\mathbf{k})\cdot\mathbf{r} - ij\phi_\xi]\,J_j^*(\zeta,\eta,\phi_\xi) \tag{11}$$

in (7) and (10) [23]. These large-photon-number solutions correspond to the energy $E_e=P^2/2\,m_e + \Delta$, where $\Delta= Z\omega$ is given by (2). The generalized Bessel functions

$$J_j(\zeta,\eta,\phi_\xi) = \sum_{m=-\infty}^{+\infty} J_{-j-2m}(\zeta)\,J_m(\eta)\exp(2\,im\phi_\xi) \tag{12}$$

depend on three parameters, $\zeta=(2|e|\Lambda/m_e\omega)|\mathbf{P}\cdot\boldsymbol{\varepsilon}| = 2|\alpha_o||\mathbf{P}\cdot\boldsymbol{\varepsilon}|$, $\eta=(Z/2)\cos\xi$, and $\phi_\xi=\arg(\mathbf{P}\cdot\boldsymbol{\varepsilon})$ in the general case of elliptical polarization.

4. RELATIONSHIP BETWEEN MULTIPHOTON IONIZATION AND ELECTRON CAPTURE

It is customary to consider MPI as a reaction in which the removal of the electron inside the pulse can be treated separately from the last stage of the process (see Fig. 1) during which it escapes the spatial region of the field [11,14]. In this approximation MPI can be viewed as a pick-up reaction, i.e. the electron is captured by the field in analogy to EC in ion-atom collisions, where it is captured by the projectile.

The theory is that of a three-body rearrangement collision and it can be summarized for EC [24] with reference to Fig. 3 as follows. We consider an electron in the field of the projectile (P) and target

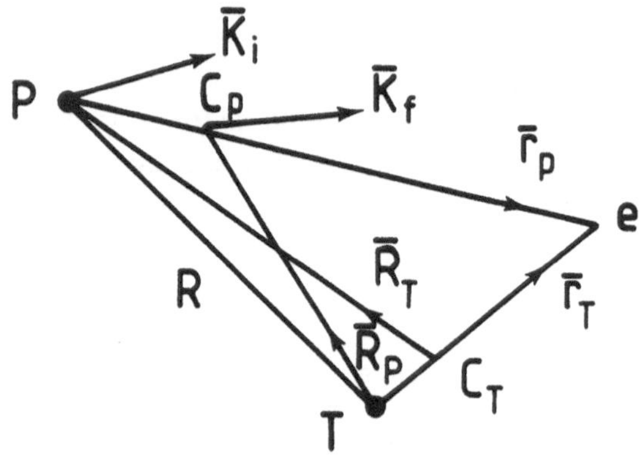

Fig. 3. Coordinate system defining the vectors in (13) and (14). The center-of-mass of T(P)+e is denoted by $C_{T(P)}$.

(T) nuclei. The electron is initially bound to T by the potential V_T and becomes bound to P by V_P. The total Hamiltonian is $H=T_N+T_e+V_T+V_P+V_{PT}$, where T_N and T_e are the nuclear and electronic kinetic energy operators, respectively and where V_{PT} is the repulsive Coulomb potential between P and T. The asymptotic energies are given by $E=\varepsilon_i+(1/2\ \mu_i)K_i^2= \varepsilon_f+ +(1/2\ \mu_f)K_f^2$, where μ_i (μ_f) is the initial (final) reduced mass of the whole three-body complex. According to Taulbjerg et al [24] the on-the-energy-shell **T** matrix element is

$$T_{fi} = \langle \Phi_f(\mathbf{r}_p)\, \chi^{-}{}_{K_f}(\mathbf{R}_P)\, |U| \Psi_i^{+} \rangle \tag{13}$$

where

$$\Psi_i^{+} = (1 + G^{+}V) |\, \Phi_i(\mathbf{r}_T)\, \chi^{+}{}_{K_f}(\mathbf{R}_T) \rangle \tag{14}$$

describes the disintegration of the target due to P while it is scattered by T. In the potential $V=V_P+V_{PT}-U_i$, $U_i=U_i(\mathbf{R}_T)$ is the initial-channel distortion potential and in $U=V_T+V_{PT}-U_f$, $U_f=U_f(\mathbf{R}_P)$ is the final-channel distortion potential. They account for elastic scattering of P before and after the EC event. The corresponding wave functions are χ^+ in (14) and χ^- in (13), which are associated with the initial and final electronic wavefunctions Φ_i and Φ_f, respectively.

If we substitute P by the photon field, then U becomes the atomic potential and V the photoelectron interaction $V=i(e/m_e)A(-\mathbf{k}\cdot\mathbf{r})\cdot\nabla + (e^2/2\ m_e)\ A^2(-\mathbf{k}\cdot\mathbf{r})$ in (3). The scattering wave function $|\Phi_i\, \chi^+\rangle$

in (14) must be replaced by the direct product $|\Phi_i, l>$ of Φ_i and the number state $|l>$ since the projectile is initially <u>bare</u>. The wavefunction $|\Phi_f \chi^->$ in (13) must however be replaced by the Volkov solution Φ_v which is obtained from (7) by substituting the coefficients (11). The reason is that in EC the electron is now dressed by P to all orders since $\chi^- = \chi^-_{K_f}(R_P)$ depends on R_P which according to Fig. 3 refers to the center-of mass motion of the projectile <u>atom</u>. We have

$$T_{fi}^{PICK} = <\Psi_v|U|\Psi_i^+> = T_{fi}^{PICK} + <\Psi_v|UG^+V|\Phi_i, l> \tag{15}$$

where the Green's operator $G^+ = (E - H + i\varepsilon)^{-1}$ refers to H in (3). If U is a short-range potential it is acceptable to replace T_{fi}^{PICK} by

$$T_{fi}^{PICK} = <\Psi_v|U|\Phi_i, l> = <\Psi_v|V|\Phi_i, l> \tag{16}$$

This is the famous and debated [25] strong-field approximation by Keldysh [26] which has been further developed by Faisal [27] and in particular by Reiss [25,28]. It has usually been treated within the framework of the semiclassical time-dependent S-matrix theory, and corresponds in this picture to an adiabatic switching-on of the pulse.

Whether the formal analogy between the one-electron theory of EC and MPI has any deeper meaning depends on whether the projectile-electron interaction can be simulated by a linear superposition of transient EM fields acting on the electron with various frequencies and amplitudes in the pertinent part of the phase space of the collision. The correlation of collisions with radiation is not a new problem [29], and some fragmentary results exist in the case of EC vs. MPI [30]. Here we shall only point at some profound similarities between EC and MPI.

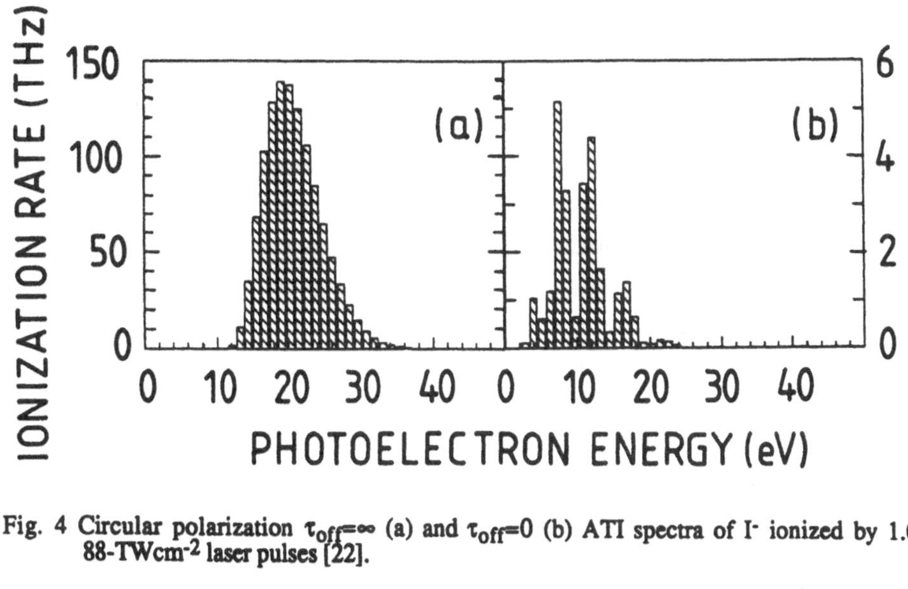

Fig. 4 Circular polarization $\tau_{off}=\infty$ (a) and $\tau_{off}=0$ (b) ATI spectra of I^- ionized by 1.064 -μm 88-TWcm^{-2} laser pulses [22].

First, there is a similarity between ATI distributions and final-state distributions in charge transfer from Rydberg atoms. Figure 4a shows a calculated ATI spectrum for MPI of I^- by circularly polarized Nd-YAG light at I=88 TWcm^{-2} [22]. The calculation is based on (16). The spectrum is plotted as a function of the photoelectron energy but could equally well be given as a function of the number of absorbed photons which in this case is about 20 at the maximum. A distribution like this is strikingly similar to distributions [31] of final n_f states in reactions like $Na^+ + Na(n_il_i) \rightarrow Na(n_fl_f)+Na^+$, where typically n_i=23-35 (l_i=0,2) and the kinetic energy of Na^+ ranges from 0.5 keV to 2 keV. The reason may be that the spread of n_f occurs during the initial and final phases of the collision through the projectile-field induced Stark mixing whereas the EC itself occurs resonantly at some distance of closest approach [32].

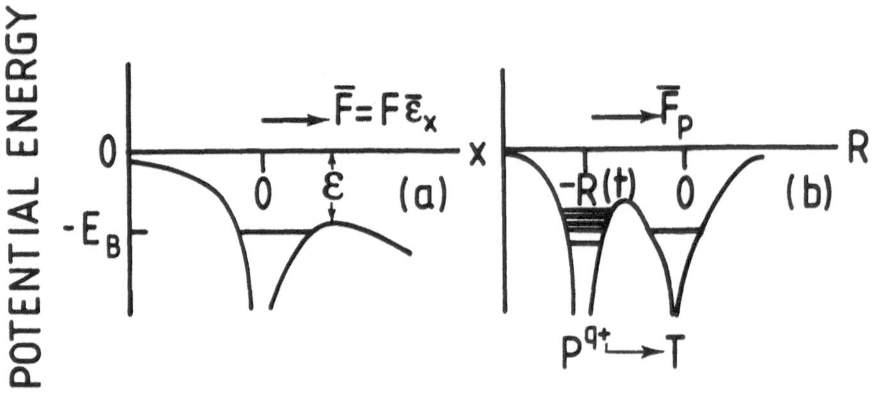

Fig. 5. Mechanism of tunneling or overbarrier transitions in (a) MPI and (b) EC.

Second, there is some similarity [33] between charge-state distributions produced by MPI and by slow collisions of the type

$$P^{q+} + T \rightarrow P^{(q-k)+} + T^{n+} + (n-k)e \qquad (17)$$

in which the target atom T may loose more electrons (n>k) than is captured by P. In the limit of low frequencies MPI may be associated with either tunneling [34] or an overbarrier transition [35] as shown in Fig. 5a. In the former case multiple ionization is associated with the condition $\varepsilon < E_B(q)$ at each stage of ionization. Here $E_B(q)$ is the Stark-shifted binding energy of the sequentally stripped ion of charge $q(\geq 0)$. In the latter case the criterion $\varepsilon = E_B(q)$ together with the assumption $W= U+V=-qe^2/4\pi\varepsilon_0 x+eEx$ (see Fig. 5a) leads to a simple prediction of the appearance intensity at which the laser pulse starts to produce ions of charge q. Figure 5b shows the corresponding situation which prevails close to the internuclear axis for a high q-value in a slow reaction of type (17). Also in this case we may imagine that the electron is captured by P^{q+} to a high Rydberg state or continuum either by tunneling [36] or by an overbarrier transition [37]. In the latter case the same criterion as applied to MPI leads to a critical internuclear distance R_c at which the electron may cross over the barrier from T to P^{q+}. With decreasing charge of P and increasing n in reaction (17) the corresponding R_c's become smaller. We have a sequence of R_c's on the way in and also out [38] which lead to multiple capture and recapture followed by autoionization.

Third, with increasing frequency and intensity we reach in MPI a nonperturbative region in which ionization is severely suppressed [39]. As shown by Gavrila et al [40] there is a limit of stabilization which manifests itself in a deformation of the time-averaged charge distribution rather than in ionization. Calculations for H(1s) [40] shows that the distribution is stretched and shifted in the direction of the linearly polarized field between 0 and $2|\alpha_0|$. This situation should be compared with the low probability of EC at high projectile velocities. It is attributed to energy and momentum matching which has to be fulfilled before the electron can jump from T to P^{q+} [41].

There is thus a correspondence between MPI and EC in the sense that these processes are phenomenologically similar when the laser frequency is comparable to the cut-off Fourier frequency $\omega_0 \cong v\, b^{-1}$. Here v is the projectile velocity and b a representative impact parameter of the collision under scrutiny [29].

5. THE ESCAPE PROBLEM

So far we have neglected the last stage of MPI which is the escape of the photoelectron from the spatial region of the laser pulse into the detector (see Fig. 1). According to [15] we have

$$E_{obs} = j\omega - \kappa\,\Delta - E_B - \Sigma \tag{18}$$

where E_{obs} is the kinetic energy of the photoelectron as observed by the detector. Here j is the number of photons that ionize the electron inside the field. The coefficient κ describes the fraction of the quiver energy Δ [see (2)] that is converted into photons when the electron is detached from the field. It depends primarily on the time τ_{off} it takes for the pulse to decay after the electron has been released, but also on j, ω, and I. The difference between the initial- and final-state ac Stark shifts is denoted by Σ and it is usually small. Figure 6 illustrates the escape process in the case of a monomode field. According to Sec. 3 the electron energy E_e inside the field consists of the kinetic energy $E_{drif} = P^2/2\, m_e$ and the "potential" energy Δ.

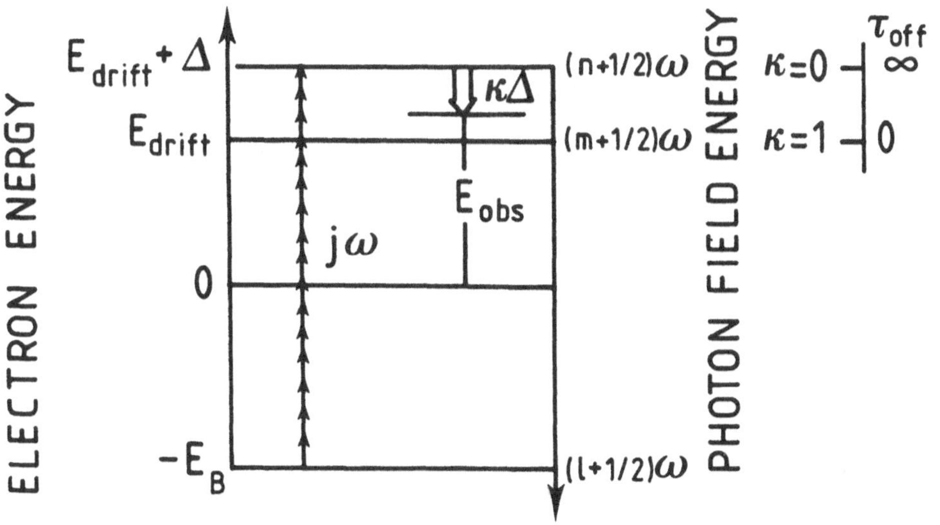

Fig. 6. Energy-balance diagram for the switching-off of a monomode laser.

We have two limiting cases depending on τ_{off}: (i) If $\tau_{off} = \infty$, then κ=0 and Δ is entirely converted into kinetic energy of the electron. In classical terms the elctron is expelled by grad Δ, where Δ is weakly dependent on time but spatially inhomogeneous due to I[15]. Consequently $E_{obs} = E_{drif} + \Delta \cong j\omega - E_B$. Figure 4a shows a typical ATI spectrum in a situation like this. (ii) If $\tau_{off}=0$, then Δ is entirely converted into photons. Consequently κ=1 and $E_{obs}=E_{drift} \cong j\omega - E_B - \Delta$. As indicated in Fig. 6 $\Delta = Z\omega$ must be an integer number of ω for this to happen if a strictly monochromatic field is forced to suddenly vanish. Figure 4b shows that the corresponding ATI spectrum is not only red-shifted with respect to the $\tau_{off} = \infty$ ATI but may exhibit interference effects as well. The interference pattern stems from the overlap factor $[J|^2$ in

$$(dw/d\Omega)_j^{BREAK} = |J_Z(\zeta,\eta,\xi)|^2 \, (dw/d\Omega)_j^{PICK} \qquad (19)$$

where the BREAK rate is obtained by assuming that the electron is neither interacting with the atom (ion) nor the field in the remote future [23]. The PICK rate for j photon absorbtion corresponds to (15), and $J_Z \propto \langle \psi_v | \chi \rangle$ is given by [12]. It represents the sudden escape of the electron in accordance with sudden approximation [22]. Equation (19) is based on the assumption of an adiabatic switching-on of the pulse. Depending on the polarization this assumption has limited validity in the short-pulse regime which may reveal both Stark-induced resonances [42] and satellite structures [43] in the ATI spectrum.

In conclusion, the main purpose of this rather selective review has been to point out some similarities between multiphoton ionization and electron capture. Whether the uncovered relationship leads to a deeper understanding of these processes remains to be seen.

ACKNOWLEDGEMENTS

I would like to thank my co-workers B Crasemann, D-S Guo, X-D Mu and J Ruscheinski at University of Oregon for their contributions to this study. This work has been supported by the Academy of Finland.

REFERENCES

1. L Rosenberg, Adv. At. Mol. Phys. **18**, 1 (1982).
2. M H Mittleman, *Introduction to the Theory of Laser-Atom Interactions* (Plenum, New York 1982).
3. F H M Faisal, *Theory of Multiphoton Processes* (Plenum, New York 1987).
4. P Francken and C J Joachain, J. Opt. Soc. Am. B **7**, 554 (1990).
5. P H Bucksbaum, M Bashkansky, and T J McIlrath, Phys. Rev. Lett **58**, 349 (1987).
6. N M Kroll and K M Watson, Phys. Rev. A **8**, 804 (1973).
7. P S Kristic and D B Milosevic, J. Phys. B **20**, 3487 (1987).
8. S Geltman and A Maquet, J. Phys. B **22**, L419 (1989); see also B N Chichkov, J. Phys. B **23**, L333 (1990).
9. A Weingartshofer, J K Holmes, J. Sabbagh, and S L Chin, J. Phys. B **16**, 1805 (1983); see also Ref. 4.

10. B Wallbank, J K Holmes, and A Weingartshofer, J. Phys. B **22**, L615 (1989).
11. P H Bucksbaum in *Atomic Physics 11*, eds. S Haroche, J C Gay, and G Grynberg (World Science, Singapore 1989) p. 311.
12. P Agostini, F Fabre, G Mainfray, G Petite and K N Rahman, Phys. Rev. Lett. **42**, 1127 (1979); P Kruit, J Kimman, and M J Van der Wiel J. Phys. B **14**, L597 (1981); see also Ref. 11 and M H Mittleman in *The Physics of Electronic and Atomic Collisions*, eds. A Dalgarno, R S Freund, P M Koch, M S Lubell, and T B Lucarto, AIP Conf. Proc. 205 (AIP, New York 1990), p. 184 .
13. A L'Huillier, L A Lompré, G Mainfray, and C Manus, Phys. Rev Lett **48**, 1814 (1982); see M Crance, Phys. Rep **144**, 117 (1987) for a review.
14. X Mu, Phys. Rev. A **42**, 2944 (1990).
15. J Kupersztych, Phys. Scr. **42**, 51 (1990).
16. L Pan, K T Taylor, and C W Clark, Phys. Rev. Lett **61**, 2673 (1988).
17. H R Reiss in *Atoms in Strong Fields*, ed. C A Nicolaides, C W Clark, and M H Nayfen (Plenum, New York 1990) p. 425.
18. M H Mittleman in Ref. 12.
19. J S Briggs, Rep. Progr. Phys. **39**, 217 (1976).
20. M L Goldberger and K M Watson, *Collision Theory* (Wiley, New York 1964).
21. See e.g. *Semiclassical descriptions of atomic and nuclear collisions*, eds. J Bang and J de Boer (North-Holland, Amsterdam 1985).
22. T Åberg, D S Guo, J. Ruscheinski, and B Crasemann, to be published.
23. D S Guo, T Åberg, and B Crasemann, Phys. Rev. A **40**, 4997 (1989).
24. K. Taulbjerg, R O Barrachina, and J H Macek, Phys. Rev. A **41**, 207 (1990); K. Taulbjerg, Phys. Scr. **42**, 205 (1990).
25. H R Reiss, Phys. Rev. A **42**, 1476 (1990).
26. L V Keldysh, Zh. Eksp. Teor. Fiz. **47**, 1965 (1964) [Sov. Phys.- JETP **20**, 1307 (1965)].
27. F H M Faisal, J. Phys. B **6**, L89 (1973).
28. H R Reiss, Phys. Rev. A **22**, 1786 (1980).
29. E J Williams, Kgl. Dan. Vid. Selsk. Mat. Fys. Medd. **13**, No. 4 (1935).
30. E L Duman, L I Men'shikov, and B M Smirnov, Zh. Exsp. Teor. Fiz. **76**, 516 (1979) [Sov. Phys.-JETP **49**, 260 (1979)].
31. R G Rolfes and K B MacAdam, J. Phys. B **15**, 4591 (1982); K B MacAdam, L G Gray, and R G Rolfes, to be published.
32. T Åberg, A Blomberg, and K B MacAdam, J. Phys. B **20**, 4795 (1987).
33. T Åberg, A. Blomberg, J. Tulkki, and O Goscinski, Phys. Rev. Lett. **52**, 1207 (1984).
34. K Codling, L J Frasinski, and P A Hatherly, J. Phys. B **22**, L321 (1989).
35. S.August, D. Strickland, D D Meyerhofer, S L Chin, and J H Eberly, Phys. Rev. Lett. **63**, 2212 (1989); see also R Shakeshaft, R M Potvliege, M Dörr, and W E Cooke, Phys. Rev. A **42**, 1656 (1990).
36. T P Grozdanov and R K Janev, Phys. Rev. A **17**, 880 (1978).
37. A Bárány, G Astner, H Cederquist, H Danared, S. Huldt, P. Hvelplund, A Johnson, H Knudsen, L Liljeby, and K-G Rensfelt, Nucl. Instrum. Meth. B **9**, 397 (1985).
38. A Niehaus, J. Phys. B **19**, 2925 (1986).

39. Q Su, J H Eberly, and J Javanainen, Phys. Rev. Lett. **64**, 862 (1990).
40. M Pont, N R Walet, and M Gavrila, Phys. Rev. A **41**, 477 (1990); see also M Gavrila in *Atoms in Unusual Situations*, ed. J-P Briand (Plenum, New York 1986), p. 225.
41. J S Briggs in Ref. 21, p. 183.
42. R R Freeman, P B Bucksbaum, H Milchberg, S Darack, D Schumacher, and M E Geusic, Phys. Rev. Lett. **59**, 1092 (1987); P. Agostini, P Breger, A L'Huillier, H G Muller, G Petite, A Antonetti,and A Migus, Phys. Rev. Lett. **63**, 2208 (1989).
43. V C Reed and K. Burnett, Phys. Rev. A **42**, 3152 (1990).

Search for an alignment in double capture processes between multicharged ions and helium below 8 keV/amu.

A.Bordenave-Montesquieu, P.Moretto-Capelle, P.Benoit-Cattin, and A.Gleizes
Centre de Physique Atomique, URA CNRS 770, Université P.Sabatier, 118 Route de Narbonne, 31062 Toulouse Cedex, France

At low collision energy the double capture (DC) process occurs in the quasimolecule $(AHe)^{q+}$, where A^{q+} is the incident multicharged ion with charge q. Mainly Σ and Π diexcited exit channels are populated around the crossings with the Σ entrance channel through radial and rotational couplings (may be Δ ones can be produced through an intermediate level); they are correlated at infinite internuclear distance to the atomic $|M_L|$=0,1,(2) sublevels respectively. Depending on the experimental conditions the relative importance of these primary couplings can induce different populations of the sublevels. Moreover, 3l3l' and 3l4l' high angular momentum states involve values of $|M_L|$ up to 4 and 5 respectively, which cannot be populated in this way. Therefore an anisotropic angular distribution of the electrons ejected by autoionisation of these states is expected. These schematic considerations are somewhat obscured by secondary couplings at large internuclear distance which redistribute the initial populations among all the sublevels. Then it appears that the electron angular distributions must be measured to give some insight into the mechanisms of double capture; of course, the integration of the angular distributions also allows to deduce accurate total DC cross sections.

Up to now the $|M_L|$ population of states produced by electron transfer on multicharged ions has not attracted attention of experimentalists in this energy range. The very scarce data available at this time concern single capture processes; they have been obtained by a measurement of the polarisation of UV and X radiations [1, 2, 3]; even in this simplest case theory [4] fails to explain the X-ray result [2]. In the present work we

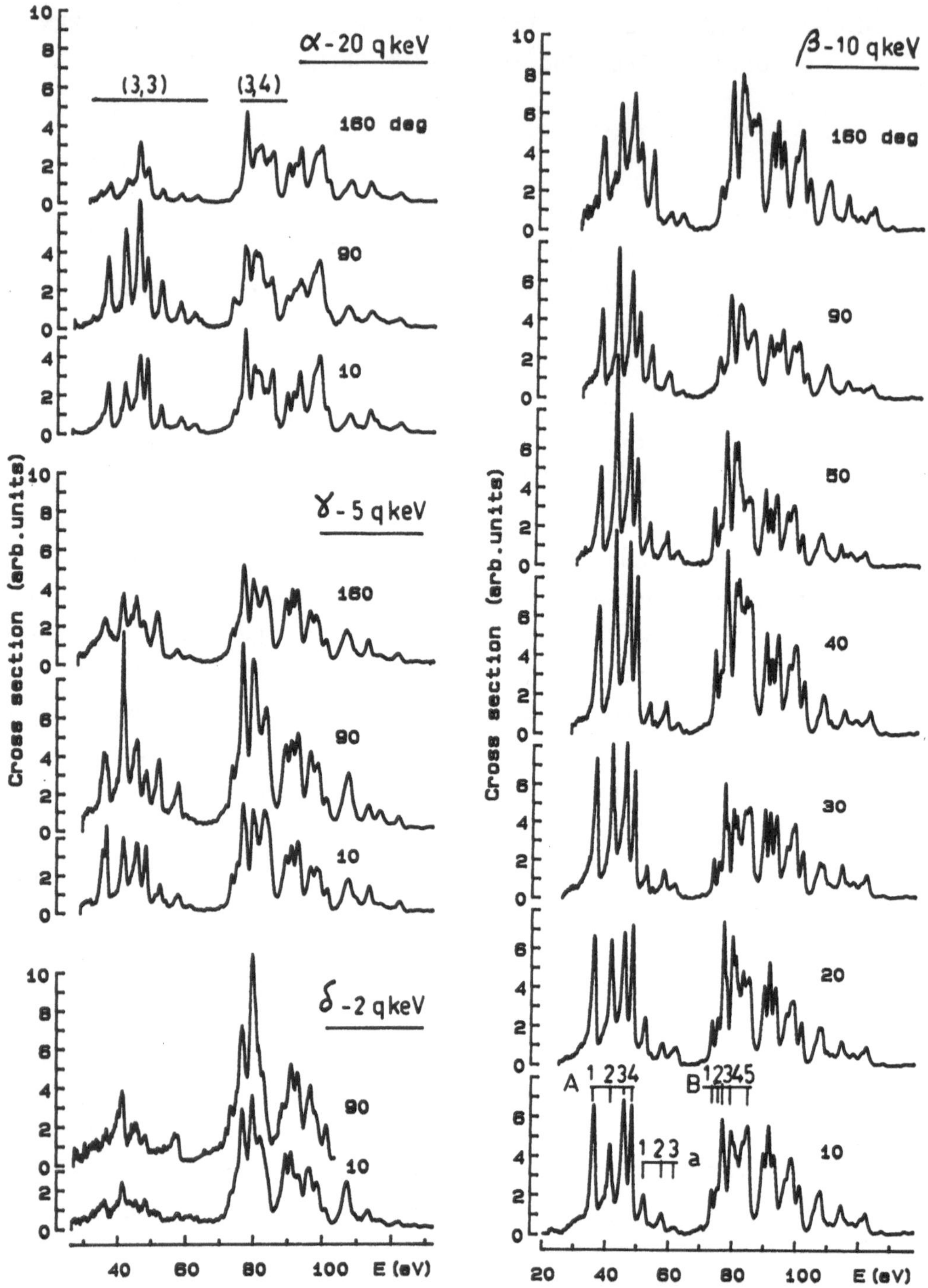

Fig. 1. Angular dependence of the electron spectra (emitter frame) in Ne^{8+}+ He for different collision energies. α: 20 qkeV; β: 10 qkeV; γ: 5qkeV; δ: 2qkeV. The doubly differential cross sections, given in arbitrary units, can be compared with each other.

have looked at a possible alignment of diexcited states in $Ne^{8+}(1s^2)$-He in the 0.8-8 keV/amu energy range.

Results for the Ne^{8+} + He system

Double capture into autoionising states obeys the following general scheme for this system:

$$Ne^{8+}(1s^2)^1S + He(1s^2)^1S \rightarrow Ne^{6+}(1s^2 3lnl')^1L + He^{++} \rightarrow Ne^{7+}(1s^2 2l) + e^- + He^{++}$$

Some examples of angular dependence of the electron spectra (in the emitter frame) are shown in figures 1α to 1δ at 20, 10, 5 and 2 qkeV collision energy respectively (angles are given in the laboratory frame); only the 10 qkeV data have been obtained with a 10° angle step. These results show an obvious and strong anisotropy for many (3,3) and (3,4) lines (noted A,a and B,b in Fig.1β when the decay occurs towards the $1s^2 2p$ and $1s^2 2s$ continua respectively); at 10 qkeV, even large modifications of the relative line intensities are seen when the angle is changed by only 10°. Unhappily these results also reveal an asymmetrical behaviour with respect to 90°: at 10° and 160° compare lines A_4 and a_1 at 10qkeV (fig.1β) and all the (3,3) lines at other collision energies (fig 1α,γ,δ). These observations cannot be explained by an alignment of the states alone; interferences between neighbouring states certainly affect the electron spectra, as already demonstrated for another system [5]. We illustrate more quantitatively the angular dependence of the 10 qkeV spectra in fig.2.

In order to reduce the error when changing the angle we only give in fig.2 *relative* cross sections normalized to the sum of the two groups of lines seen between 104 and 120 eV which can be supposed to be isotropic. Indeed, the *ratio* of the groups at 104-113 eV and 113-120 eV was found to be independent of the angle within ±8% indicating that they do not have very different angular dependences; since they result from the decay of many n=5,6,7,...states it can be assumed that they behave isotropically. The (3,n) cross sections (corresponding to a group of lines) are obtained by direct integration whereas a fitting procedure already described elsewhere (without allowing for interferences; see [5] and references given therein) is necessary to determine the angular dependence of the n=3 or 4 blended

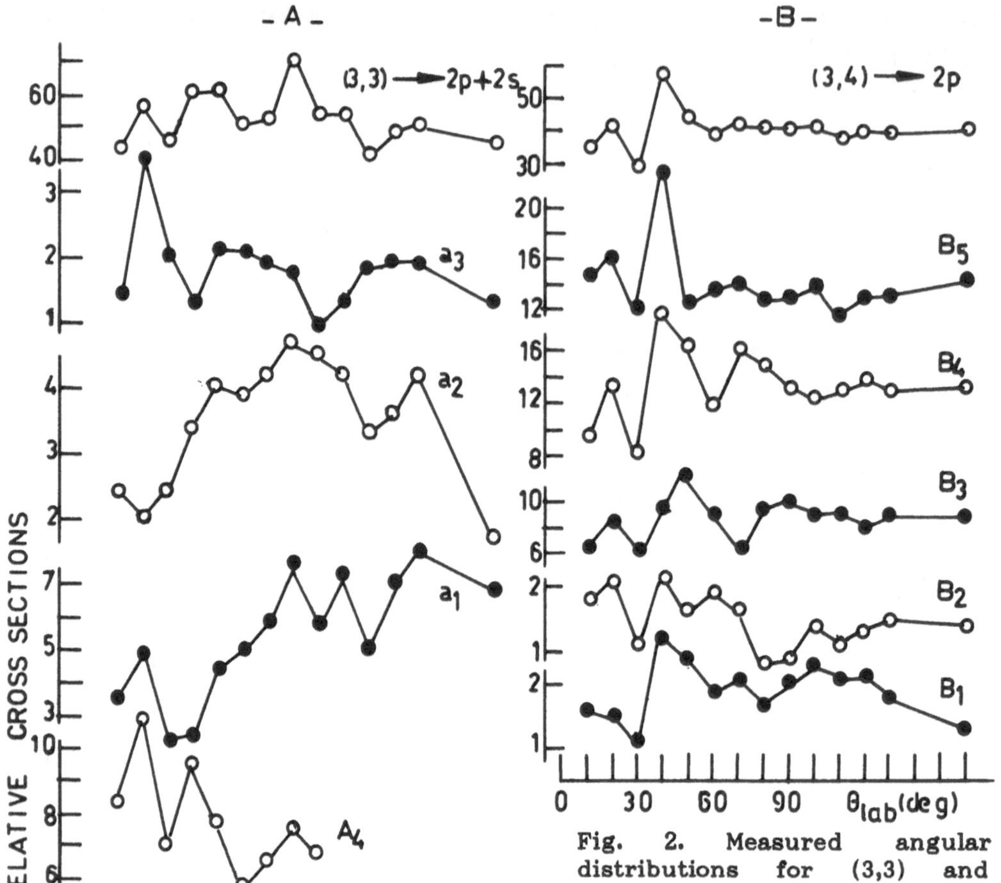

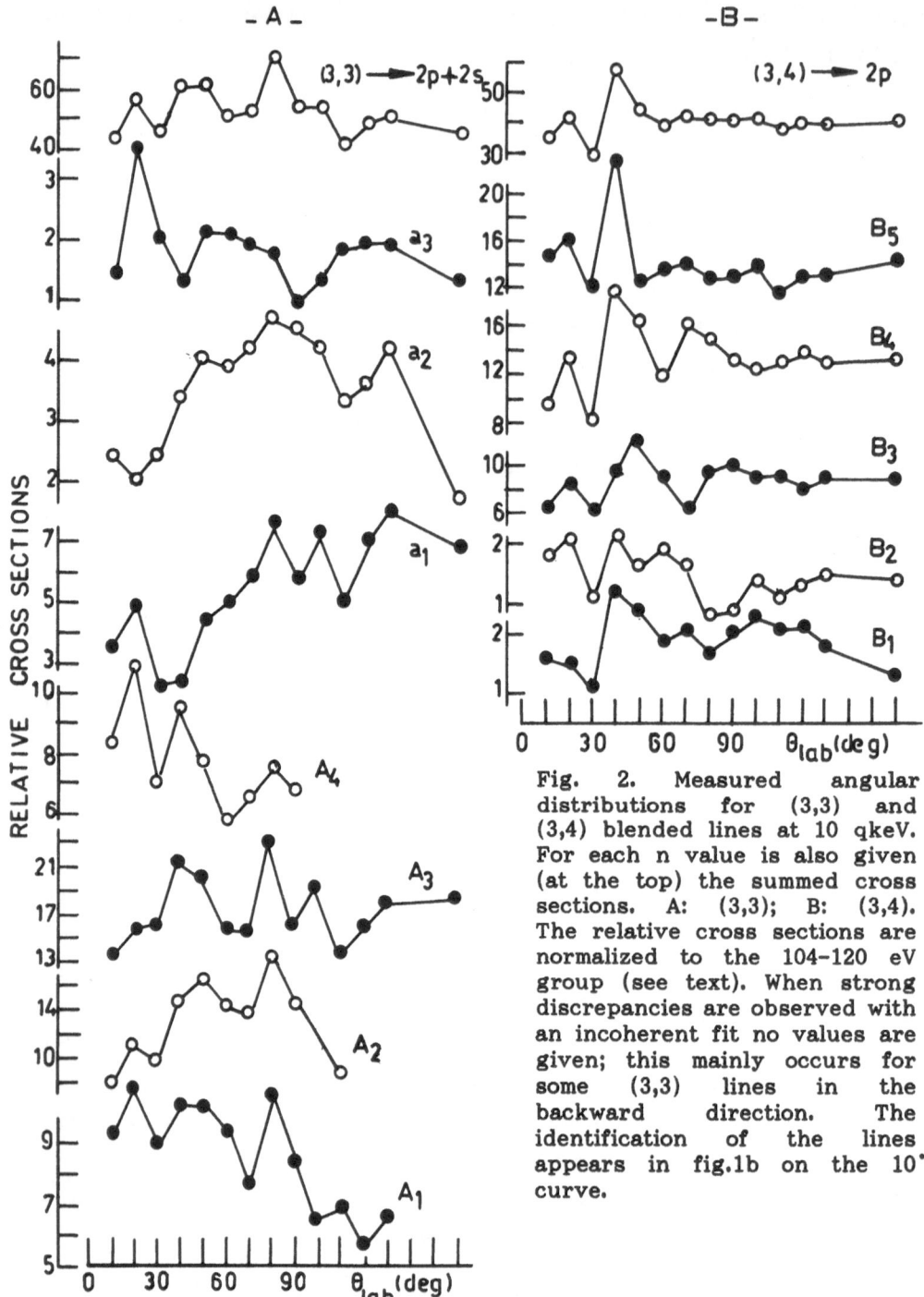

Fig. 2. Measured angular distributions for (3,3) and (3,4) blended lines at 10 qkeV. For each n value is also given (at the top) the summed cross sections. A: (3,3); B: (3,4). The relative cross sections are normalized to the 104-120 eV group (see text). When strong discrepancies are observed with an incoherent fit no values are given; this mainly occurs for some (3,3) lines in the backward direction. The identification of the lines appears in fig.1b on the 10° curve.

lines (the identification of n=3 states is given in [5]). The data displayed in fig.2 present a very complicated angular dependence which cannot be analysed more thoroughly at the present stage. Variations within a factor larger than 2 are often observed for the blended lines; they largely exceed the uncertainty coming from the fit procedure. Even the summed (3,3) and (3,4) groups present a significant oscillatory angular dependence; the error on these relative values is smaller than ±10%.

Conclusion.

In the collisional system investigated in the present work the angular distributions have been found to be so strongly perturbed that the question of the alignment in double capture processes remains open. Our measurements show that the relative population of the 3l3l' and 3l4l' states as well as a reasonable estimate of the total cross section cannot be obtained *without measuring a detailed angular distribution.* Whenever these conditions are not fulfilled a quantitative comparison of the experimental and theoretical data becomes somewhat meaningless.

This work has been performed at LAGRIPPA, a common laboratory of the CNRS and of the CEA.

References.

1. R.Baptist, J.J.Bonnet, G.Chauvet, J.P.Desclaux, S.Dousson, D.Hitz, J.Phys.B 17, L417 (1984)

2. D.Vernhet, A.Chetioui, K.Wohrer, J.P.Rozet, P.Piquemal, D.Hitz, S.Dousson, A.Salin, C.Stephan, Phys.Rev.A 32, 1256 (1985)

3. R.Hoekstra, M.G.Suraud, F.J. de Heer, R.Morgenstern, J.de Physique, Colloque C1, 50, 387 (1989)

4. C.D.Lin, J.H.Macek, Phys.Rev.A 35, 5005 (1987)

5. M.Boudjema, P.Moretto-Capelle, A.Bordenave-Montesquieu, P.Benoit-Cattin, A.Gleizes, H.Bachau, P.Galan, F.Martín, A.Riera, M.Yañez, J.Phys.B 22, L121 (1989)

INFLUENCE OF ROTATING ELECTRON WAVE FUNCTIONS ON L-SHELL IONIZATION AND ALIGNMENT*

I.C. Legrand[1)], R. Dörner[2)], H. Schmidt-Böcking[2)], V. Zoran[1)]

1) Institute of Atomic Physics, Bucharest, Romania

2) Institut für Kernphysik, Universität Frankfurt/Main, Germany

Abstract

In adiabatic ion-atom collisions, the target electron distribution will follow the rotation of the internuclear axis resulting in a change of the subshell ionization probabilities as well as of the alignment tensor as compared with the case of frozen electron wave functions. In the present work we compare SCA calculations based on frozen and rotating relativistic electron wave functions with experimental L-subshell probabilities and L_3-subshell differential alignment measured by Dörner et al. for 1 MeV protons on Samarium. In this case, where the intrashell couplings play a minor role, the agreement between calculations and experiment is significantly improved assuming a partial rotation of the target electron distribution during the collision, possibly depending on substate and impact parameter.

First order perturbation theories such as the semiclassical approximation (SCA)[1)] provide a good description of the inner-shell ionization by light ion impact as far as K-shell ionization probabilities and total cross sections are concerned[2,3)]. According to these theories, the ionization probability depends on the orientation of the angular momentum of the innershell electronic state with respect to the beam direction. Thus, the atom ionized in non-spherical states turns out to be aligned with respect to the direction of the beam. The degree of alignment is measured by the alignment tensor A_{kj}[4)]. Generally, SCA theory yields good agreement also for the

*Supported in part by Intern. Büro, Kernforschungszentrum Karlsruhe and BMFT

"total" alignment tensor component A_{20}, when the L_3 subshell is ionized by protons[5]. However, significant differences were found for differential, i.e. impact parameter dependent, L-shell ionization probabilities P(b) or the alignment tensor component $A_{20}(b)$[6,7] in spite of the fact that the conditions for the validity of the semiclassical approach should be reasonably well fulfilled in such cases.

Recently Dörner[8] performed accurate measurements of the L-subshell ionization probabilities $P_{Li}(b)$ i = 1,2,3 and the alignment tensor components A_{2k} (b) for the collision system 1 MeV p - Sm. Again, depending on the impact parameter, differences up to a factor 2 have been found between the experiment and the SCA calculations. The aim of the present work is to clarify the origin of these discrepancies.

A current explanation for the disagreement between theory and experiment, even in rather asymmetric collisions is that 1. order perturbation theory is not adequate. Thus, second order SCA calculations improved the description of the L-subshell ionization data for light projectiles[9,10]. For heavier projectile coupled channel calculations become necessary (see e.g. ref. 11,12 and references therein). The main process responsible for the higher order effects is the vacancy sharing within target L-shell during the collision: first the vacancy is created in any of the target L-subshells and then it is transferred to another subshell in the same collision.

We performed coupled channel calculations in the coupled subshell approximation (CSA)[11] for 1 MeV protons in Samarium. Except for the L_1 subshell intrashell couplings bring only marginal improvements relative to the SCA calculations, the couplings remaining weak for such an extremely asymmetric collision[12].

Another effect, not included in the SCA, which might explain the deviations has been suggested by the Frankfurt - Heidelberg group[7,8]. In adiabatic ion-atom collisions, the target electron distribution could in principle follow the rotation of the internuclear axis, resulting in a change of the subshell ionization probabilities as well as of the alignment tensors as compared with the case of frozen electron wave functions. In view of the coupled channel results discussed above, the p - Sm collision system might be a suitable case to study the rotation effect in more detail. This will be done in following: A natural frame to describe wave function rotation effects is a rotating coordinate system having the internuclear vector as quantization axis. The collision takes place in the xz plane. Since we consider an initially occupied shell, it is more convenient to regard vacancies instead of electrons as the active particles in the collision. The scattering wave function is expanded in hydrogenic Dirac wave functions of the target atom which are rotating with the above coordinate system. Neglecting couplings among the bound states, the time-dependent Dirac equation is equivalent to the following system of coupled differential equations for the amplitudes a_{nf} (t,E) describing the presence of the vacancy in the target bound state $|n\rangle$:

$$\dot{a}_{nf}(t,E) = i \langle \psi_n(\vec{r}) \mid \frac{Z_p e^2}{|\vec{r} - \vec{R}(t)|} \mid \psi_f(\vec{r}) \rangle \; e^{i(E-\varepsilon n)t} + \dot{\Theta} \sum_m \langle \psi_n(\vec{r}) \mid \hat{\vartheta}_y \mid \psi_m(\vec{r}) \rangle \qquad (1)$$

Here n and m label the various L-subshells and within a subshell the various magnetic substates f labels the initial (continuum) vacancy states. ϑ_y is the component of total angular momentum along the y axis[13] and θ the rotation angle. The first term in (1) is similar to the SCA, while the second term describes the rotational coupling among the different magnetic substates. The system (1) is integrated along each Rutherford trajectory for the available initial states f at various continuum energies E. The ionization probability is given by

$$P_n(b) = \sum_f \int_E |a_{nf}(+\infty,E)|^2 \, dE \qquad (2)$$

The results of the above calculations are shown in Fig. 1, for L_3 subshell ionization probabilities as a function of the impact parameter. The disagreement between experiment and the calculations with adiabatic rotation of the target wave functions is at least as serious as for the SCA calculations with frozen wave functions.

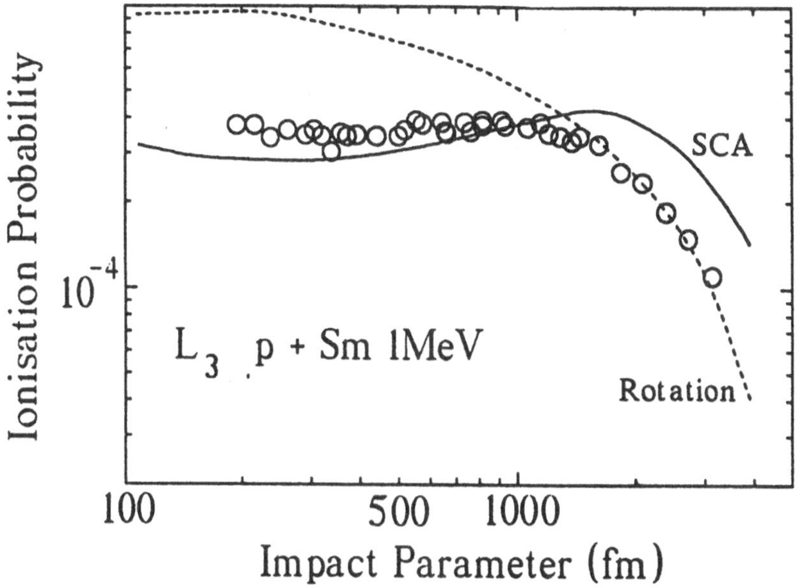

Fig. 1: The ionization probability of L_3-subshell of Sm by 1 MeV protons as a function of impact parameter. Theory: full line, SCA; dashed line, SCA with adiabatic rotation of target wave functions. Data: ref. 8. Calculations: this work.

The fact that the adiabatic rotation overcompensates the deficiencies of the classical SCA calculations suggests that the electron distribution legs behind the internuclear axis as the collision evolves. This phenomenon has been revealed e.g. by the calculations of the capture process to the 2p states in B^{9+} - He collisions[14)] and the quoted authors coined the name "slippage" for it. We remade the caculation of the L-subshell ionization probabilities allowing for slippage. The simplest way to do it is to consider a "uniform slippage", i.e. the slippage angle is $\rho = \delta\theta$ where the parameter δ does not depend on time (Fig. 2). This way the dynamics of wave function rotation is simulated by a kinematical effect. While $\delta = 0$ corresponds to full rotation, for $\delta = 1$ the problem reduces to the classical SCA in a rotation coordinate system. Since the subshell ionization probabilities should not depend on the coordinate system, the case $\delta = 1$ has been used to check the numerical procedures.

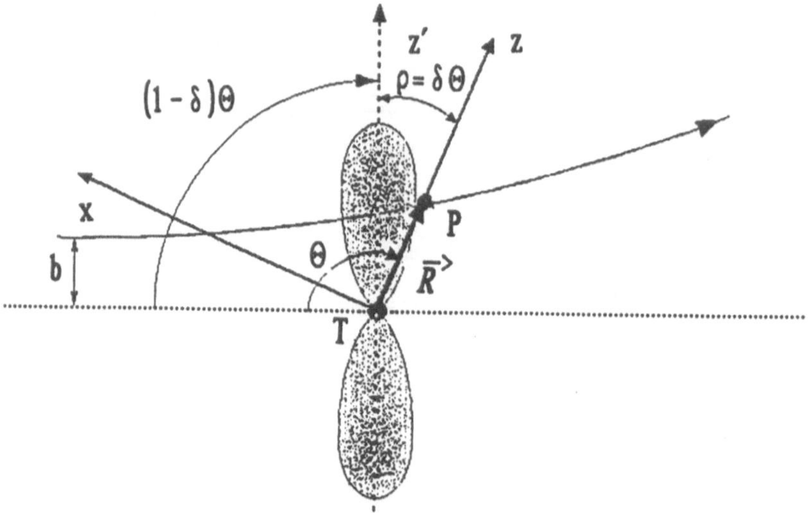

Fig. 2: Schematic of the electron wave function slippage relative to the internuclear vector $\vec{R}$. Notations: T target, P projectile, θ rotation angle of $\vec{R}$, $(1-\delta)\theta$, rotation angle of the major symmetry axis of the electron distribution.

The results of our calculations are shown in Figs. 3 - 6 for several values of δ. Interesting enough, the spherical L_1 subshell (Fig. 3) is also affected by rotation at large impact parameters, because there the contribution to ionization from non-spherical continuum states is incr asing due to the increased weight of the higher multipoles in the potential expansion. Concerning the L_2 and L_3 subshells, there is a relatively narrow range of δ values for which the overall agreement (in magnitude and shape) to the experimental ionization probabilities is improved. The discrepancies for L_1 and L_2 subshells at small b, while L_3 remains practically unchanged.

The A_{20} component of the alignment tensor is readily expressed in terms of the substate ionization probabilities (2):

$$A_{20}(b) = \frac{P_{3/2,3/2} - P_{3/2,1/2}}{P_{3/2,3/2} + P_{3/2,1/2}} \quad (3)$$

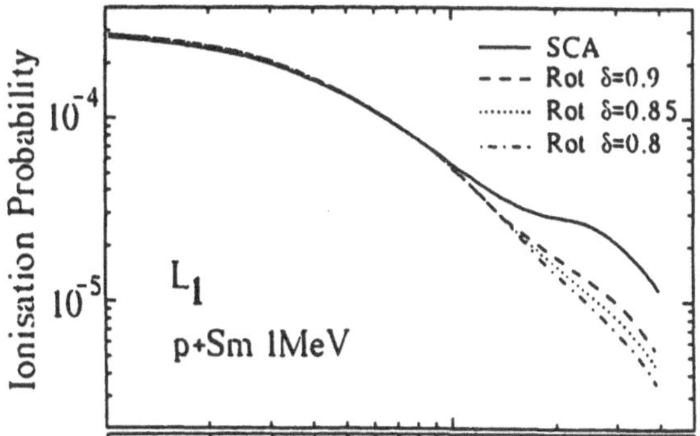

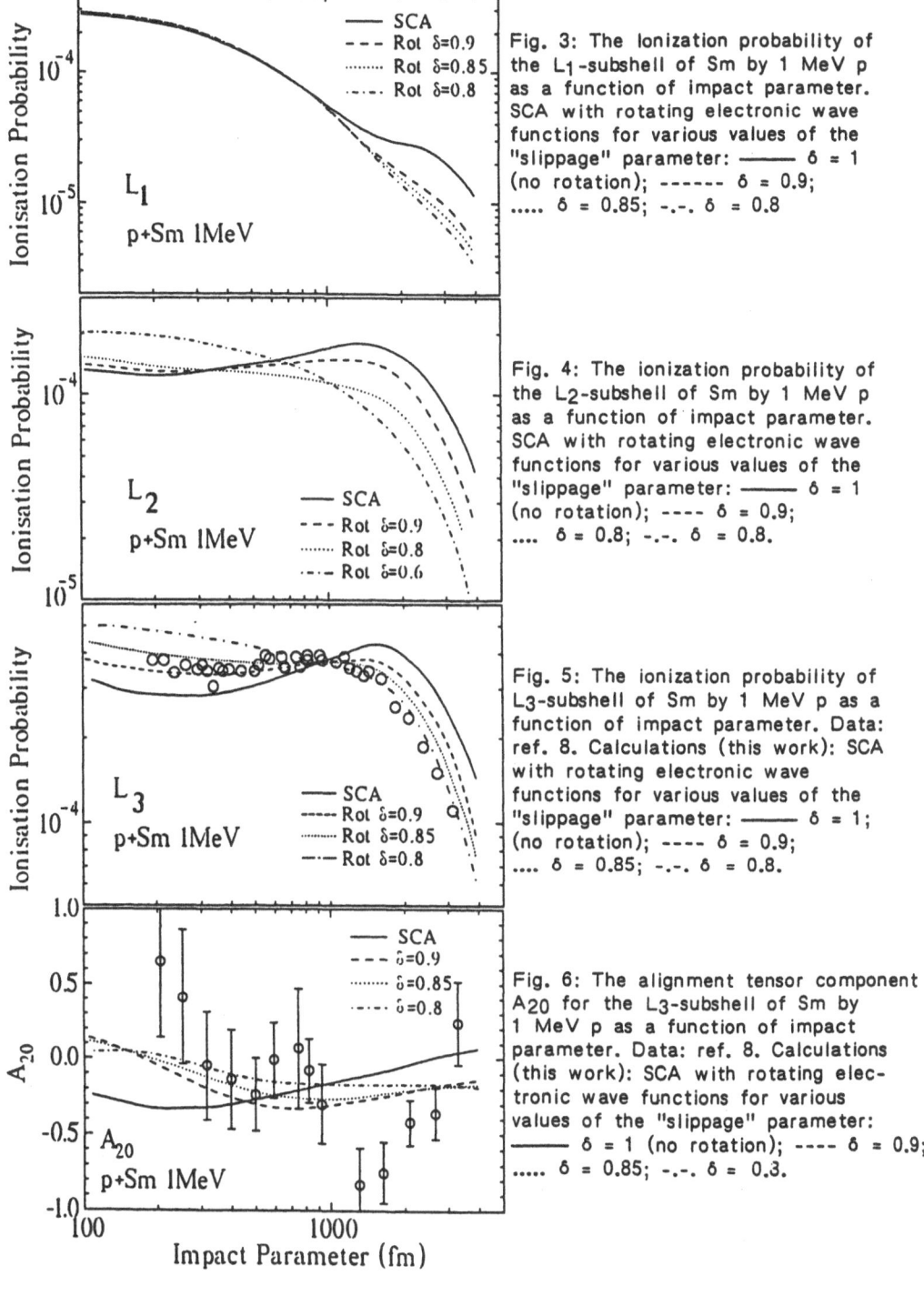

Fig. 3: The ionization probability of the L_1-subshell of Sm by 1 MeV p as a function of impact parameter. SCA with rotating electronic wave functions for various values of the "slippage" parameter: ——— δ = 1 (no rotation); ------ δ = 0.9; ..... δ = 0.85; -.-. δ = 0.8

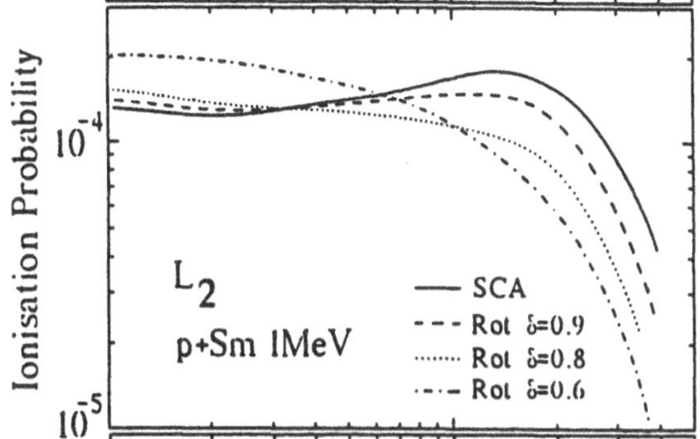

Fig. 4: The ionization probability of the L_2-subshell of Sm by 1 MeV p as a function of impact parameter. SCA with rotating electronic wave functions for various values of the "slippage" parameter: ——— δ = 1 (no rotation); ---- δ = 0.9; .... δ = 0.8; -.-. δ = 0.8.

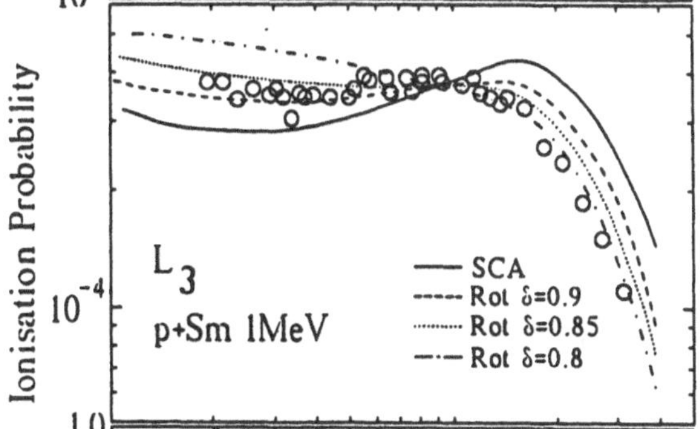

Fig. 5: The ionization probability of L_3-subshell of Sm by 1 MeV p as a function of impact parameter. Data: ref. 8. Calculations (this work): SCA with rotating electronic wave functions for various values of the "slippage" parameter: ——— δ = 1; (no rotation); ---- δ = 0.9; .... δ = 0.85; -.-. δ = 0.8.

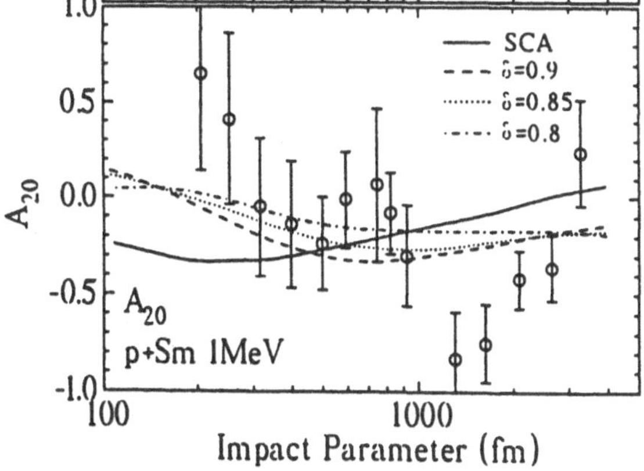

Fig. 6: The alignment tensor component A_{20} for the L_3-subshell of Sm by 1 MeV p as a function of impact parameter. Data: ref. 8. Calculations (this work): SCA with rotating electronic wave functions for various values of the "slippage" parameter: ——— δ = 1 (no rotation); ---- δ = 0.9; ..... δ = 0.85; -.-. δ = 0.3.

The alignment is particularly sensitive to kinematic effects[15]. Unfortunately, the differential alignment data have large error bars due to experimental difficulties. Nevertheless, the trend of A_{20} as a function of b for p + Sm at 1 MeV is better described by our calculations including the wave function rotation as compared to SCA calculations with frozen electronic wave function. It is gratifying that both $P_{L3}(b)$ and $A_{20}(b)$ are best described by a common value of the slippage parameter around $\delta = 0.85$.

To conclude, the data and calculations presented in this work support the hypotheses of the electronic wave function rotation during asymmetric ion-atom collisions. Although in the case studied it represents only about 15% of the internuclear axis rotation, its influence on the ionization probabilities may reach a variation up to factor 2. In view of this finding, the good agreement between the SCA calculations and experimental total integral cross section for L subshell ionization with protons seems to be accidentally achieved by the mutual compensation of the disagreements for small and large impact parameters. On the other hand, the present approach is limited by the assumption of uniform slippage. Nevertheless, the way towards a dynamic theory may be paved by a closer look to possible dependence of δ on impact parameter and energy, for various collision systems.

References

1. J.M. Hansteen, Advances in Atomic and Molecular Physics (edited by D.R. Bates and B. Benson) Vol. 11, p. 299 Academic Press, New York (1975) and L. Kochbach, J.M. Hansteen, R. Gundersen Nucl. Instr. Meth. 169 (1980) 281
2. H. Paul, Proc. 2nd Workshop on Inner Shell Ionization by Light Ions (Linz) 1981, Nucl. Instr. Meth. 162 (1982) 1
3. W. Jitschin, R. Hippler, R. Shanker, H. Kleinpoppen, R. Schuch, H.O. Lutz, J. Phys. B16 (1980) 1417
4. K. Blum, Density Matrix Theorie and Applications, Plenum Press, New York (1981)

5. W. Jitschin, A. Kaschuba, H. Kleinpoppen, H.O. Lutz
Z. Phys. A304 (1982) 69
6. J. Konrad, R. Schuch, R. Hoffmann, H. Schmidt-Böcking
Phys. Rev. Lett. 52 (1984) 188
7. S. Zehendner, G.B. Baptista, R. Dörner, E. Justiniano,
J. Konrad, H. Schmidt-Böcking, R. Schuch
Z. Phys. D4 (1987) 243
8. R. Dörner, Thesis, University of Frankfurt (unpublished)
and R. Dörner et al. to be published
9. J. Pàlinkas, L. Sarkadi, B. Schlenk, I. Török,
Gy. Kalman, C. Bauer, K. Brankoff, D. Grambole, C. Heiser,
W. Rudolph, H.J. Thomas,
IEEE Trans. Nucl. Sci., Vol. NS-30, 970, 1983
10. J. Pàlinkas, L. Sarkadi, B. Schlenk, I. Török,
Gy. Kalman, C. Bauer, K. Brankoff, D. Grambole, C. Heiser,
W. Rudolph, H.J. Thomas; J. Phys. B17 (1983) 131
11. P.A. Amundsen, D.H. Jakubassa-Amundsen, J.Phys. B21
(1988) L99
12. I.C. Legrand et al., to be published
13. M.E. Rose, Relativistic Electron Theory, John Wiley 1961
14. J.P. Hansen, L. Kocbach, A. Dubois, S.E. Nielsen
Phys. Rev. Lett. 64 (1989) 2491
15. A. Berinde, V.E. Iacob, I.C. Legrand, I. Piticu, V. Zoran
J. Phys. B18 (1985) L229

I Other Actual Issues (Hot Topics)

Chairman: S.T. MANSON

I.2 Invited Contributions

T. MUKOYAMA and K. SHIMA
Rearrangement of Electronic States of Heavy Ions Passing Through Solid Targets

N.P. POPOV
Coulomb Deexcitation of Muonic Hydrogen in Collision Processes

P.H. MOKLER
First Experiments in the Heavy Ion Cooler Ring ESR

I.3 Contributions (Posters)

J.M. HANSTEEN and J.M. BANG
On e^+e^- Pair Production in Heavy Ion Collisions: Possible Electronic Coherence Effects

S. BHATTACHARYYA
Single Ionisation of H_2 by High Energy Electron – A QED Approach

V.V. BALASHOV, V.K. DOLINOV and V.A. SHAKIROV
Collisionally Induced Transitions between Excited States of He_μ^+-Ion in Muon-Catalyzed DT- and DD-Fusion

Rearrangement of Electronic States of Heavy Ions Passing Through Solid Targets

Takeshi Mukoyama [1] and Kunihiro Shima [2]

[1]Institute for Chemical Research, Kyoto University, Uji, Kyoto, 611 Japan
[2]Accelerator Center, University of Tsukuba, Tsukuba, Ibaraki, 305 Japan

Abstract: Electron rearrangement of fast heavy ions passing through solid targets has been studied by the Monte Carlo simulation. Starting from the electron configuration of ions in solids, the change in electronic states due to successive radiative and Auger transitions was traced from the time when the ion emerges from the solid surface to the time when it is detected. The calculations were made for 63-MeV Cu ions emerging from carbon target. It is found that the charge increase due to vacancy cascade plays a minor role for the mean charge difference between gases and solids. The origin of the shell effect in the equilibrium charge distribution of ions is discussed.

1 Introduction

The charge distribution of fast ions passing through matter is of fundamental importance from early days of atomic physics. When heavy ions enter the medium, their charge state fluctuates by loss and capture of electrons. The form of charge distribution is determined by the velocity of the ion and the properties of the medium. A number of reviews on this phenomenon have been published [1-5].

It is well known experimentally that the mean charge of ions emerging from solid targets is higher than that from gas targets. Several phenomenological models to explain this effect, called the *density effect*, have been proposed. In passing through matter, the projectile ion loses its electron and captures atomic electrons into excited states. The excitation of electrons in ions also takes place. In gases, the time between successive collisions of ions with atoms is longer than the lifetime of excited states. The excited ion loses its energy through radiative or Auger transitions and goes to the ground-state configuration. On the other hand, the collision time is shorter in solids and the excited states play an important role on charge state distribution of ions.

Bohr and Lindhard (BL) [6] assumed that the ions in an excited state lose electrons more easily than in the ground state because of decrease in the binding energy, while the probability of electron capture for an excited state is small. This suggests that in solids the electron-loss cross section is higher and the electron-capture cross section is lower than that in gases. In their model, called the BL model, the main part of the increase in the mean charge of ions takes place inside solids.

On the other hand, Betz and Grodzins (BG) [7] assumed that in a single collision the electrons of ions are not so highly-excited as in the BL model, but a number of electrons are excited in successive collisions. When these excited ions emerge from the solid surface into vacuum, their mean charge increases substantially through the emission of Auger electrons. Therefore, in the BG model the main reason for the increase in the mean charge is due to the Auger transitions of ions after passing through solids.

Based on these two models, a lot of arguments have been done on the charge state of heavy ions in solids. However, it is not clear until now which model is more useful to describe the charge distribution of ions inside solids. In the present work, we perform the Monte Carlo simulation to estimate the effect of electron rearrangement of ions after passing through solid targets on the charge distribution of ions experimentally measured at a certain distance from the target. The calculated results are used to investigate the origin of the shell effect in the equilibrium charge state of ions.

2 Method of Computation

The method to calculate the rearrangement of electronic states is, in principle, the same as that used for vacancy cascade following inner-shell ionization [8-10]. The electronic state of an ion at the time t after emerging from the solid is simulated by the Monte Carlo method. Two processes, x-ray emission and Auger transition, are considered. The initial electron configuration of ions is estimated from the experimental data for electronic states of ions in solids.

For a given initial configuration, the computer program calculates x-ray emission rates and Auger rates. The sum of both transition rates gives the total transition probability per unit time and the time of the transition is determined from an exponential random number. If the time elapsed is shorter than the time necessary for ions to travel from the target to the detector, the transition is determined whether radiative or not. For radiative transitions, the next position where the vacancy moves is selected by a random number. In the Auger transitions, two new vacancy states are chosen and the number of ionic charge is increased by one. This procedure is repeated until all inner-shell vacancies are filled and no transition takes place or until the time elapsed is larger than the flight time of the ion between the target and detector. Then the final vacancy distribution is recorded and the computer program generates a new history. After 10000 histories, the electron configuration of ions at the detector is printed out.

3 Results and Discussion

The radiative transition rates for various shells are taken from the tabulated Hartree-Fock-Slater values of Manson and Kennedy [11]. We use the Auger transition rates calculated by Kostroun *et al.* [12] for K shell and those for L subshells are obtained from the theoretical results of McGuire [13]. For M subshells, the Auger rates are taken from the values of McGuire [14]. We treat N shell as one shell and the transitions between N subshells are not considered. When there are no tabulated data for special elements, we estimate the values by interpolation for atomic numbers. These atomic transition rates correspond to the values for a singly ionized atom. For multiply-charged ions, we modify these values to be proportional to the number of electrons available to a particular transition, following the method of Larkins [15].

In order to test the present computer program, we have calculated the charge distribution for 42-MeV Br ions passing through a carbon foil. Betz [16] estimated the electron configuration of Br ions in carbon target based on the BL and BG models. In both case, K and L shells are fully occupied. The M- and N-shell configuration is $(3s)^2(3p)^4(3d)^2(4s)^2(4p)^4(4d)^3(4f)^1$ for the BG model and $(3s)^2(3p)^6(4p)^1$ for the BL model. For gas targets, M shell is also considered to be fully occupied and higher shells are vacant. We assume that the detector is located at the distance of 3 m from the target. In this geometry, the flight time of ions between the target and the detector is much longer than the lifetime of atomic states and all the electrons in the excited states decay to the ground state when the ion is detected.

In the case of the BG model, the Monte Carlo calculation indicates that the ionic charge increases by five or six due to Auger cascade and the difference in mean charge between the gas and solid targets is $\Delta q = 5.6$. First, two $3p$ shells are filled by $3p3d$N Auger transitions and then four $3d$NN Auger transitions take place. In this case, six electrons are emitted. There is another decay channel, in which one $3p$NN Auger transition occurs instead of $3p3d$N transition. In the final stage of this decay channel, only one electron remains in N shell and the $3d$-shell vacancy is filled by a radiative transition. The increase in the ion charge in this channel is $\Delta q = 5$. On the other hand, in the BL model only single electron exists in N shell and the $4p \to 3d$ radiative transition takes place. This yields also $\Delta q = 6$ and is in agreement with the result of the BG model.

As a next example, we calculate the charge distribution of 63-MeV Cu ions emerging from carbon target, since the number of K-, L- and M-shell electrons in solids is roughly known from K-x-ray energy shifts and $K\beta/K\alpha$ x-ray intensity ratios [17]. According to the experimental results of Shima *et al.* [17], the mean number of electrons inside the carbon foil is 2 for K shell, 5.6 ± 0.4 for L shell, and $3.3 \sim 7.4$ for M shell, respectively. For simplicity, we use an electron configuration $(1s)^2(2s)^2(2p)^4(3s)^2(3p)^3$ plus two N-shell electrons and investigate the charge distribution at the detector for various target-to-detector distances. The calculated results are shown in Fig. 1. It can be seen that the atomic transition takes place at very small distances and in the present case the charge distribution does not change for distances larger than 0.01 cm.

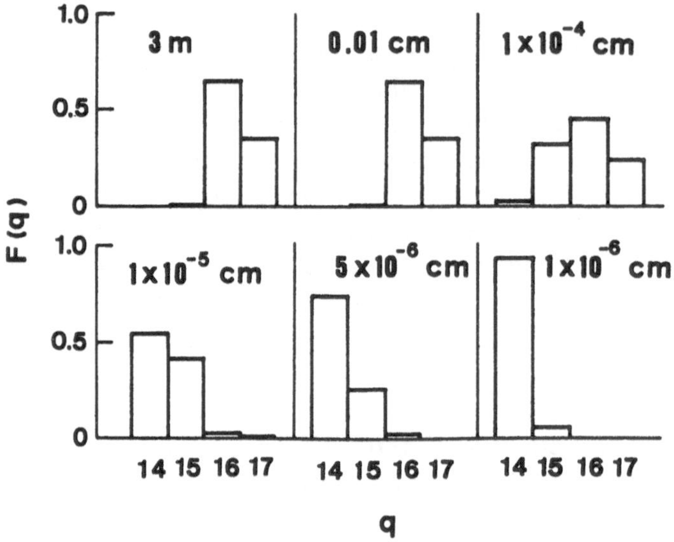

Fig. 1. Charge state distribution $F(q)$ of 63-MeV Cu ions with charge state q after passing through carbon target for various target-to-detector distances

For charge distributions of ions observed behind solid targets, there is the phenomenon called the *shell effect*, i.e. the charge distribution consists of two Gaussian components and the boundary charge state between these two Gaussians correspond to the ion whose outer-shell electron configuration is in a closed shell. The shell effect was first observed by Moak *et al.* for Br and I ions in carbon foils [18] and has been systematically investigated by Shima *et al.* for various ions passing through carbon foils [19]. A clear demonstration of the shell effect is obtained when the ratio of charge fractions of ions between two adjacent charge states, $F(q+1)/F(q)$, is plotted against the charge state q in a semi-logarithmic scale. There is a sharp change in the slope of the curve at the boundary charge states, for example at Cu^{19+} in the case of Cu ions.

The origin of the shell effect is not clear. In the BL model, the shell effect can be ascribed to the collision processes in the target. On the other hand, in the BG model it is possible that the shell effect arises during the decay process through Auger cascade. In order to investigate the origin of the shell effect in the equilibrium charge state distributions, we calculate the change in charge states of 63-MeV Cu ions after emerging from the carbon foil. The distance between the foil and the detector is taken to be 3 m as in the actual experimental condition.

Since we know only the mean number of electrons for each shell in the foil and have no information of distribution of configurations, the initial configuration of Cu ions emerging from the foil is chosen as a combination of (n_K, n_L, n_M, n_N), where n_i is the number of electrons in the i-th shell. From the experimental data [17], each n_i value is taken as $n_K = 2$, $n_L = 2,3,4,5,6,7,8$, $n_M = 5$, and $n_N = 0,1,2,3,4$. For fixed n_i, all configurations among the subshells are considered. Each configuration is assumed to occur with the same probability.

In Fig. 2, the calculated result for the charge distribution $F(q)$ and the charge fraction $F(q+1)/F(q)$ is plotted against q [20] and compared with the experimental fraction ratios of Shima *et al.* [19]. It is clear that there is no indication of the shell effect in the calculated result. This fact means that the origin of the shell effect can be attributed to the scattering processes of ions inside solid targets and the Auger cascade plays a minor role.

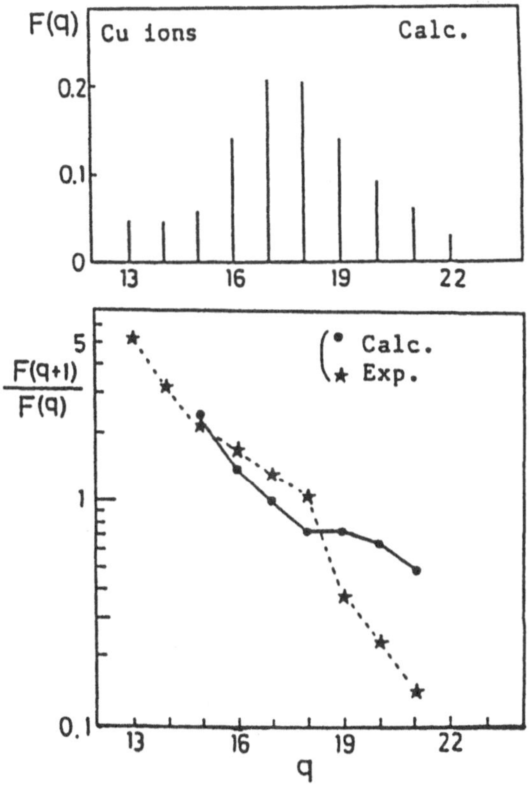

Fig. 2. Calculated result of the charge distribution $F(q)$ and charge fraction ratio $F(q+1)/F(q)$ for 63-MeV Cu ions after passing through carbon target as a function of the charge state q. The experimental data for the charge fraction ratio (Ref. 19) are also shown.

Furthermore, it is found from the present calculations that in most cases the number of Auger electrons emitted from ions after passing through the target is less than two. This is consistent with the experimental results of Baragiola *et al.* [21], Lennard and Phillips [22], and Della-Negra *et al.* [23]. The present result suggests that the charge state of ions in solids is higher than in gases and is in favour of the BL model.

References

1. S.K. Allison: Rev. Mod. Phys. **30** 1137 (1958)
2. H.D. Betz: Rev. Mod. Phys. **44** 465 (1972)
3. H.D. Betz: In *Methods of Experimental Physics, Vol. 17*, ed. by P. Richard (Academic, New York, 1980), pp. 73–148
4. M.A. Kumakhov, F.F. Komarov: *Energy Loss and Ion Ranges in Solids* (Gordon and Breach, New York, 1981)
5. H.D. Betz: In *Applied Atomic Collision Physics, Vol. 4*, ed. by S. Datz (Academic, New York, 1983), pp. 1–42
6. N. Bohr, J. Lindhard: Kgl. Danske Vidensk. Selsk. Mat.-Fys. Medd. **28** No. 7 (1954)
7. H.D. Betz, L. Grodzin: Phys. Rev. Lett. **25** 211 (1970)
8. T. Mukoyama: Bull. Inst. Chem. Res., Kyoto Univ. **63** 373 (1985)
9. T. Mukoyama: J. Phys. Soc. Japan **55** 3054 (1986)
10. T. Mukoyama, T. Tonuma, A. Yagishita, H. Shibata, T. Koizumi, T. Matsuo, K. Shima, H. Tawara: J. Phys. B **20** 4453 (1987)
11. S.T. Manson, D.J. Kennedy: At. Data Nucl. Data Tables **14** 111 (1975)
12. V.O. Kostroun, M.H. Chen, B. Crasemann: Phys. Rev. A **3** 533 (1971)
13. E.J. McGuire: Phys. Rev. A **3** 1801 (1971)
14. E.J. McGuire: Phys. Rev. A **5** 1043 (1972)
15. F.P. Larkins: J. Phys. B **4** L29 (1971)
16. H.D. Betz: Nucl. Instr. Meth. **132** 19 (1976)
17. K. Shima, S. Fujioka, Y. Tajima, T. Ishihara, M. Yamanouchi: Nucl. Instr. Meth. **A262** 132 (1987)
18. C.D. Moak, H.O. Lutz, L.B. Bridwell, L.C. Northcliffe, S. Datz: Phys. Rev. Lett. **18** 41 (1967)
19. K. Shima, N. Kuno, M. Yamanouchi: Phys. Rev. A **40** 3557 (1989)
20. K. Shima, T. Mukoyama: To be published.
21. R.A. Baragiola, P. Ziem, N. Stolterfoht: J. Phys. B **9** L447 (1973)
22. W.N. Lennard, D. Phillips: Phys. Rev. Lett. **45** 176 (1980)
23. S. Della-Negra, Y. Le Beyec, B. Monart, K. Standing, K. Wien: Phys. Rev. Lett. **58** 17 (1987)

Muonic Atom – Atom Collision Processes

N.P. Popov

Leningrad Nuclear Physics Institute, Gatchina, Leningrad 188350, USSR

Abstract: Elastic and inelastic processes of interaction of mesohydrogen in mixture of hydrogen isotopes with light nuclei are considered. The role of the electron screening in the elastic scattering is shown to be very important. Isotopic exchange reactions as well as muon transfer to light nuclei are considered. Kinetics of the excited mesohydrogen in the mixture of hydrogen isotopes are considered.

The destiny of a muonic hydrogen isotope atom in the mixture of hydrogen isotopes or in the mixture of hydrogen isotopes with light atoms is determined by many competing processes. The excited muonic hydrogen atom is slowing down due to elastic collisions, it is accelerating in the process of Auger or Coulomb deexcitation, and, besides this, its muon can be transferred to another (heavier) nucleus. The characteristic time of these processes is determined by the cascade deexcitation time $\sim 10^{-11}$ s at liquid hydrogen (LHD). The time scale for ground state muonic hydrogen is usually much larger and is characterized by the time > 1 ns (with the exception of muon transfer from the ground state muonic hydrogen to nuclei of $Z > 3$ with a characteristic time $\sim 10^{-11}$ s at liquid hydrogen LHD). Finally, muonic hydrogen dies either due to muon decay for the time $\sim 1 \mu s$ or due to the nuclear synthesis in the hydrogen mesomolecules, which are formed for time $\sim 10^{-8} - 10^{-6}$ s. The last process is the heart of the problem of the muon catalyzed fusion which is now under active investigation. In the mixture of hydrogen isotopes with nuclei of $Z > 1$ muon transfer to these nuclei leads, as a rule, to the weak nuclear capture by this nucleus.

The processes which determine the kinetics of mesic hydrogen must be taken into account when studying the muon capture by protons [1] and in the muon catalysis [2]. The study of these processes, however, is interesting in itself, allowing one to check the energies of the bound states with an accuracy inaccessible in atomic physics [3].

1 Elastic Scattering

Elastic scattering of muonic hydrogen atoms on hydrogen isotopes nuclei and on light elements nuclei was considered in a number of papers. Due to the high rate of the spin-flip reaction for muonic protium [4] and the large value of the hyperfine splitting (~0.2 eV) the experimental study of the elastic scattering of mesoprotium is confined to the lower singlet state of hyperfine structure. Recent experimental data

$$\sigma_{\rm pp} = (1.5 \pm 0.1) \times 10^{-20}{\rm cm}^2 \qquad [1]$$

$$\sigma_{\rm pp} = (2.1 \pm 0.4) \times 10^{-20}{\rm cm}^2 \qquad [5]$$

calculated from the results on $p\mu$ scattering on H_2 for the scattering on the bound proton in a reasonable agreement with the theoretical result $\sigma_{\rm pp}^{\rm theor} = 1.9 \times 10^{-20}{\rm cm}^2$, obtained by numerical calculation of the multichannel scattering problem in the framework of the adiabatic representation of the three-body problem, without taking into account electron screening effects [6]. The account of the electron screening increase the calculated cross sections [7]. A further increase comes if one takes into account the molecular structure which leads to a disagreement with experimental results.

One cannot exclude, however, the change in the calculated data if one extends the basis and includes some additional channels.

For muonic deuterium the hyperfine splitting energy is comparable with the kinetic energy of muonic atoms at room temperature (~ 0.04 eV), so the relative population of the quartet and doublet hyperfine structure states is not fixed definitely. However, as follows from calculations, elastic scattering cross sections for different hyperfine structure states are approximately equal to each other, so the population of the states seems to be not very important. At liquid deuterium temperature the transition between hyperfine structure states is irreducible. Experimental value of the elastic cross section

$$\sigma_{\rm dd} = (0.8 \pm 0.2) \times 10^{-19}{\rm cm}^2 \qquad [1]$$

is in a good agreement with calculated one, averaged over hyperfine structure states, assuming their statistical population, and with electron screening taken into account [7,9].

The role of the electron screening increases for low collision energies. It is seen the most clearly when calculating the scattering lengths [7,10]. The same situation holds also for the scattering of muonic hydrogen on helium and lithium [11], which was not yet measured experimentally. Also there is no experimental results, as well as reliable theoretical calculations, on the elastic scattering of the excited muonic hydrogen. At the same time there are some experimental indications [12,13] or rather large energy (~ 1.8 eV) of deexcited muonic deuterium atoms at room temperature, which could be an evidence of either the acceleration of muonic atoms during deexcitation process [14] or Ramsauer-effect in the elastic scattering.

2 Deexcitation

Muonic atoms of hydrogen isotopes deexcite themselves via target molecules dissociation as well as via external Auger process on target atoms (molecules), radiative and Coulomb deexcitation [15,18]. Auger deexcitation of mesic hydrogen was considered in a Born approximation in [16] and in the eikonal one in [17]. The most interesting now is the probable competition of various deexcitation processes at early stage of the cascade. As it was shown in [14], even in the process of Auger deexcitation, in which the most part of deexcitation energy is taken away by a light particle, a considerable kinetic energy (~ 1 eV) of the residual ground state muonic hydrogen is possible. Muonic atom acceleration during deexcitation process will be much stronger, if one takes into account Coulomb deexcitation reaction, in which all the energy excess is realized in the kinetic energy of mesic hydrogen and recoil nucleus. This process can really compete with Auger deexcitation one at energy early stage of the cascade. It is of importance to obtain the energy distribution functions for excited and ground state mesic atoms, as well as the population of various states. This is an urgent problem for muon catalyzed fusion research, for which the energy distributions and ground state population q_{1s} with the account of the charge exchange of the excited mesic hydrogen are very important.

3 Charge exchange

Muon transfer from muonic hydrogen to the nucleus of heavier hydrogen isotope or of an element with $Z > 1$ plays an important role in the kinetics of the processes in the mixtures mentioned. The rates of muon transfer from the excited states of mesic hydrogen (with principal quantum number $n > 1$) to nuclei with $Z \geq 1$ and from the ground state mesic hydrogen ($n = 1$) to nuclei with $Z > 3$ are large ($\sim 10^{11}s^{-1}$ for LHD). These processes can be treated in a framework of WKB approximation and the rates in question are determined by nonadiabatic transitions in the crossing points of the terms, corresponding to the initial and final states of the reaction. Muon transfer from the ground state mesic hydrogen to nuclei with $Z \leq 3$ is determined by quantum processes and suppressed by three orders of magnitude as compared with transfer mentioned above. More detailed treatment of the transfer process is given in the review paper [19]. Here just note, that the most reliable information, both experimental and theoretical, exists for the rates of the isotopic exchange reactions for the ground state mesic atom in d-t and p-d mixtures, as well as for muon transfer rates from the ground state mesic hydrogen to nuclei with $Z > 3$. One needs a theoretical explanation of some peculiarities, observed in muon transfer to argon, neon and oxygen [20]. As to muon transfer in hydrogen-helium mixture, experimentally proved is the molecular mechanism for transfer from the ground state mesic hydrogen. For p-α mixture additional measurements are necessary due to contradictions between experimental data available. For t-α mixture there are no measurements up to now. Finally, the further investigations, both theoretical and experimental, of the muon transfer from the excited mesic hydrogen would be very important, because experimental data on q_{1s} in d-t mixture cannot be explained yet in the framework of the models used.

References

1. A. Bertin, A. Vitale: Fifty Years of Weak-Interaction Physics eds. A. Bertin, R.A. Ricci, A. Vitale, Bologna, Italian Physical Society, p. 130 (1984)
2. L.I. Ponomarev: Muon Catalyzed Fusion **3** 629 (1988)
3. G. Fiorentini: Nucl. Phys. **A374** 607c (1982)
4. S.S. Gershtein: Sov. Phys. - JETP **7** 318 (1958)
5. V.M. Bystritsky, V.P. Dzhelepov, V. Petrukhin, A.I. Rudenko, V.M. Suvorov, V.V. Filchenkov, N.N. Khovansky, B.A. Khomenko: Zh. Eksp. Teor. Fiz. **87** 384 (1984)
6. V.S. Melezhik, L.I. Ponomarev, M.P. Faifman: Sov. Phys. - JETP **58** 254 (1983)
7. A. Adamczak, V.S. Melezhik, L.I. Menshikov: Z. Phys. **D4** 153 (1986)
8. A.I. Mikhailov, V.I. Fomichev: Muon Catalyzed Fusion, **2** 137 (1988)
9. N.P. Popov: Muon Catalyzed Fusion, **2** 207 (1988)
10. A.V. Kravtsov, A.I. Mikhailov, N.P. Popov: Phys. Lett. **A116** 180 (1986)
11. A.V. Kravtsov, A.I. Mikhailov, N.P. Popov: J. Phys. B: At. Mol. Phys. **19** 1323 (1986)
12. W.H. Breunlich, P. Kammel, J.S. Cohen, M. Leon: Ann. Rev. Nucl. Part. Sci. **39** 311 (1989)
13. J.B. Kraiman, G. Chen, P.P. Guss et al.: Proc. Int. Conf. on Muon Catalyzed Fusion (Vienna, May 27 - June 1, 1990), to be published.
14. L.I. Menshikov: Muon Catalyzed Fusion **2** 173 (1988)
15. W. Czaplinski, A. Gula, A. Kravtsov, A. Mikhailov, N. Popov: This issue
16. M. Leon, H.A. Bethe: Phys. Rev. **127** 636 (1962)
17. A.P. Bukhvostov, N.P. Popov: Sov. Phys. - JETP **55** 13 (1982)
18. L. Bracci, G. Fiorentini: Nuovo Cim. **43A** 9 (1978)
19. N.P. Popov: Proc. Workshop "Exotic Atoms in Condensed Matter" (Erice, 19-25 May, 1990), to be published.
20. H. Schneuwly: Muon Catalyzed Fusion **4** 87 (1989)

Closing

Chairman: B. FASTRUP

Concluding Remarks by S. DATZ

Concluding Remarks

Sheldon Datz

Oak Ridge National Laboratory, Oak Ridge TN, USA

and Manne Siegbahn Institute of Physics, Stockholm, Sweden

At this workshop we have been presented with a melange, or perhaps I should say a *gulyás*, of extremely interesting results. Although the subject matter covered a broad range it all properly fit under the rubrik "High-Energy Ion-Atom Collisions," and a considerable amount of cross fertilization has occurred. The invited papers, for the most part, have been pedagogically correct and the discussions have, in some cases, been spirited. To try to touch on all the subjects presented would be presumptuous and redundant. However, if one were to try to sum up this workshop in a few sentences they would be: A truly remarkable panoply of results has been paraded before us. In the past three years some areas of research have been nearly completed, some old problems remain unsolved, some have just been newly and strongly developed and some are just beginning to open up. *Some* examples in these various areas are given below.

As an example of an almost completed area we take resonant transfer and excitation, RTE. The formation of the dielectronically excited intermediate state has been studied by coincidences between charge capture and X-ray emission (RTEX), by coincidence between two X-rays for H-like Ions (RTEXX) and by Auger emission from the doubly excited state (RTEA).

The non-ion-atom collision analogue of RTE, i.e. dielectronic recombination (DR), has come under extensive study with merged electron-ion beam experiments in heavy ion storage rings and in single pass experiments, and hybrid DR/RTE experiments have been carried out using a crystal channel as a dense source of loosely bound electrons. For K-shell excitation in fast collisions (i.e. where the impulse approximation is strictly valid) the method of substituting a lightly bound for a free electron has proven very valuable indeed. The arch example of this which was discussed by Tanis is the recent and superb study of RTE in $U^{90+} + H_2$ collisions where the effect is clearly visible in the total capture cross section alone. Some difficulties remain in the exact energy positions of the resonances and some difficulties also remain in the interpretation of L-shell RTE.

What of the future? One possibility is to use RTE/DR to prepare specific intermediate states. In the case of channeling these short-lived states are created in dense electron media and thus electron collisions with excited states can come under study. Another future possibility lies in the study of collisions where the impulse approximation is not valid i.e. where the exciting electron's orbital velocity is comparable to the collision velocity and the electron becomes "heavier", to put it into the interesting but controvesial terms introduced by N. Stolterfoht in these proceedings.

In the category of those problems which remain unsolved even after many years, one can think of the dilemma posed by the large disparity between charge state distributions of swift heavy ions emerging from solids as compared to gaseous target in contrast with the rather small differences in stopping power in the two media. A search for electrons which were theorized to screen the moving ions charge in the solid and then be released by the ion at its exit was carried out by the author of this theory (H.D. Betz) and is reported in these proceedings. They were not found. Instead an interesting study of Rydberg electrons accompanying the exiting ion resulted. It showed unusually long lifetimes for Stark stripping of the Rydberg states in electric fields. So the attempt to solve one mystery leads to another. In this regard a very interesting method for the quantitative study of Rydberg electron production was introduced at this conference in a talk by J. Wagner et al. In spite of the confusion which still exists in this field, one must say that the theoretical approach used by Burgdörfer for treating the states of ions moving in solids has yielded remarkably good results and its extension to the numerous experimental results on convoy and Rydberg electrons should do much to clarify the picture.

The field which is now being strongly developed is the use of anti-particles $\bar{p}$ and e^+ and charged mesons (μ^+ and μ^-) in atomic collision physics. As Walter Meyerhof points out "Collisions with antiparticles are particularly useful in transitions between the same initial and final states of an atomic system when amplitudes must add coherently. Typically these amplitudes are of the form $A = a_1 Z_1 + a_2 Z^2$ (where Z_1 is the projectile atomic number) so that in any cross section (integral or differential) $\alpha |A|^2$ there exist interference terms. The coefficients a_1 and a_2 can express

(1) first and second order terms in theory (critical intermediate states)
(2) one and two step mechanisms (real intermediate states)
(3) different mechanisms (due to electron and nucleus).

The prime example for (3) is e^-, e^+, $\bar{p}$, p on He where double ionization by various mechanisms can occur. Using various projectiles, one can distinguish empirically the various contributions to $|A|^2$ (Morenzoni) and check them against theory (Ford). Particularly striking is the fact that the ratio of double to single ionization is independent of Z_1 at high energy.

An example of (2) is excitation e.g. from the ^1S_o to $^1D_1$ state of He^0 by single (non-dipole) and double excitation, the latter via the $^1P_1$ state (Balashov).

An example of (1) is K-shell ionization where Coulomb ionization with or without distortion of the wave function ("polarization effect") can occur. Differences occur in the cross section with p and $\bar{p}$ projectiles (charge effect) as well as e^+ and e^- (mass effect) (Morenzoni)."

The future of work with anti-protons depends strongly upon the future existance of the LEAR (Low Energy Anti-Proton Ring) at CERN which may come into question in a few years because of financial problems. However, two projects aimed at the production of anti-hydrogen are presently being pursued. One involving a charge transfer collision

$$\bar{p} + Ps \rightarrow \bar{p}e^+ + e^-$$

is being developed by a group from Aarhus, University College London and Los Alamos. the second involves three body recombination (Gabrielse) at very low temperatures in an ion trap $\bar{p} + 2e^+ \rightarrow \bar{p}e^+ + e^+$ (note $\bar{p}e^+ = \bar{H}$). We shall have to wait for the next conference to find out the results of these efforts.

The largest recorded effort in the history of high energy ion-atom collisions has been the study of very high Z (e.g. U + U) collisions in the 5–10 MeV/amu region and the observation of positrons created by dynamic Coulombic interactions and with, perhaps, the production of positrons from filling of K vacancies from the negative continuum ("dive positrons"). In addition, peaks in the positron spectra, which were first reported in 1981, are now well characterized but not at all understood. The mystery persists and it may well be that the explanation lies in atomic physics. A contribution by Hansteen and Bang presented as a poster at the workshop may represent a possible road to a solution and deserves further development.

At sufficiently high energies (γ) and nuclear charge perturbative treatments for the production of $e^{\pm}$ pairs should clearly fail. Thus higher order QED effects should be observable and therefore testable under these conditions. Experiments on 200 GeV/amu S^{16+} on Au have already been mounted at CERN and it is expected that 200 GeV/amu Pb beams which will further stretch the limits of QED will be available in 1993.

We can expect that much excitement will be generated by the time of the next workshop, and given the same ambiance and superb conditions for exchange which our hosts have supplied for this workshop, we can anticipate another fruitful conference in 1993.

Author Index

List of Participants

4th Workshop on High-Energy
Ion-Atom Collision Processes
Debrecen, September 17–19, 1990

T. ÅBERG
Laboratory of Physics
Helsinki University of Technology
SF-02150 Espoo
Finland

S. ANDRIAMONJE
Université de Bordeaux 1
C.E.N.B.G.
Le Haut-Vignean
F-33170 Gradignan
France

V.V. BALASHOV
Institute of Nuclear Physics
Moscow State University
Moscow 119899
USSR

M. BARAT
Laboratoire des Collisions Atomiques
et Moleculaires (CNRS)
Batiment 351
Université Paris-Sud-91405
Orsay Cedex
France

O. BENKA
Institut für Experimental-Physik
J. Kepler Universität
A-4040 Linz/Auhof
Austria

E.M. BERNSTEIN
Western Michigan University
Kalamazoo, Michigan 49008-5151
USA

D. BERÉNYI
Institute of Nuclear Research
of the Hungarian Academy of Sciences
P.O.Box 51
H-4001 Debrecen
Hungary

H.D. BETZ
Sektion Physik
Universität München
Am Coulombwall 1
D-8046 Garching
BRD

S. BHATTACHARYYA
Gokhale College
370/1 N.S.C.Bose Road
Calcutta 700047
India

A. BORDENAVE-MONTESQUIEU
Centre de Physique Atomique
URA CNRS 770,Université P.Sabatier
118 Route de Narbonne
Toulouse Cedex 31062
France

J. BURGDÖRFER
Dept. of Physics and Astronomy
University of Tennessee
Knoxville, Tennessee 37996
USA

C. CISNEROS
Instituto de Fisica
Laboratorio de Cuernavaca
62191 Cuernavaca
Morelos
Mexico

C.L. COCKE
Department of Physics
Kansas State University
Manhattan, Kansas 66506
USA

A. CSAPÓ
I. Bródy Research Centre,
TUNGSRAM Co. Ltd.
Váci u 77
H-1340 Budapest
Hungary

I. CSERNY
Institute of Nuclear Research
of the Hungarian Academy of Sciences
P.O.Box 51
H-4001 Debrecen
Hungary

S. DATZ
Oak Ridge National Laboratory
P.O.Box 2008
Oak Ridge, Tennessee 37831
USA

R.D. DuBOIS
Battelle Pacific
Northwest Laboratories
P.O.Box 999
Richland, Washington 99352
USA

A.K. EDWARDS
Dept. of Physics and Astronomy
University of Georgia
Athens, GA 30602
USA

S.B. ELSTON
Dept. of Physics and Astronomy
University of Tennessee
Knoxville, Tennessee 37996
USA

B. FASTRUP
Institute of Physics
University of Aarhus
DK-8000 Aarhus C
Denmark

D. FLUERAŞU
Central Institute of Physics
P.O.Box MG-6
Bucharest
Romania

A.L. FORD
Department of Physics
Texas A & M University
College Station, Texas 77843
USA

C.R. GARIBOTTI
Centro Atómico Bariloche
Comision Nacional de Energia Atómica
Bariloche, R.N. 8400
Argentina

R. GAYET
Université de Bordeaux I.
Laboratorie de Coll. Atomiques
351 Com. de la Liberation
Talence Cedex 33405
France

M. GERETSCHLÄGER
Institut für Experimental-Physik
J. Kepler Universität
A-4040 Linz/Auhof
Austria

K.-O. GROENEVELD
Institut für Kernphysik
J.W. Goethe Universität
A-Euler Str. 6.
D-6000 Frankfurt/M 90.
BRD

L. GULYÁS
Institute of Nuclear Research
of the Hungarian Academy of Sciences
P.O.Box 51
H-4001 Debrecen
Hungary

J.M. HANSTEEN
Department of Physics
University of Bergen
Allegaten 55.
N-5007 Bergen U
Norway

R. HIPPLER
Fakultät für Physik
Universität Bielefeld
Universitätsstrasse
D-4800 Bielefeld 1
BRD

G. HOCK
Institute of Nuclear Research
of the Hungarian Academy of Sciences
P.O.Box 51
H-4001 Debrecen
Hungary

T. KAMBARA
Institute of Physical and
Chemical Research (RIKEN)
Hirosawa, Wako-shi
Saitama 351-01
Japan

E. KOLTAY
Institute of Nuclear Research
of the Hungarian Academy of Sciences
P.O.Box 51
H-4001 Debrecen
Hungary

H. KUIPER
Physikalisches Institut der
Universität Erlangen-Nürnberg
Erwin-Rommel Str.1
D-8520 Erlangen
BRD

Á. KÖVÉR
Institute of Nuclear Research
of the Hungarian Academy of Sciences
P.O.Box 51
H-4001 Debrecen
Hungary

L. KÖVÉR
Institute of Nuclear Research
of the Hungarian Academy of Sciences
P.O.Box 51
H-4001 Debrecen
Hungary

I. KÁDÁR
Institute of Nuclear Research
of the Hungarian Academy of Sciences
P.O.Box 51
H-4001 Debrecen
Hungary

I.C. LEGRAND
Central Institute of Physics
P.O.Box MG-6
Bucharest
Romania

M.W. LUCAS
School of Matem. and Phys. Sci.
University of Sussex
Falmer, Brighton BN1 9QH
England

R. MAIER
Institut für Kernphysik
J.W. Goethe Universität
August-Euler Str. 6
D-6000 Frankfurt/M
BRD

S.T. MANSON
Dept.of Physics and Astronomy
Georgia State University
Atlanta, Georgia 30303
USA

I. MARISCSÁK
Department of Physics
University of Miskolc
H-3515 Miskolc-Egyetemváros
Hungary

W. MEHLHORN
Fakultät für Physik
A-Ludwigs Universität
H-Herder Str. 3.
D-7800 Freiburg/Br
BRD

M.G. MENENDEZ
Dept. of Physics and Astronomy
University of Georgia
Athens, Georgia 30602
USA

W.E. MEYERHOF
Dept. of Physics
Stanford University
Stanford, CA 94305-4060
USA

B.L. MOISEIWITSCH
Dept. of Applied Mathematics
Queen's University Belfast
Belfast BT7 1NN
UK

P.H. MOKLER
GSI Darmstadt
Postfach 110541
D-6100 Darmstadt 11
BRD

E. MORENZONI
Paul Scherrer Institut
Würenlingen und Villigen
CH-5232 Villigen
Switzerland

T. MUKOYAMA
Institute for Chemical Research
Kyoto University
Kyoto 606
Japan

R.E. OLSON
Dept. of Physics
University of Missouri
Rolla, MO 65401-0249
USA

W. OSWALD
Section Physik
Universität München
Am Coulombwall 1
D-8046 Garching
BRD

T. PAPP
Institute of Nuclear Research
of the Hungarian Academy of Sciences
P.O.Box 51
H-4001 Debrecen
Hungary

B. PARIPÁS
Department of Physics
University of Miskolc
H-3515 Miskolc-Egyetemváros
Hungary

J. PÁLINKÁS
Institute of Nuclear Research
of the Hungarian Academy of Sciences
P.O.Box 51
H-4001 Debrecen
Hungary

S. RICZ
Institute of Nuclear Research
of the Hungarian Academy of Sciences
P.O.Box 51
H-4001 Debrecen
Hungary

P. RYMUZA
Institute for Nuclear Studies
Department P-II
05-400 Swierk
Poland

L. SARKADI
Institute of Nuclear Research
of the Hungarian Academy of Sciences
P.O.Box 51
H-4001 Debrecen
Hungary

A.S. SCHLACHTER
Lawrence Berkeley Laboratory
Berkeley, California
USA

H. SCHMIDT-BÖCKING
Institut für Kernphysik
J.W. Goethe Universität
August-Euler Str. 6
D-6000 Frankfurt/M 90
BRD

I.A. SELLIN
Dept. of Physics and Astronomy
University of Tennessee
Knoxville, Tennessee 37996
USA

O.B. SHPENIK
Inst. for Nuclear Research of the
Ukrainian SSR, Acad. of Sciences
Uzhgorod Department
Uzhgorod
USSR

N. STOLTERFOHT
Hahn-Meitner-Institut
Glienicker Str.100
D-1000 West-Berlin 39

B. SULIK
Institute of Nuclear Research
of the Hungarian Academy of Sciences
P.O.Box 51
H-4001 Debrecen
Hungary

Gy. SZABÓ
Institute of Nuclear Research
of the Hungarian Academy of Sciences
P.O.Box 51
H-4001 Debrecen
Hungary

E. SZMOLA
I. Bródy Research Centre,
TUNGSRAM Co. Ltd.
Váci u 77
H-1340 Budapest
Hungary

L. SZÓTÉR
Department of Physics
University of Miskolc
H-3515 Miskolc-Egyetemváros
Hungary

J. SÁRKÖZI
I. Bródy Research Centre,
TUNGSRAM Co. Ltd.
Váci u 77
H-1340 Budapest
Hungary

E. TAKÁCS
Institute of Nuclear Research
of the Hungarian Academy of Sciences
P.O.Box 51
H-4001 Debrecen
Hungary

J.A. TANIS
Department of Physics
Western Michigan University
Kalamazoo, Michigan 49008
USA

M. TERASAWA
Department of Physics
Himeji Institute of Technology
21 Shora Himeji-Shi
Hyogo, 671-22
Japan

K. TŐKÉSI
Institute of Nuclear Research
of the Hungarian Academy of Sciences
P.O.Box 51
H-4001 Debrecen
Hungary

I. TÖRÖK
Institute of Nuclear Research
of the Hungarian Academy of Sciences
P.O.Box 51
H-4001 Debrecen
Hungary

J. TÓTH
Institute of Nuclear Research
of the Hungarian Academy of Sciences
P.O.Box 51
H-4001 Debrecen
Hungary

Z. TÓTH
I. Bródy Research Centre,
TUNGSRAM Co. Ltd.
Váci u 77
H-1340 Budapest
Hungary

C.H. TRIKALINOS
Physics Department
Solid State Section
Athens University
Solons Str. 104.
Athens 10680
Greece

J. ULLRICH
GSI Darmstadt
Planck str. 1
D-6100 Darmstadt
BRD

T. VAJNAI
Department of Physics
University of Miskolc
H-3515 Miskolc-Egyetemváros
Hungary

D. VARGA
Institute of Nuclear Research
of the Hungarian Academy of Sciences
P.O.Box 51
H-4001 Debrecen
Hungary

J. VÉGH
Institute of Nuclear Research
of the Hungarian Academy of Sciences
P.O.Box 51
H-4001 Debrecen
Hungary

L. VÉGH
Institute of Nuclear Research
of the Hungarian Academy of Sciences
P.O.Box 51
H-4001 Debrecen
Hungary

T.J.M. ZOUROS
Physics Department
University of Crete and
Research Centre of Crete
Iraklion
Greece